I0064006

Caution!
Risk of electric shock.

The circuits and subjects discussed in this book operate from or involve dangerously high voltages. Do not attempt to build any of the circuits described here, or work on live equipment, without proper training. It is the reader's responsibility to ensure that any equipment constructed or modified using the information given in this book, is safe to use.

Designing Power Supplies for Valve Amplifiers

Second Edition

by
Merlin Blencowe, MSc

2022

This work is presented "as is" and no expressed or implied warranties are offered by the author, printer or publisher. The author, printer and publisher specifically disclaim any and all liability and responsibility to any person, property or entity for loss or damage caused, or alleged to have been caused, directly or indirectly, by the use, misuse, negligent use, or inability to use the information presented within this book. No warranty may be created or extended by sales or promotional materials. Any brand name or trademark expressed herein is used only for identification and explanation without intent to infringe. Information and diagrams reproduced herein do not imply freedom from patent rights.

Copyright © 2022, Merlin Blencowe. All rights reserved

No part of this publication may be reproduced in any material form (including photocopying or storing in any medium by electronic means) without the written permission of the author, except in accordance with the provisions of the Copyright, Designs and Patents Act 1988.

Contents

3. Voltage Multipliers   65

4. Smoothing Filters   83

5. Fusing, Switching, and Other Refinements 108

6. Shunt Stabilisers and Regulators 128

7. Series Stabilisers and Regulators   162

Preface

If you are reading this then perhaps you are one of the growing number of enthusiasts who embrace the anachronism that is valve audio technology. I certainly belong in that category, and despite the protestations of modernists, it is not too difficult to appreciate the appeal of valve amplifiers beyond mere nostalgia. Valve circuits tend to be simple, which makes them easy to understand, modify, and repair, and gives them an appealing longevity that is sadly missing from our modern age of disposable consumer electronics and software obsolescence.

The intention of this book is to provide a deep dive into linear power supplies, particularly as they apply to valve amplifiers. Nevertheless, much of the information will be found to be generally applicable to other areas of electronics, too. Some of the information is original, and some has been collected through research of sources going back more than a century. Many of the referenced sources are now available for free online, and interested readers are encouraged to consume them for even more detail, see for example:
https://worldradiohistory.com/
http://www.tubebooks.org/

Although there are hundreds of electronics textbooks from the 1930s onwards which cover valve circuit theory thoroughly, power supplies often occupy just one chapter, or even less. Books dedicated to specifics like transformer design or regulator design are not too difficult to find, but to my knowledge there is no other book dedicated to linear power supply design as a broad topic. It was for this reason that I wrote the first edition in 2010. Apparently I was not alone in wanting to see this information collected into one place, as the book was extremely well received despite its shortcomings, and the clamour for a new edition took me by surprise. Therefore, for this edition, errors have been corrected, key topics expanded, and superfluous material ruthlessly pruned.

Be warned, this book is not a beginner's guide. Nevertheless, I appreciate that many readers will be enthusiastic amateurs who may be self-taught and may have gaps in their general electronics knowledge, so this book attempts to occupy the middle ground between pure, armchair theory, and real-world practical circuits. I have not shied away from the mathematics, but neither do I want to beat the reader over the head with it. Most of it is algebra, not calculus, and is explained step by step. Other basic topics are covered incidentally as part of more general discussion without –I hope– patronising experienced readers. If it helps you even a little bit, I shall consider it a success!

Acknowledgements

The author would particularly like to thank Paul Fawcett and Paul Reid for the Herculean effort they put into proof reading the manuscript, and Stephen Keller for his apparently supernatural ability to source reference material. In addition, this book would not have been possible without the help, encouragement and co-operation of AMS Neve Ltd., David Ivan James, Mallory Nicholls, Holly Smith, Lee Taylor, The University of Leeds Library, and The University of York Library.

Chapter 1: Mains Power and Transformers

"And out of old books, in good faith,
Cometh all this new science that men learn."
Chaucer, 14th century.

Power supplies can be divided into two main types: linear power supplies and switch-mode power supplies (SMPSs). Linear supplies are the traditional kind using expensive and heavy transformers with big capacitors and hot, brute-force regulation techniques. By contrast, SMPSs are cheap, lightweight, and ubiquitous in modern appliances. Both types can provide galvanic isolation from the mains by using transformers, and this is absolutely necessary for audio equipment since the user regularly comes into physical contact with the audio circuit, e.g. when touching audio connectors. The difference is that SMPSs can use much smaller transformers because they operate at much higher frequency.

Linear supplies are conceptually simple and generally robust; repair is usually straightforward. SMPSs are more complicated, and while they are not necessarily less reliable, the motivation to make them small and cheap inevitably leads to borderline heat sinking and sweating capacitors, so they have become associated with shorter lifetimes. Repair is often a specialist job. In principle, SMPSs are more efficient than linear supplies, and this is particularly true at the low-voltages and high-currents needed by most modern appliances. But at higher voltages and modest currents the difference is sometimes exaggerated; SMPSs *can* be made more efficient but it requires considerable optimisation. Often they are criticised for being noisy, but the state of the art has progressed to the point where they can now compete with linear supplies in this regard. However, they do require more close attention to shielding and high-frequency filtering if a product is to meet RFI and EMI regulations. The design of SMPSs is an engineering specialism by itself (about which this author does not pretend to be an expert), which is not an inviting prospect for the hobbyist, and SMPSs are presumably the sworn enemy of those audio peculiarists who eschew silicon of all kinds. This book therefore deals mainly with linear supplies.

The power transformer is one of the most expensive and awkward items to obtain for a valve amplifier. There is an understandably limited range of off-the-shelf devices tailored to valve circuits these days, particularly outside north America. For the designer on a budget it is often more prudent to begin by considering what power transformer is actually available, and what might be done with it, than to conjure a grand amplifier design and then try to find a power supply to suit it.

1.1: Mains Electricity

The official mains frequency in Europe and most of the world is 50Hz. In the US/Canada and parts of Japan it is 60Hz. In developed countries the frequency is closely regulated to much better than ±1%.

For the last few decades the official mains voltage across Europe has been harmonised at 230V ±10%. However, most places in Britain still receive the old figure of 240V, while many parts of mainland Europe still receive 220V. Both of these are within the official 10% tolerance of the new standard.

The official mains voltage in the US/Canada is 120V and most populous areas receive this to within ±5%. However, in various regions the voltage has been as low as 110V in the past, and older equipment may suffer from over-voltage problems when used on the modern 120V supply. The use of two-conductor (no earth) wall sockets is still fairly common, but must be avoided with any equipment that is not double insulated, as the chassis must remain connected to earth for safety (section 1.2.1).

1.1.1: Mains Distribution

In Europe, domestic electricity enters the building through two conductors called the Live[*] (brown) and Neutral (blue). The Neutral conductor is bonded to ground (soil / dirt / planet Earth) at multiple locations along its route, thereby making it nominally zero volts relative to the ground beneath your feet. The Live conductor swings all the voltage of $230V_{rms}$ or $325V_{pk}$. After entering the building, the mains conductors will pass through the utility meter and a box called the consumer unit, which contains circuit breakers. In an old building these may be simple wire fuses, whereas new installations contain more sophisticated miniature circuit breakers (MCBs) and residual current devices (RCDs). The MCBs detect overcurrent and the RCDs monitor the difference in current between the Live and Neutral, tripping if there is a significant imbalance which would imply a fault to earth somewhere.

In the US/Canada domestic supplies normally include two Hot conductors (black and red), plus the Neutral (white or grey). The two Hot conductors are both $120V_{rms}$ but 180° out of phase with one another (or 120° for some buildings). The Neutral conductor is bonded to planet Earth at the service panel (equivalent to the European consumer unit) so is nominally at zero volts. High power appliances like ovens are connected across the two Hot phases so they receive a total of 240V (or 208V). However, most appliances –including audio amplifiers– will be connected between a single Hot conductor and neutral. We will therefore forget about split-phase power for the rest of this book.

---

[*] Electricians insist the preferred terminology is 'line' rather than 'live', which this author finds underwhelming and chooses to ignore.

2

Just because the Neutral is *nominally* at zero volts does not mean it can be considered fully safe; a break, or some cause of high resistance along its length, will cause one side of the break to rise to the same voltage as the Live/Hot if there happens to be a circuit connected across the supply at the time. For this reason, an Earth or Ground conductor is also needed. In Europe this is green-yellow striped, and in America it is green, although regulations currently allow either colour variation to be used within appliances for export.

The Earth/Ground is bonded to planet Earth but, unlike the Neutral conductor, it is not part of any normally functioning circuit. Under normal conditions it carries no current and should be thought of simply as an extension of the ground beneath your feet. It is there to ensure that any metal object –such as the metal casing of an appliance– which could accidentally come into contact with the Live/Hot or Neutral, is always and absolutely connected to ground. The metal remains at ground potential at all times and is therefore always safe to touch by anyone who also happens to be standing on the ground, which is where most people spend their time. If a main conductor does accidentally touch an earthed object it will trip the circuit breaker.

In old British houses the Earth conductor may enter the building from outside; it is bonded to planet Earth at the local substation or via a ground rod. In new installations the Earth conductor exists only within the building and is bonded both to Neutral and to planet Earth via the

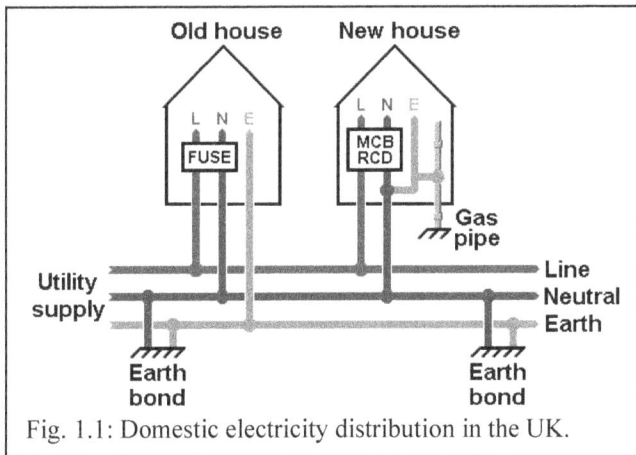

Fig. 1.1: Domestic electricity distribution in the UK.

domestic gas/water pipes or a local ground rod, as illustrated in fig. 1.1. New American installations are essentially the same, with the additional Hot conductor.

1.2: Mains Appliance Wiring

At the heart of a linear power supply is the power transformer. This takes the AC mains (wall) voltage and converts it into one or more AC voltages that are more suitable for our needs. For a valve amp this usually means a low voltage supply for the heaters and a high voltage supply for the anodes, at the very least. As well as providing the voltages we want, the transformer provides safety isolation from the mains wall supply. Although the secondary voltages in a valve amp might well be higher than the wall voltage, the secondaries are inherently current limited by the source impedance of the transformer. Touch the wrong thing and the wall supply can

3

dump nearly unlimited energy into your body, whereas a transformer cannot. The chances of getting a severe burn, or stopping the heart, are lower when a transformer sits between you and your energy provider.

1.2.1: Earthing and Safety

A valve amplifier typically needs a low voltage supply for the heaters, and a high voltage supply for the anodes, called the HT or B+. Most hobbyists have therefore stopped to wonder, why not get the HT directly from the mains wall voltage, with no need for an expensive transformer? The reason not to do this has everything to do with safety and the law.

Until at least the 1960s it was common to find valve equipment using an arrangement something like fig. 1.2.[1,2] The heaters for all valves in the set were connected in series and supplied directly from the wall through a dropping resistor; the Philips/Mullard 'U' series of valves were manufactured specially for this type of operation. The HT was obtained by half-wave rectifying the mains[*] (the small inductor and capacitor shown faint were sometimes included to suppress modulation hum in radio receivers). The chassis was connected to Neutral which was bonded to earth somewhere outside the building, so in theory it should be

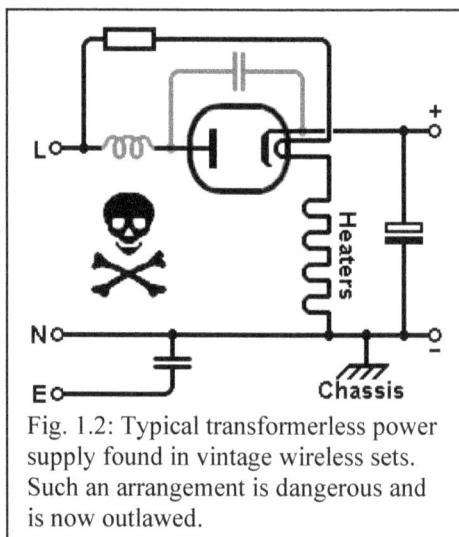

Fig. 1.2: Typical transformerless power supply found in vintage wireless sets. Such an arrangement is dangerous and is now outlawed.

at zero volts. The chassis might also be connected to a local earth through a capacitor to bypass radio frequencies more effectively.

Unfortunately, it is the chassis-to-neutral connection that is the main problem here. If an old-fashioned reversible mains plug is used, or it happens to be miswired, or if there is a break in the Neutral conductor somewhere in the building, the chassis can become live. This is an extraordinary risk to the user, so this practice is now outlawed. Any appliance built into a metal chassis which can be touched by the user *must* have that chassis connected directly to earth. A corollary of this requirement is that the user must be galvanically isolated from the mains supply using either a

---

[1] Bulley, E. G. (1956). Constructing AC/DC Equipment. *Practical Wireless*, February, pp113-4.
[2] Nash, L. N. (1962). Power Rectifier Circuits, *Practical Wireless*, June, pp128-31.
[*] Full-wave rectification of the mains was not possible since the neutral is bonded to earth, and the receiver/audio circuit also needs an earth reference, which would therefore short out a full-wave rectifier.

4

traditional power transformer or an isolating SMPS. In the world of safety regulations, earthed appliances like this are called Class-I appliances, which includes virtually all audio equipment that is powered from the wall.

Where the mains cable enters the chassis, usually via an IEC inlet or sometimes a captive cable, the incoming mains earth must be connected to the chassis through a suitably heavy wire. Regulations

Fig. 1.3: The earth bond should be made to a dedicated screw, close to the mains inlet.

currently permit push-on connectors to be used, although common sense would suggest a permanent soldered connection is more reliable (provided the wire is also mechanically 'hooked' onto the terminal). Where this wire is bolted to the chassis is known as the **main earth bond**. It should be a dedicated stud or screw, as in fig. 1.3, not a screw that is used to fix some other piece of hardware which might loosen over time. The Earth wire should be soldered or crimped to a tag and fixed to the stud with a nyloc™ nut or a shake-proof washer and two nuts. Any paint or oxidation on the chassis must first be scraped off to ensure a good electrical connection, and a star washer should be placed between the chassis and solder tag. The purpose of the star washer is to bite into the metal, creating gas-tight contacts which should be immune to oxidation. Other earth connections can be stacked on top of the first earth bond if necessary, always using green or green-and-yellow striped wire.

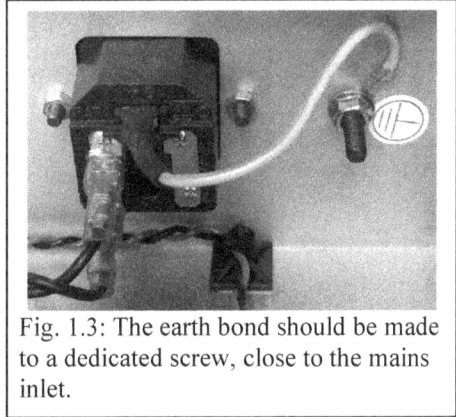

For commercial products the main earth bond must be labelled with the earth symbol, but any other earthed terminals must *not*. If the bond is made to an ordinary screw then the earth symbol must be visible on the outside of the chassis, to remind anyone not to unscrew that one. If a welded or pressed stud is used instead, then the symbol should be on the inside, for the benefit of the technician. You can buy adhesive stickers for this purpose.

The earth bond is legally required and is for safety only; it plays no part in circuit operation and no current flows in it except under fault conditions. It should be regarded as just another part of the chassis. Once it has been firmly connected to mains earth you can forget about the chassis as far as circuit grounding is concerned. Although the terms 'earth' and 'ground' are often used interchangeably, the audio circuit ground does not *necessarily* have to be connected to earth at all. In practice, however, we do connect the audio circuit to chassis (ideally at a single point near the audio input) since this ensures the amplifier's working voltages are properly defined with respect to zero volts and therefore with respect to any other equipment we might connect to it.

Some appliances are double-insulated meaning they have at least two layers of insulation between the user and any hazardous conductors. If a metal chassis is used

5

then one of these layers will be some kind of plastic casing which completely covers the metal so it cannot be touched. Such products do not need an earth connection to be safe, and these are called Class-II appliances. However, it is considerably more difficult to build a compliant Class-II appliance, particularly for audio appliances where metal shielding is needed almost everywhere for hum and RF suppression, so double-insulation is not considered in this book.

1.3: Basic Electromagnetics

Now that we understand why an isolating power transformer is necessary, we can move on to how transformers actually work. It might be supposed that we don't really need to know *how* a transformer works, only what it does, but knowing the basic principles of how something works makes the limitations of the device more obvious and intuitive, and makes us more sympathetic to its hardship. Since the power transformer is usually the limiting factor in the whole power supply, a good understanding of its functioning is deserved. Nevertheless, we will try to keep the lesson broad and qualitative; this is not a physics textbook, after all.

1.3.1: The Ohm's Law of Magnetism

It is a Law of Nature that magnetic fields are created by moving electric charges such as electrons.* When current travels along a wire, a magnetic force field is generated around the wire. We use the symbol H for magnetic force, so it is called the H-field (also called the magnetic field intensity in some books). This force in turn causes **magnetic flux** to 'flow' around the wire, which is called the B-field. And beware, some books use the term 'magnetic field' ambiguously for both the H- and B-fields, which can be quite maddening. Additionally, since the technology of magnetic components has not really changed in more than a century, old scientific units are still often encountered.

The notion of magnetic flux as a kind of invisible magnetic current or fluid is only a conceptual model. There is no physical matter truly flowing around the wire, it is only a complementary magnetic field. The H-field creates or 'drives' the B-field. It is also essential to grasp that B is flux *density*. Flux itself –the fictitious magnetic fluid– is given the symbol Φ. How much flux we get depends on how large the area is where the B-field is created; we will return to this later.

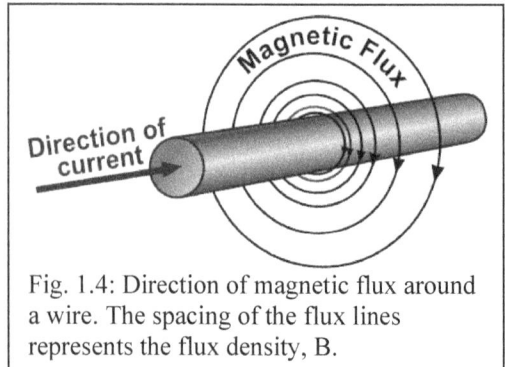

Fig. 1.4: Direction of magnetic flux around a wire. The spacing of the flux lines represents the flux density, B.

---

*Even a permanent magnet contains moving electrons, spinning endlessly around atomic nuclei, each generating a magnetic field.

The greater the electric current, the more charge is being moved per second, and therefore the stronger the magnetic force (H-field) grows. The greater the magnetic force, the more flux (Φ) it pushes around, thus increasing the flux density too (B-field). The H-field is strongest at the surface of the wire, close to all that moving charge, and it decays exponentially with distance. Therefore, there is also more flux flowing close to the surface of the wire than at a great distance. This can be represented on paper by drawing the path of the flux as a collection of lines as in fig. 1.4, where the spacing of the lines represents the density of the flux.

The direction of the flux can be figured using the 'corkscrew rule'.[*] When a corkscrew is turned clockwise it moves forwards into the cork. Likewise, if current (positive charge)[†] is moving in that direction, into the cork as it were, flux will flow clockwise around it.

The flux density B, which results from a given magnetic force H, depends on the material in which the field is set up. The property of a material that allows flux to flow within it is called **magnetic permeability** –much like how everyday permeability is a measure of how easy it is for water to pass through the material. The formula which relates the two fields is:

$$B = \mu H \qquad \text{teslas} \tag{1.1}$$

Where:
B = flux density in webers per square metre, or teslas for short;
μ = permeability in henries per metre;
H = magnetic field strength in amps per metre.

This formula is sometimes called the 'Ohm's law of magnetism'. It represents a magnetic force H, acting across a material with permeability μ, which pushes some flux through its volume resulting in a flux density B inside the material. This is somewhat equivalent to a voltage acting across a conductance, pushing electric current through it.

Nearly all materials have practically the same permeability as air / vacuum, which is extremely small at 1.26 microhenries per metre. It takes a huge magnetic force to produce even a little flux in air, and to a magnetic H-field almost everything looks like air. Except, that is, for iron, nickel and cobalt, and their various alloys. These metals are unusual in having permeability that is hundreds to thousands of times greater than air. These special materials have come to be known as **ferromagnetic**, from the Latin *ferrum* meaning iron, the most significant of the trio, known since ancient times to have magnetic properties.

---

[*] You may have been taught the 'right-hand grip rule' in school, involving an outstretched thumb for current and curled fingers for flux. But there are so many 'hand rules' in physics that I find them more of a hindrance than a help.
[†] Electrons carry negative charge of course, in the opposite direction to positive current. Isn't physics fun?

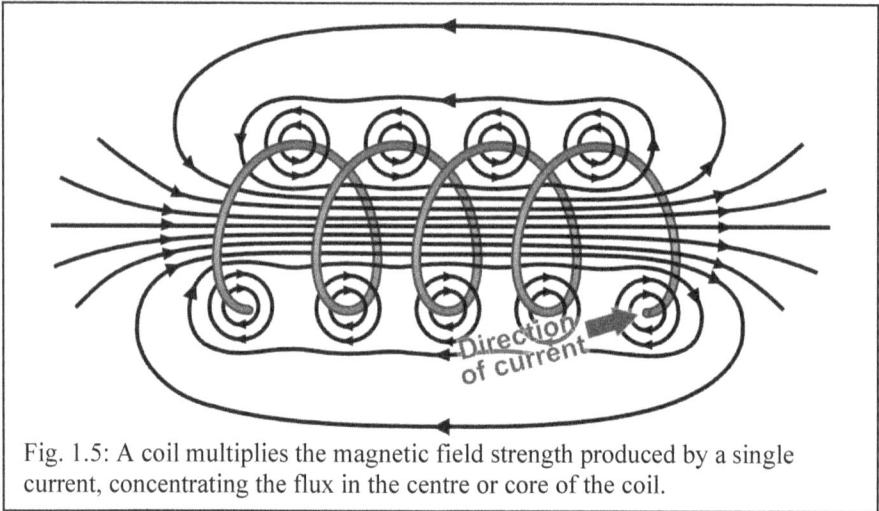

Fig. 1.5: A coil multiplies the magnetic field strength produced by a single current, concentrating the flux in the centre or core of the coil.

Ferromagnetism is caused by tiny 'domains' within the metal which behave like microscopic permanent magnets. Putting a chunk of iron inside a pre-existing magnetic H-field will force its domains to line up with the field like compass needles, so their own self-generated flux adds to the flux already flowing through that chunk of space, thereby boosting the overall flux density B. In other words, by putting some iron within the vicinity of our current-carrying wire, we get more flux for our money. And the closer we place the iron to the wire, where the H-field is strongest, the better the deal.

We can go further still. By winding the wire into a coil we effectively multiply up the H-field. Since each turn of wire is carrying the same current, each turn produces a similar H-field. If the turns are very close together then these H-fields will almost perfectly overlap, adding together and thus supercharging the amount of H-field in a given volume of space. If we now draw the lines of flux we will find it is most concentrated in the centre of the coils, as illustrated in fig. 1.5. Here will be the best place to put some iron.

Notice, however, that there are some small circular flux paths that do not flow through or 'link' with any neighbouring turns. In a practical inductor this is known as **leakage flux**, and it is usually minimised by winding the turns very neatly, as close together as possible.

1.3.2: Faraday's Law of Inductance

Where things get really interesting is when there is a *change* in a magnetic field. An alternating current is constantly changing and therefore produces a constantly changing magnetic field, in direct sympathy with it (H in phase with i). Now we encounter another Law of Nature. Any change in the amount of magnetic flux will induce a voltage across any conductor that happens to be within the changing field at the time, including the conductor that created the field in the first place. This is expressed by Faraday's law of induction:

$$E = -N\frac{d\Phi}{dt} \qquad \text{volts} \qquad (1.2)$$

Where:
N = number of turns;
$d\Phi/dt$ = rate of change of flux in webers per second.

The induced voltage may be called a **back-EMF** or **flyback voltage**, because it is always induced with such a polarity that it tries to drive a current whose magnetic field *opposes* the original change. This principle is called Lenz's law, which is incorporated into Faraday's law by the minus sign; two laws for the price of one. Thus if the B-field is declining, the induced EMF will attempt to make more current flow, to generate more flux and halt the decline. Contrarily, if the flux is increasing then the EMF will be induced in the opposite direction, in an attempt to reduce the current and choke off the increasing flux. Put simply, the magnetic field does not *want* to change, so it generates an opposing voltage in a desperate attempt to keep the current flowing just the same as it was a moment ago. In a practical circuit this means the back-EMF opposes the voltage we apply across the wire or coil.

Note that the induced voltage is not proportional to the flux itself but to the *rate of change* of flux, and therefore correspondingly to the *rate of change* of current. We relate changing current to changing flux by introducing a constant called the **self inductance** or simply the **inductance** of the wire, measured in henries.

$$E = -N\frac{d\Phi}{dt} = -L\frac{di}{dt} \qquad (1.3)$$

Where:
N = number of turns;
L = inductance in henries;
E = EMF in volts;
di/dt = rate of change of current in amps per second.

Although we are mostly interested in coils, even a straight wire has some inductance, typically one microhenry per metre for average hookup wire.

1.4: Basic Transformer Theory

At the most fundamental level, a transformer is two or more coils/inductors placed close to one another, so they intimately share each other's magnetic fields. A voltage applied across on of them will cause current to flow, and this current in turn produces a magnetic field. Because the coils are in close proximity, the flux produced by the first coil will unavoidably flow through or 'link with' the turns of the other coil too. If the current is alternating then the flux must follow likewise, and this will in turn induce EMFs into *both* coils. Transformers therefore only do something useful when energised with AC, not DC. This is why the world adopted AC mains electricity in the first place –because it can be transformed.

The coil that we drive is called the **primary** and the other coil/s which we draw upon are called **secondaries**. Ideally, we would like the tightest possible coupling between the coils so very little energy is wasted. This is usually accomplished by winding them very close together on a single ferromagnetic core. Nearly all the flux will flow preferentially in the high permeability core material rather than in the low permeability air outside, so nearly all the flux will be 'guided' through each and every turn of wire.

If the sinusoidal mains wall voltage is applied across the primary of the transformer, an alternating current will flow which in turn produces a corresponding H-field in the core. The H-field, when multiplied by the permeability of the core material, produces a certain flux density or B-field in the core. The alternating flux flowing in the core in turn produces a back-EMF which opposes the applied wall voltage. Without this opposing EMF the primary current would be limited only by the wire resistance. In other words, the primary has *inductance* which limits the current to a steady value known as the **magnetizing current**. This is equal to the AC voltage divided by the inductive reactance:

$$I_{mag} = \frac{V_{pri}}{2\pi fL}$$

Magnetising current does nothing useful as far as transforming is concerned; it is an unwanted inefficiency. The transformer designer will minimise it by winding hundreds or thousands of primary turns to achieve a fairly high inductance, limiting this current to perhaps a few tens of milliamps in the sort of transformers we're interested in.

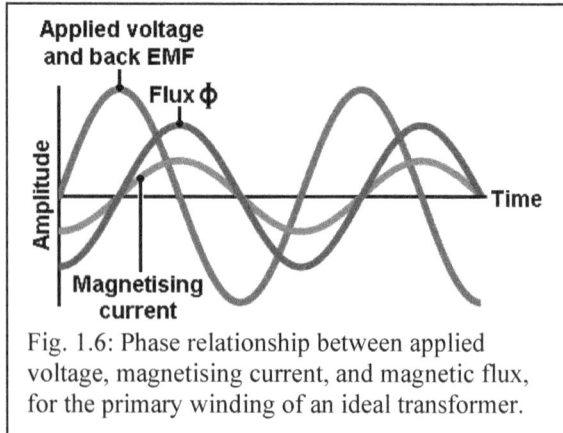

Fig. 1.6: Phase relationship between applied voltage, magnetising current, and magnetic flux, for the primary winding of an ideal transformer.

Fig. 1.6 shows the phase relationship between voltage, flux, and magnetising current in a transformer primary. The flux is in phase with the current since it is conjured by it. But the back-EMF (which exactly opposes the applied voltage and therefore lies directly on top of the same trace) is proportional to the *rate of change* of the flux, in accordance with faraday's law. The flux is changing fastest at its zero crossings, so it is here that the EMF reaches its peaks. This is the physical explanation for why the voltage across a simple inductor always leads the current through it by 90°.

The flux which links with a secondary coil will generate an EMF across the secondary too, in proportion to the ratio of the number of turns on the primary to the number of turns on the secondary, which is called the **turns ratio**. If *all* the flux is perfectly linked then the voltage generated across the secondary coil will be:

$$V_{sec} = V_{pri} \frac{N_{sec}}{N_{pri}} \qquad (1.4)$$

Where:
N_{sec} = number of the secondary turns;
N_{pri} = number of the primary turns.

Since the secondary EMF is also proportional to the rate of change of flux, and the flux is already 90° out of phase with the primary voltage, the secondary voltage will be in phase with the primary voltage (or 180° out of phase, depending on which way we connect the voltmeter). If we now connect a resistive load across the secondary, this EMF with drive a current around the secondary circuit, in phase with the secondary voltage. Electrical power is therefore being drawn from the secondary and, since we can't break the law of energy conservation, an equal amount of power must also flow into the primary. The current which consequently flows into the primary is in phase with the primary voltage and is called the working current. The total primary current is now the vector sum of the magnetising current –which is always present– and the working current which varies according to our needs.

The secondary load current and the primary working current both necessarily generate their own H-fields in the core. However, this does *not* create additional flux in the core. After all, more flux would mean larger back-EMFs than the voltages already present, creating an impossible paradox. Instead, the two H-fields are always in precise opposition, cancelling each other out, so the core flux density remains the same even if the secondary is a short circuit. In this way the power supplied from the wall is not dissipated in the transformer core but is cleverly converted into magnetic energy, and then immediately back into electrical energy in the load. To reiterate, secondary load current does *not* cause core saturation; we will return to this later.

1.4.1: Phasor Diagrams

For readers who prefer phasor diagrams, suppose we have a transformer with 2000 turns on the primary and 55 turns on the secondary. This is a turns ratio of: 55/2000 = 0.0275. Therefore, whatever voltage we apply to the primary should produce 0.0275 times as much voltage across the secondary, assuming no energy is lost along the way.

Now suppose we apply $230V_{ac}/50Hz$ to the primary and discover the magnetizing current to be $50mA_{ac}$. No current is yet being drawn from the secondary. The magnetizing current I_{mag} lags the primary voltage V_{pri} by 90°, so no power is actually dissipated in the primary, it is only sloshing back and forth every quarter cycle. The magnetizing current is held at this steady value by an opposing EMF, E_{pri}, which is equal-but-opposite to V_{pri}, as shown in fig. 1.7.

An EMF is also generated across the secondary coil according to the turns ratio:

$$E_{sec} = E_{pri}\frac{N_{sec}}{N_{pri}} = 230 \times \frac{55}{2000} = 6.3V$$

This has been drawn in phase with E_{pri} but it could as easily be drawn in phase with V_{pri} if the transformer leads were reversed.

Since the secondary voltage is less than the primary voltage we say it has been **stepped down**, or that we have a step-down transformer. We can of course apply more turns to the secondary than to the primary, in which case the secondary voltage would be **stepped up** instead.

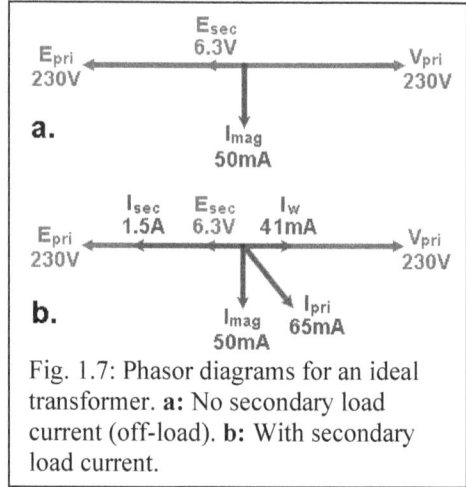

Fig. 1.7: Phasor diagrams for an ideal transformer. **a:** No secondary load current (off-load). **b:** With secondary load current.

Now suppose we draw 1.5A of current from the secondary. This amounts to: $6.3 \times 1.5 = 9.45W$ of power being consumed. An equal amount of power must be supplied to the primary, meaning the working current I_w in the primary must be: $9.45 / 230 = 41mA$ *in phase* with the primary voltage.

However, from fig. 1.7b it can be seen that the *total* primary current I_{pri} is the vector sum of the working current and the magnetizing current:

$$I_{pri} = \sqrt{I_w^2 + I_{mag}^2} = \sqrt{41^2 + 50^2} = 65mA \text{ and lags the applied voltage by:}$$

$$\tan^{-1}(50/41) = 51°$$

From this we see that when voltages are stepped down, currents are stepped up by the same proportion, and vice versa. The power supplied must at least equal the power consumed, and in reality the primary current will be somewhat larger still, owing to various inefficiencies.

Using the previous example, we can also calculate from Ohm's law that the load resistance on the secondary was $6.3V/1.5A = 4.2\Omega$, but the resistance –or more accurately the *impedance*– of the primary winding was $230V/0.065A = 3538\Omega$. We can, therefore, regard the transformer alternatively as an *impedance* transformer. Any impedance placed across one winding is 'reflected' across to all other windings and will appear larger or smaller when 'looking into' those windings.* Voltages are

---

* Valve amplifiers normally use an *output* transformer to step-up the low impedance of a loudspeaker to a higher value that is more compatible with the power valves. The transformer principle is the same, but the quality of manufacture is different.

12

transformed by the turns ratio, but impedances are transformed by the *square* of the turns ratio:

$$Z_{pri} = \left(\frac{V_{pri}}{V_{sec}} \right)^2 Z_{sec}$$

Where:
Z_{pri} = impedance looking into the primary, in ohms;
Z_{sec} = impedance placed across the secondary, in ohms.

In the previous example the secondary resistance appeared from the primary to be:

$$\left(\frac{230V}{6.3V} \right)^2 \times 4.2\Omega = 5598\Omega$$

This is what we expect from the primary voltage divided by the working primary current. However, there was also the unwanted magnetising current caused by the primary inductance, which is effectively in parallel with whatever we see reflected across from the secondary. Knowing the magnetising current was 50mA the primary *inductive* reactance was:

$$Z_L = \frac{V_{pri}}{I_{mag}} = \frac{230}{0.05} = 4600\Omega$$

The total impedance of an inductance in parallel with a resistance is found using vector addition: $1/\sqrt{(1/4600)^2 + (1/5598)^2} = 3554\Omega$. This agrees with the earlier calculation, except for a little rounding error.

1.4.2: Transformer Losses

Practical transformers cannot be one hundred percent efficient. Losses which are due to shortcomings of the core are collectively referred to as the **core losses** or **iron losses**. Losses due to shortcomings of the windings are called the **copper losses** or **I²R** losses. A typical EI power transformer might be around 85% to 90% efficient, and toroidal transformers can do a little better, which is quite good for nineteenth-century technology.

Since the iron core is conductive, EMFs will be generated in the core itself by any flux that does not successfully link with all turns. These core EMFs will drive **eddy currents** around the core, generating heat and wasting energy. To minimise this effect, transformer cores are designed to have high electrical resistance. This may be done by using high-resistance silicon steel, stamped into thin laminations which are naturally insulated from each other by surface oxide, or by making the core hollow and filling it with iron dust.

Energy is further lost in forcing the magnetic dipoles in the core material to change direction with each AC cycle –a sort of friction at the molecular level. This is called **hysteresis**. Hysteresis also results in a core 'memory effect' which can be seen in fig. 1.9 later.

13

However, the greatest loss in a power transformer is usually due to simple resistive heating of the copper windings, hence calling it I^2R loss. Ordinary mains transformers are usually limited to a maximum winding temperature of around 70 to 100°C, above which the winding insulation may be compromised (some have built-in thermal fuses).

A transformer that handles only small currents can use thin wire which therefore takes up little winding space –called the winding window. This means the core itself will be small too. A more powerful transformer must use thicker wire to handle the larger currents, which means a lager winding window is needed, leading to a larger core, even though more iron it isn't strictly needed for magnetic reasons.

Most modern power transformers use very similar materials and construction techniques to one another. Core sizes are mostly standardised, so transformers from different manufacturers but with similar ratings will usually be very similar or even identical in size. With a little experience and by comparison with known transformers, it is usually possible to estimate the power rating of an unknown transformer with reasonable accuracy –see section 1.6.

1.4.3: The BH Curve

Ferromagnetic materials have very non-linear permeability, that is, doubling the magnetic force field H does not necessarily double the flux density B. This can be appreciated by plotting a curve showing how much B is produced as H is varied. This is called the BH curve of the material, and a working knowledge of its characteristics is useful when dealing with transformers and chokes of all kinds.

It is quite easy to view the BH curve of a transformer, using an oscilloscope with XY capability. Fig. 1.8 shows a practical circuit for doing so. The primary current is sampled with a small resistor R_2; this is proportional to the magnetising force H and therefore provides the horizontal (X) deflection. R_1 and C_1 form a simple integrator; the resulting

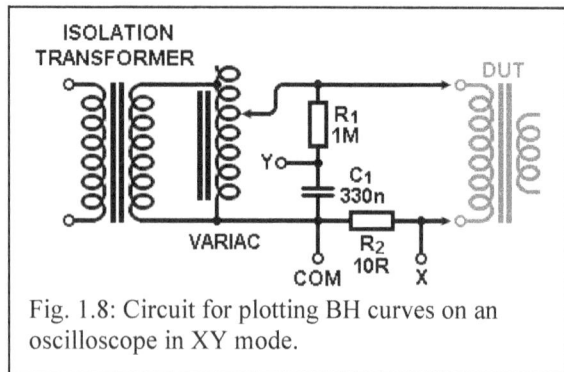

Fig. 1.8: Circuit for plotting BH curves on an oscilloscope in XY mode.

voltage is proportional to the flux density B and provides the vertical (Y) deflection. The variac allows the primary voltage to be adjusted, and the isolation transformer is needed only because the common terminal of an oscilloscope is connected internally to mains earth.

Fig. 1.9 shows the BH curves for a toroidal transformer measured by the author using the circuit of fig. 1.8. Three images have been superimposed, showing how the BH loop grows as the applied voltage is increased from 30V$_{rms}$ up to the nominal wall voltage of 240V$_{rms}$. The small arrowheads show the anticlockwise path taken by the core operating point. As field strength H increases in the positive direction, flux increases too, albeit nonlinearly. But when the

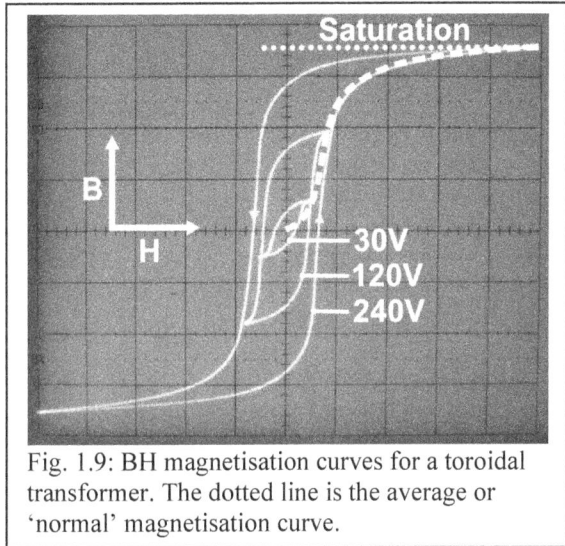

Fig. 1.9: BH magnetisation curves for a toroidal transformer. The dotted line is the average or 'normal' magnetisation curve.

field strength relaxes again the flux does not decay back along the same path. Indeed, even when H falls back to zero there is still some flux remaining in the core, which is called **remanent flux** (if we instantly disconnected the voltage at this moment the core would remain magnetised). The flux only drops back to zero when the force swings somewhat negative. This is the 'memory' phenomenon exhibited by ferromagnetic materials –an effect more properly called **hysteresis**.

Since we must expend energy to grow the flux (by aligning domains), and still more energy to bring it back down to its original level (flipping the domains around again), power is dissipated in the core as heat, thereby contributing to transformer inefficiency. Each AC cycle traces out one complete BH curve or **hysteresis loop**, and the area of the loop is proportional to the power lost. It is perhaps also worth pointing out that the BH curve may change with different applied frequencies, but we are only interested in 50/60Hz here.

From the figure it can be seen how increasing the applied voltage causes the BH loops to expand in length and width until the voltage is large enough to cause saturation. Once saturated, the core is 'full up' with flux, so further increase in H causes practically no increase in B. The loop can no longer grow vertically; the 'tails' stretch out horizontally instead. In this case it is clear that the transformer has been designed to reach saturation with the nominal 240V$_{rms}$ wall voltage applied. This is standard practice for a power transformer since it makes full use of the (relatively expensive) core material. If it had been designed for 120V mains instead then it would still be deliberately engineered to reach saturation –by using half as many primary turns.

The core permeability is equal to the slope of the BH curve:

$$\mu = \frac{dB}{dH} \quad \text{henries per metre} \tag{1.5}$$

This of course varies enormously depending on where you care to measure it on the hysteresis loop. To make things a little simpler it is usual to forget about hysteresis and plot only the path traced out by the tip of each loop, which is known as the **normal magnetisation curve**, shown dashed in fig. 1.9. This is effectively the gross average BH curve for the core material. A transformer designer would then further approximate this curve as a straight line over the region actually being used, and use this to calculate a gross average permeability:

$$\mu = \frac{B}{H}$$

We have arrived back at equation 1.1, the Ohm's law of magnetism, though it took several big approximations to do it.

Can the BH curve tell us anything about inductance? Inductance is, after all, more immediately useful to us as circuit designers than abstractions like magnetic fields. Recall equation 1.3:

$$E = -N\frac{d\Phi}{dt} = -L\frac{di}{dt}$$

Therefore:

$$L = N\frac{d\Phi}{di}$$

To put this into words, inductance is a measure of how much flux is created in the core by a given magnetising current. Now, since the core has a fixed size, flux must be proportional to flux density B. Meanwhile, the magnetising current directly causes –and is therefore proportional to– H. From this we deduce that inductance must be proportional to dB/dH, which

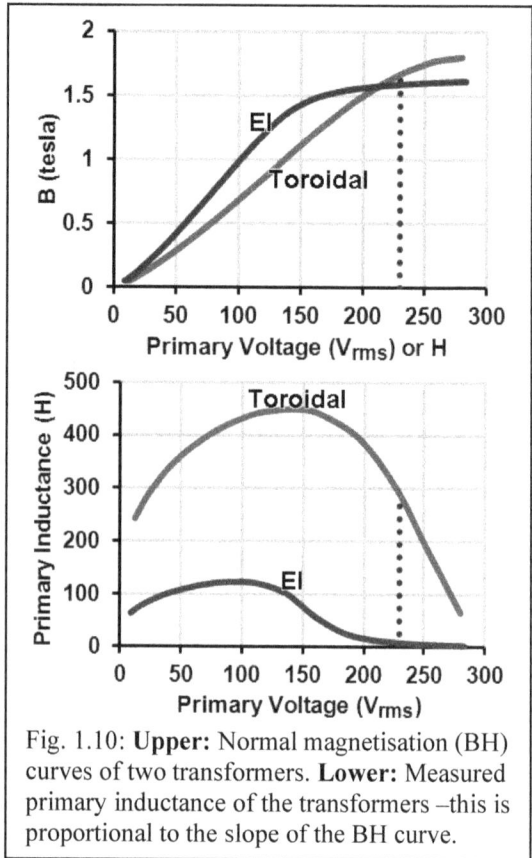

Fig. 1.10: **Upper:** Normal magnetisation (BH) curves of two transformers. **Lower:** Measured primary inductance of the transformers –this is proportional to the slope of the BH curve.

is permeability. A plot of permeability would therefore look exactly the same as a plot of inductance, except for the units on the vertical axis.

This is demonstrated by fig. 1.10 which shows the effective primary inductance of the 50VA toroidal transformer used to produce fig. 1.9, together with a typical 50VA EI transformer for comparison. By integrating these inductance curves, the normal magnetisation BH curves have been derived, as shown in the figure. In other words, the inductance curves correspond to the *slopes* of the BH curves. The vertical flux density scale has been estimated from published information (EI cores tend to saturate around 1.6 tesla whereas toroidal cores typically reach 1.8 tesla, but it is difficult to measure true values).

From the figure we can see that with small applied voltages the inductance starts small because we are operating near the 'bottom bend' of the BH curve. Medium voltages enjoy the highest inductance because we are then operating on the steepest part of the BH curve. Large voltages approach saturation where inductance falls again. This behaviour is a great annoyance when designing audio transformers since they must operate over a wide range of voltages, but power transformers only care about one voltage –the wall voltage– indicated by the dotted line.

In this case the designer of the toroidal transformer has used an unusually large number of turns (about 2350 by the author's calculations) to achieve a very high nominal inductance of 296H. This limits the magnetising current to just 2.5mA at rated wall voltage. By contrast, the designer of the EI transformer has decided to allow a more typical magnetising current of about 80mA, by winding for only 9H of inductance at rated wall voltage. The EI transformer also operates much further into its saturation region. And remember, $230V_{rms}$ has a *peak* value of 325V, so the operating point will periodically sweep well off the right-hand side of the graph.

Note that ordinary inductance meters are likely to give meaningless readings when trying to measure primary inductance, since they apply a test voltage that is too small and a frequency that is too high. A more reliable method is to apply a suitable 50/60Hz voltage to the primary and measure the magnetising current through it, e.g. by monitoring the voltage across a 10Ω resistor. The inductance can then be calculated from:

$$L = \frac{V_{pri}}{2\pi f I_{pri}}$$ See also section 4.3.3.

1.4.4: Saturation

For a given type of core material there is a limit to the amount of flux that it can hold. Once all its magnetic domains have been fully aligned with the applied H-field there is nothing more it can do; no additional flux can be contributed by the core and it is said to have saturated. Any further increase in the field strength can only create extra flux in the air outside, where it is usually unwanted. Beyond saturation the benefits of the core's high permeability are exhausted; we basically default back to the permeability of air. The inductance therefore rapidly drops, leading to increased primary magnetising current which increases I^2R heating. Additionally, the leakage

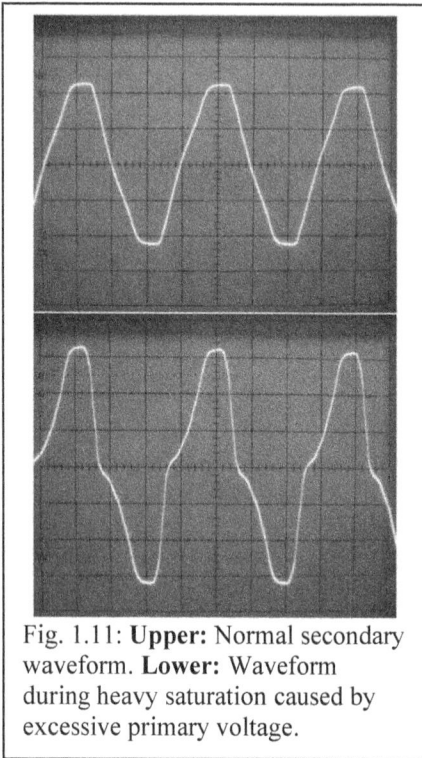

Fig. 1.11: **Upper:** Normal secondary waveform. **Lower:** Waveform during heavy saturation caused by excessive primary voltage.

field in the air induces eddy currents which further heat the core, not to mention the electrical interference it may induce into nearby circuitry. If the heating is sustained then eventually the transformer will burn out. Long term saturation in transformers must therefore be avoided.

Apart from excessive hum, buzz, or overheating, the effect of saturation can also be witnessed on the transformer voltage waveforms. The upper image in fig. 1.11 is an oscillogram of the secondary voltage produced by an unloaded power transformer when the primary voltage is the correct rated value. The clipped wave peaks are already present on the wall supply –the wall voltage is never a clean sinewave– but the waveform is relatively uniform nonetheless.

The lower image shows the waveform when the primary voltage is increased by a further 50%. The peaks of the primary voltage drive the core into saturation, but since the magnetising current is 90° out of phase, it is at the zero-crossings of the voltage waveform where the primary inductance drops. This causes disproportionately more current to flow at the zero crossings of the voltage waveform, which in turn results in a voltage drop across the primary resistance, so the waveform develops a distinctive kink. Since the transformer was unloaded during the test, this is an accurate image of the wave shape of the *current* flowing in the primary, which clearly contains severe harmonics. The difference between the rising and falling slopes is due to the asymmetry of the BH characteristic of the core material. Incidentally, the transformer quickly became hot and buzzed loudly during the test. The upper image actually shows a hint of crossover kink too, indicating that the transformer has been designed to operate on the verge of saturation under normal conditions. This observation can be used as a diagnostic tool when testing transformers (see later).

Load current, it should be remembered, does not cause any increase in core flux, so does not cause saturation. Indeed, the extra working current in the primary will cause the effective primary voltage to drop slightly due to wire resistance, which will actually pull the transformer slightly *away* from saturation. Therefore, if we view the secondary voltage waveform of a power transformer, it may exhibit a crossover kink that shrinks as the load is increased. The exception to this is when a net *DC* current is allowed to flow in any winding, such as when a half-wave rectifier is used. DC current will create DC flux in the core which cannot be cancelled out by an opposing field, since we can only transform AC. Since a standard power transformer normally operates more-or-less at the point of saturation anyway, adding additional DC flux can only push it deeper into saturation. What's worse, direct and alternating magnetising current *combined* result in lower effective permeability than AC alone, which reduces the primary inductances and leads to yet more AC magnetising current. Even a few tens of milliamps of DC (primary referred) can be enough to cause a power transformer to saturate, if it is not specially designed to handle it.

A corollary of this is that lower frequencies also push a transformer more towards saturation. Lower frequencies mean more time for current to flow in a given direction, which means greater flux growth. For a power transformer already operating close to saturation, if we reduce the wall frequency the transformer will be pushed even deeper into saturation, which leads to overheating. A transformer designed for 50Hz mains can be used at 60Hz quite safely, but not the other way around. A transformer manufactured in America must therefore be assumed to be 60Hz only, unless it has a universal primary provided by the manufacturer.

1.5: Transformers and Their Ratings

Fig. 1.12 shows some typical power transformers. Lower-cost devices are usually of the EI type where the core is made from a stack of E- and I-shaped laminations. The coils are wound on a bobbin or former, and the E and I pieces are interleaved to minimise air gaps in the core (see fig. 1.19).

Toroidal (doughnut shaped) transformers, as in fig. 1.12d, are usually a little more expensive. Toroidal cores are made from ceramic with an iron-dust centre, or from a strip of silicon steel coiled up into a spiral. The windings are then applied evenly and continuously around the core with no need for a former.

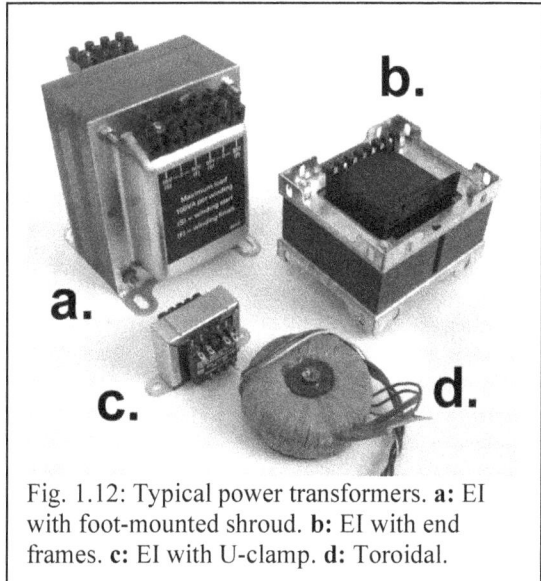

Fig. 1.12: Typical power transformers. **a:** EI with foot-mounted shroud. **b:** EI with end frames. **c:** EI with U-clamp. **d:** Toroidal.

Where core flux has to navigate around sharp corners or cross defects and joints in the core material, it will tend to leak out into the air. EI transformers therefore tend to leak more than toroids, mostly from their corners and out of the middle of the ends of the core. Leakage flux around toroidal transformers occurs mainly around the lead-out wires where the windings cannot be kept perfectly even and parallel. Consequently, toroidal transformers usually should be orientated with the lead-out wires facing away from sensitive circuitry. The leakage field is sometimes suppressed by wrapping a conductive 'belly band' (e.g. copper tape) around the outside of the whole transformer, to act as a shorted turn for the leakage flux. This drains the magnetic leakage field by converting its energy

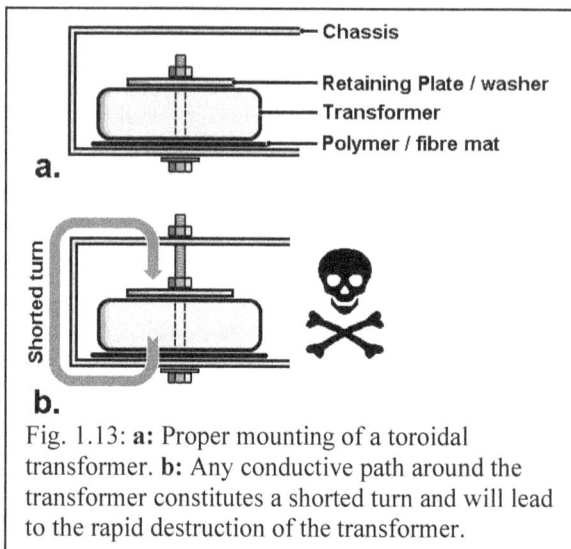

Fig. 1.13: **a:** Proper mounting of a toroidal transformer. **b:** Any conductive path around the transformer constitutes a shorted turn and will lead to the rapid destruction of the transformer.

into a harmless eddy current in the belly band. Note that this is different from the electro*static* screen sometimes wrapped around the bobbin *within* an EI core, which must *not* form a complete loop! As a related note of caution, a toroidal transformer should be mounted as shown in fig. 1.13a. Never let both ends of the central mounting bolt come into electrical contact as in b., as this will create a shorted turn, shortly followed by smoke and swearing.

1.5.1: VA (Power) Rating

The VA rating of a transformer is equal to the full-load RMS secondary voltage multiplied by the permitted full-load RMS secondary current, and is therefore the average power that that can be safely demanded from the transformer. It is given in volt-amps rather than watts because it takes into account the fact that the load may be reactive rather than purely resistive, i.e. the voltage and current may not be in phase. The most familiar example of this is a capacitor-input rectifier. The total power (in VA) that the transformer has to handle is then the magnitude of the real power dissipated in the load, combined with the reactive power flowing back and forth in the capacitance. This is explained in more detail in chapter 2.

If the device has several identical secondaries then the total VA rating is divided equally among them, so the current capacity of each winding is equal to the apportioned VA divided by the voltage of the winding in question. Transformers with several different secondary windings might give the rating of each winding in terms of VA, or as separate RMS voltage and current ratings. But just to confuse matters, power transformers intended especially for valve amps sometimes use a more old-fashioned ratings system. Here, AC RMS figures are quoted for the heater winding/s as usual, since they are assumed not to use a rectifier, but the manufacturer second-guesses the user by quoting the maximum *DC* load current for the high-voltage winding, assuming a standard full-wave rectifier and a typical (but unstated) power factor. This was no doubt useful in the days before pocket calculators and computer simulators, but today it is rather irritating. If the ratings system is in any doubt, contact the manufacturer and remind them what century it is.

1.5.2: Regulation

Some transformer datasheets will quote a 'regulation' figure. This delightful misnomer is a measure of the change in secondary voltage between no load and full load, the difference being caused mainly by the voltage drop across the winding resistance as the load current increases. The regulation is given as a percentage according to:

$$\%_{regulation} = 100 \times \frac{V_{no\ load} - V_{full\ load}}{V_{no\ load}} \qquad (1.6)$$

Fig. 1.14 shows the range of regulation to be expected from ordinary mains transformers. Why this regulation figure became the standard metric rather than the simple ratio of no-load to off-load voltages is anyone's guess. Perhaps it was handy in the days of slide rules, although it is not at all clear how. These days it would really be more helpful if manufacturers simply quoted the winding resistances and the no-load and full-load voltages. In theory, the secondary-referred winding impedance can be calculated from the regulation percentage:

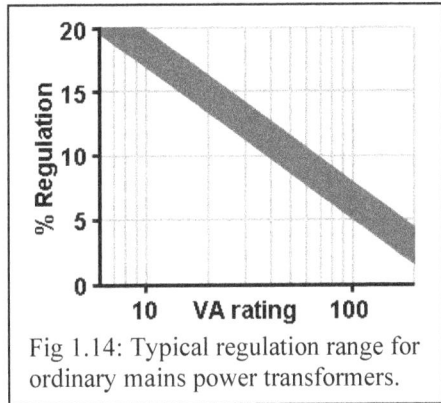

Fig 1.14: Typical regulation range for ordinary mains power transformers.

$$Z_{sec} = \frac{\%_{regulation} \times V_{full\ load}}{(100 - \%_{regulation}) \times I_{full\ load}} \qquad (1.7)$$

But since regulation percentages is always quoted in suspiciously round numbers, any calculations using them must be assumed to be approximate only.

1.5.3: Phasing

When windings are connected in series it is important to get their phasing correct so the voltages do what we expect them to. Connecting them with one phasing will causes the voltages to add together, whereas the opposite phasing will cause them to subtract or cancel out.

When windings are connected in parallel they *must* be identical and in phase. Getting this wrong will effectively short-circuit the transformer with disastrous results. Windings of different voltages must never be connected in parallel.

The relative phasings of windings is usually indicated by dots on the transformer diagram. Each dot indicates the end of the winding that is in phase with all the other dots. In other words, if the voltage happens to be increasing at one of the dots, then it is increasing at each of the other dots too.

Many transformers have dual primary windings which can be connected in series for 230/240V mains or in parallel for 115/120V mains, as shown in fig. 1.15. Many modern, off-the-shelf transformers also have dual (identical) secondary windings which can be left entirely separate, or be connected in series to obtain double the voltage or a centre-tapped supply, or be connected in parallel to obtain a single voltage with maximum current. Again, it is essential to observe the correct phasing when doing this –section 1.6.2.

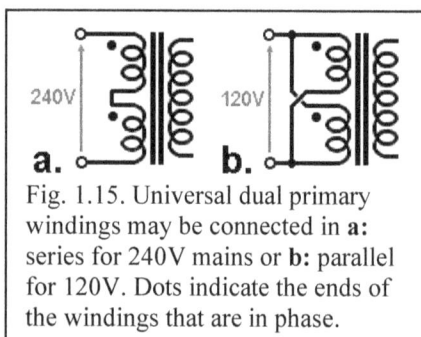

Fig. 1.15. Universal dual primary windings may be connected in **a:** series for 240V mains or **b:** parallel for 120V. Dots indicate the ends of the windings that are in phase.

1.6: Testing Power Transformers

Most hobbyists will use second-hand transformers from time to time. Over the years the self-respecting electronics enthusiast will probably accumulate an impressive collection of transformers rescued from discarded equipment and failed projects. Sometimes we will have little or no official information on the voltages and power capability of these transformers, but these things can be estimated to a reasonable degree using the following methods. It should go without saying, however, that we should be conservative in our design when using unknown transformers. Putting pre-used transformers into light retirement service should ensure reliable performance almost indefinitely. Most of the following tests are performed 'live' using mains voltage, and higher voltages may also exist on some windings, so take care to avoid electric shock.

1.6.1: Voltages

The DC resistance between each transformer connection can be measured with a multimeter, and a diagram or 'map' of the transformer built up this way. The resistance gives a rough indication of the winding ratio –higher resistance implies more turns and therefore higher voltage. From this we can usually determine which is the primary and which are the secondaries. This may also be suggested by the thickness of the wire if it is visible, e.g. step-down transformers will have thicker wire on the secondaries. With valve transformers the primary and the high-voltage secondary may be of similar resistance, but often the primary connections will be on one side of the transformer and the secondaries on the other, or else the lead-out wires may have a familiar colour scheme.

Once the primary has been identified we can apply mains voltage, ideally with a conservatively rated fuse and variac so the voltage can be turned up cautiously.

When measuring the secondary voltages off-load, remember that they will be around 5 to 20% higher than those measured under full load. It is the full-load voltage that would be advertised by the manufacturer. In other words, if you suspect one winding is *rated* for 6.3V then do not be surprised if it measures around 7V off load. Similarly, if a winding measures 274.2V off load then it is probably a 250V-rated winding, since high voltages are usually sold in round numbers.

Sometimes we may be uncertain of the correct primary voltage because the transformer is foreign or antique. One method of determining the correct primary voltage is to use a variac to increase the mains voltage gradually while viewing the secondary voltage on an oscilloscope (with no other load attached). When the secondary voltage starts to show signs of a crossover kink, like the upper image in fig. 1.11, then the transformer is approaching saturation and the mains voltage is likely to be correct.

Alternatively, if we are not sure of the primary but we *are* sure of one of the secondary voltages then we can use another transformer to drive the *secondary* with the correct voltage, i.e. use the transformer backwards, and hence measure the voltage appearing on the primary.

1.6.2: Phase

Knowing the relative phases of the windings is important if we wish to connect them in series or parallel. Phase can be very easily determined with a dual-channel oscilloscope of course, by viewing the waveforms on the different windings, but it can also be determined with a voltmeter and a little care. Draw a diagram of the transformer windings, with known voltages. Connect two secondary windings in *series*, and apply the primary voltage (no other load attached). If the voltage measured across the series-connected pair is the sum of the individual voltages then the two free ends must be out of phase with one another as in fig. 1.17a. Conversely, if the voltages have subtracted

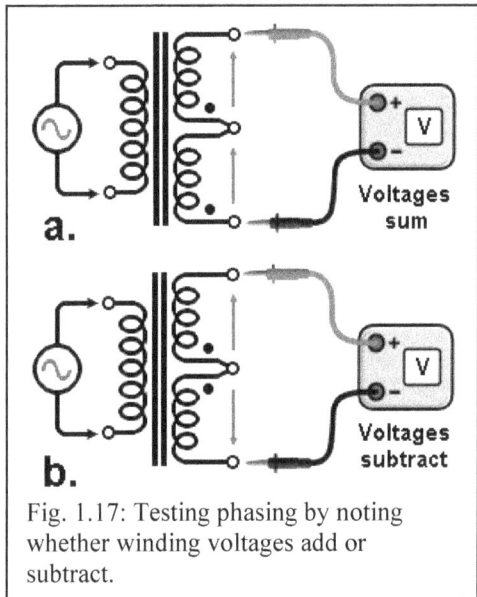

Fig. 1.17: Testing phasing by noting whether winding voltages add or subtract.

(resulting in zero volts if they are identical windings) then the two free ends must be in phase as in b. The ends can now be marked with dots on the diagram to indicate the relative phases.

If the transformer has dual primary windings and we need to know their phase then similar methods can be used. The first is simply to connect the transformer backwards as mentioned earlier, by driving one of the secondaries, and then do the phasing test described above but on the primaries.

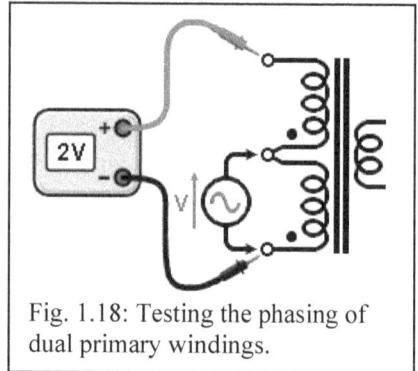

Fig. 1.18: Testing the phasing of dual primary windings.

An alternative method is to connect the two windings in series and apply the appropriate voltage (e.g. 120V) to *one* of them, and measure the total voltage across the pair, as in fig. 1.18. If the two joined ends are out of phase with each other then the measured voltage will be the double the applied voltage, otherwise they will subtract and we will measure close to zero volts, assuming the two windings are identical. The diagram can then be marked with dots accordingly. This method can also be used to determine the phase of a secondary winding relative to the primary.

1.6.3: Power and Current Capability

The overall power handling (VA rating) of a transformer can be reasonably estimated from the size of its core. Most EI transformers are made with standardised 'wasteless' laminations where the 'I' pieces are stamped out from the 'E' pieces, leaving no wasted material, as illustrated in fig. 1.19. The outer 'legs' and winding windows are all one unit wide, and the

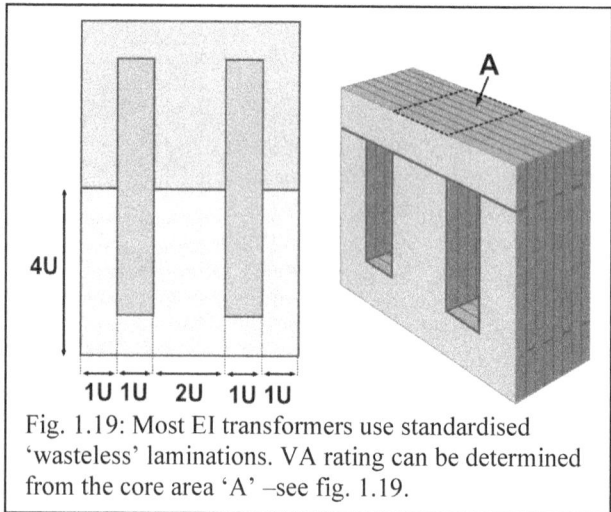

Fig. 1.19: Most EI transformers use standardised 'wasteless' laminations. VA rating can be determined from the core area 'A' –see fig. 1.19.

centre leg is two units wide. We can therefore determine the width of the centre leg of an existing transformer core from its overall length divided by three. By multiplying this by the stack thickness we find the cross-sectional area of the centre leg or 'core' proper. From this the VA rating can be determined:

$VA = (A/1.27)^2$ (50/60Hz) or

$VA = (A/1.15)^2$ (60Hz only)

Where A is in *square centimetres*. These formulas are plotted in fig. 1.20. Some vintage transformers or ones of special design such as low-profile transformers, may use laminations with different dimension ratios (this is usually obvious by inspection) and so may not obey these formulas. For toroidal transformers it is easiest to find another, known transformer with the same diameter and height and to assume the same VA rating applies. You don't necessarily need one to hand, you can simply look up the dimensions in vendors' catalogues.

If the transformer has several identical secondaries then it is reasonable to assume the total power was intended to be shared equally among them, so the current capacity of each winding will be the VA apportioned to it, divided by

Fig. 1.20: Relationship between power rating and core area in transformers using wasteless EI laminations –refer to fig. 1.19.

the rated voltage of that winding. The thickness of the secondary wire, if visible, will serve as a reality check. If there are several different secondaries then some common-sense-based guesswork is needed. If the transformer was taken from a known piece of equipment then it may be possible to estimate the capacity of each winding by looking at what it powered in the original circuit. Again, use wire gauge as a guide. It should also be noted that if one winding is run at less than its published rating then other windings can be run in excess of their published ratings *within reason*[*], provided the total VA rating of the core is not exceeded, so we may guess at several possible loading arrangements.

Once we have estimated the capacity of the windings we have little choice but to crash test it. As a first approximation, when a transformer is fully loaded so it is handling its maximum safe power level it will (eventually) feel somewhat hot to the touch, but it should not be so hot that you cannot keep your hand on it permanently.[†] This simple observation can be combined with a more reliable test of the internal core temperature, as follows.

Measure the resistance of the primary winding at room temperature. Then load every winding to its presumed full capacity and run the transformer for some time, say thirty minutes, unless it shows signs of distress, of course. Ordinary incandescent light bulbs are quite useful as power loads (the kind sold for illuminating domestic ovens are often a convenient wattage). Then switch off and immediately measure the

---

[*] Fifty per cent excess loading is probably reasonable.
[†] Some transformers are designed to run noticeably hotter than this and use special materials to do so. This is unusual, however, so we should not assume such operation from an unknown transformer.

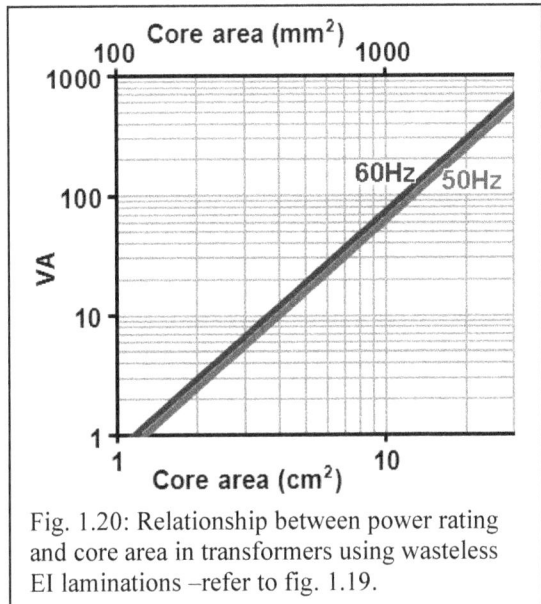

resistance of the primary before it cools down. The temperature of a wire can be closely estimated from:

$$T = \frac{R - R_0}{\alpha R_0} + T_0 \qquad \text{celsius} \qquad (1.8)$$

Where:
T = hot wire temperature
T_0 = initial temperature.
R = hot wire resistance.
R_0 = initial wire resistance.
α = temperature coefficient of resistance (0.0039/°C for copper)

From the above formula we can deduce that a 20% rise in resistance indicates a 50°C increase in copper temperature. If we take room temperature (T_0) to be 20°C then this would amount to a total core temperature of 20 + 50 = 70°C, which is the maximum safe limit for most ordinary transformers. If the resistance increases by more than 20% then the transformer is probably overloaded, but hopefully you will not let the test go that far before noticing an excessive surface temperature.

Chapter 2: Rectification

Rectification is the process of forcing alternating current to become direct current, that is, AC to DC conversion. This is the second ingredient in a linear power supply, after the power transformer. The rectifier circuit itself is non-linear, which means purely mathematical treatment can become extremely complicated. Section 2.2 does cover some of the simpler aspects of rectifier mathematics, but mostly we will concern ourselves with practical design methods.

Valve heaters and a few other circuit elements may be operated from pure AC, but most electronic circuits need DC. In many valve circuits the only DC supply we need is the high voltage for the anodes, usually called the HT (High Tension) in British textbooks, or B+ in American parlance. Most of this chapter will use the HT/B+ supply as the main example for familiarity, but the basic theory of rectification is general.

2.1: Essentials of Rectification

It is the job of the rectifier to take an AC supply, e.g. from the power transformer, and convert it into DC. There are several circuit configurations we might use.

2.1.1: Half-Wave Rectification

The simplest rectifier is the half-wave rectifier, shown in fig. 2.1. When the transformer voltage is large enough to overcome the forward voltage of the diode (about 0.7V for a silicon power diode) current is able to flow as indicated by the arrow. But when the voltage is reversed, the diode blocks current flow. The voltage appearing across the load is simply the positive halves of the input sine wave, hence the name of the circuit. The current is now DC in the sense that it is unidirectional, but it is not 'smooth' –we will return to this point later.

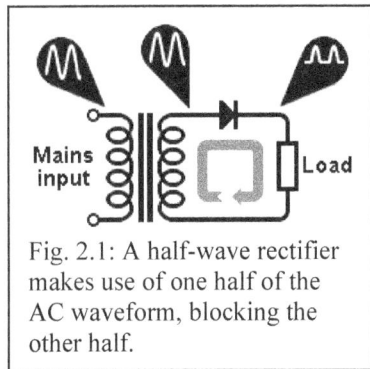

Fig. 2.1: A half-wave rectifier makes use of one half of the AC waveform, blocking the other half.

A significant problem with this rectifier is that the average current flowing in the transformer is also unidirectional, leading to some amount of net DC flux in the transformer core. Modern power transformers are normally designed to run strictly on pure AC; even a little DC flux can quickly push them into severe saturation and overheating. Half-wave rectifiers were commonly used in valve appliances years ago, but the power transformers were specially designed to cope with this DC flux. Do not fall into the trap of copying an old circuit without this proviso in mind. In a modern setting, half-wave rectifiers are only safe for very low-current supplies, or best avoided altogether.

2.1.2: Full-Wave Rectification

A better power supply can be built by taking advantage of both halves of the available AC voltage. One way to do this is to use a **bridge rectifier**. Four diodes are arranged so that alternating current in the transformer is directed into the load in only one direction, as illustrated in fig. 2.2. With each half of the AC cycle one pair of diodes is 'on' while the other pair is reverse biased or 'off', then the situation flips. Since current flows alternately and equally in both directions through the transformer, it does not suffer any net DC magnetisation, so saturation is avoided.

A bridge rectifier can be built from discrete diodes or bought as a monolithic package. The larger monolithic packages may include a hole that allows them to be bolted directly to the chassis for heat sinking. This can be handy for vintage-style point-to-point construction too, even if you don't necessarily need the heat sinking.

Another way to accomplish full-wave rectification is to use two identical but out-of-phase windings on the same power transformer, or a single winding with a centre-tap, which amounts to the same thing. Each winding is connected to its own half-wave rectifier that feeds the same shared load. The two half-wave rectifiers take turns conducting, as shown in fig. 2.3. Although current in each half-winding flows in one direction, it flows in the opposite direction to its partner, since they are out of phase. The two opposing magnetic fields therefore cancel out, which avoids any saturation problems. This is known as a **two-phase rectifier** or **two-diode full-wave rectifier**.

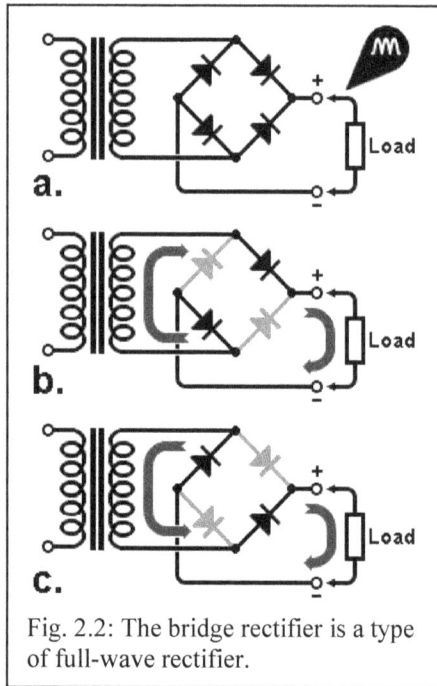

Fig. 2.2: The bridge rectifier is a type of full-wave rectifier.

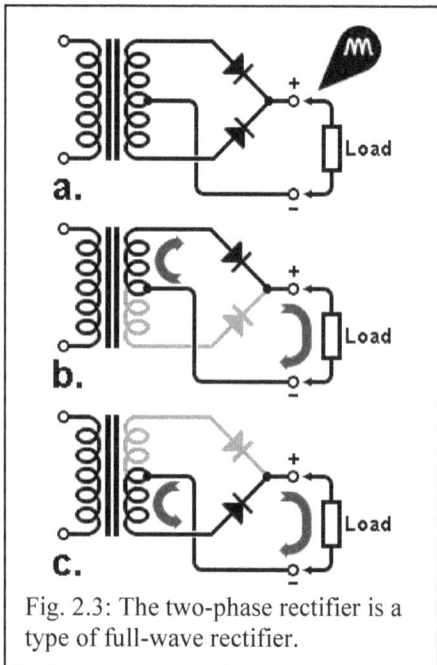

Fig. 2.3: The two-phase rectifier is a type of full-wave rectifier.

The two-phase rectifier is very common in valve equipment, for historical reasons. Valve rectifiers were expensive and needed occasional replacement, so using only two diodes economised on this. Also, the two-phase rectifier only suffers one diode drop per half-cycle, and valve diodes drop a lot of voltage which would make a bridge rectifier very lossy indeed. Even the drop across a silicon diode becomes a hindrance for low-voltage power supplies, so modern low-voltage appliances often use two-phase rectifiers for this reason. For anything more than about twelve volts, however, solid-state diode drop becomes less of a problem, making the bridge rectifier more attractive since it economises on transformer cost.

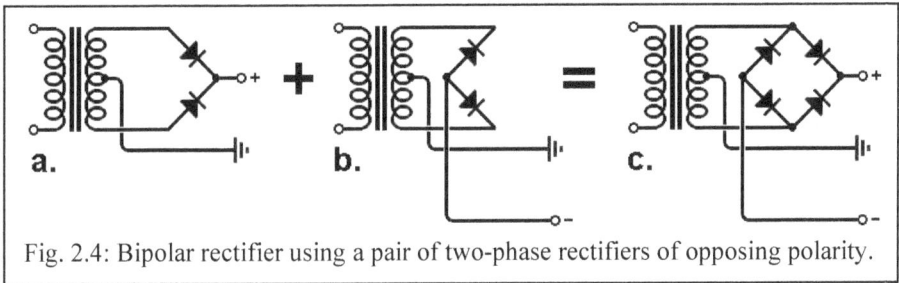

Fig. 2.4: Bipolar rectifier using a pair of two-phase rectifiers of opposing polarity.

By combining *two* two-phase rectifiers with opposing polarity we can obtain positive and negative voltages, i.e. a symmetrical or bipolar supply, as in fig. 2.4. This arrangement is commonly found in modern solid-state audio electronics. Although this circuit looks at first glance like a bridge rectifier –and is often built using a monolithic bridge-rectifier package– it is important to remember that it is really a pair of two-phase rectifiers. This distinction has important bearings on the component ratings, which differ from those of the ordinary bridge rectifier –see later.

2.1.3: The Reservoir Capacitor

The rectifier turns the alternating transformer waveform into a unidirectional pulsating waveform, but we usually need a much smoother DC supply. The most common way to do this is to connect a large capacitance across the output of the rectifier, as shown in fig. 2.5. Ideally, the rectifier dumps current into this capacitor, charging it up to the peak voltage which will be $\sqrt{2} \times V_{rms}$ (ignoring losses). Once charged, the capacitor holds a huge amount of stored energy which the load can draw upon until it is topped up again during the next rectified half cycle. In other words, it is a 'reservoir' of energy, so is called the reservoir capacitor.

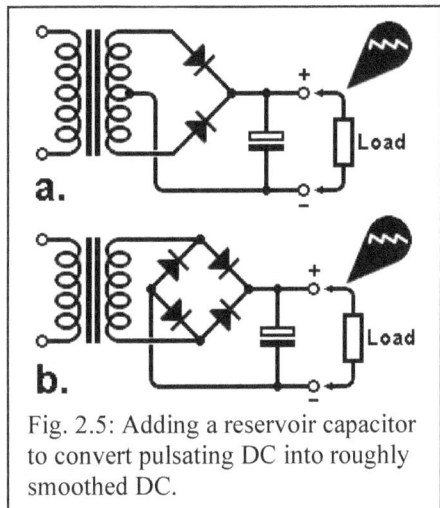

Fig. 2.5: Adding a reservoir capacitor to convert pulsating DC into roughly smoothed DC.

If we do not drain very much water from a reservoir then the water level will hardly change. Similarly, if the reservoir of energy held by the capacitor is very great compared with the amount of energy drained off by the load, the voltage across it will hardly change; we will have a fairly clean DC voltage. However, it may take an inordinately large reservoir capacitance to get *acceptably* clean DC, so further stages of filtering or smoothing may be necessary too, as illustrated by figure 2.6. Smoothing is discussed in chapter 4.

The power transformer, rectifier, and reservoir capacitor, form a holy trinity. They should be thought of as one unit performing the basic task of converting AC into roughly smoothed DC; think of them as taking the place of a somewhat noisy battery. Wherever possible they should be laid out in a physically compact

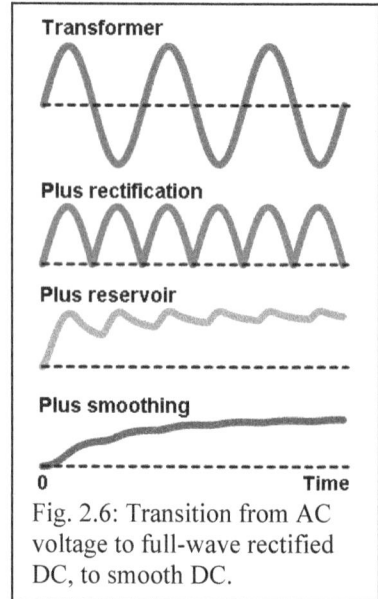

Fig. 2.6: Transition from AC voltage to full-wave rectified DC, to smooth DC.

way to minimise loop area and therefore electromagnetic emission. Once we have this fundamental circuit block taken care of then we are free to attach the rest of the power supply/amplifier as required. Ideally these 'onward' connections should be made right on the reservoir capacitor's terminals themselves.

2.2: Estimating Idealised Rectifier Performance

No textbook about power supplies would be complete without some fist-principle mathematical treatment of rectifiers. However, let us say right away that this is of minimal use for practical design work because real-world power supplies are not ideal and therefore do not conform to the ideal, manageable equations. The maths is presented here more out of obligation, and because it may be interesting to discover later just how much the real world departs from the ideal. Readers who are not mathematically inclined may skip this section without feeling too guilty. Section 2.3 covers how to predict *realistic* power supply behaviour both graphically and by rule of thumb.

2.2.1: Ripple Voltage

In a full-wave rectifier the reservoir capacitor is charged up once per half cycle, and it partially discharges in the interim periods as the load drains current from it. The DC voltage across the capacitor is therefore not perfectly smooth but ramps up and down or 'ripples'. This roughly sawtooth waveform which rides on top of the mean average voltage is therefore called the **ripple voltage**, as illustrated in fig. 2.7. This shows how the capacitor voltage tracks the rectified sine waveform as it charges up to the peak voltage, beyond which the diodes become reverse biased as the transformer voltage drops away, leaving the capacitor to discharge slowly into the load until the next pulse comes along. Because the wave is sawtooth it will contain many harmonics of twice the wall frequency, which is why power supply noise may sound more buzzy than pure AC hum.

Knowing the capacitor is initially charged up to the peak transformer voltage, we can calculate the voltage remaining

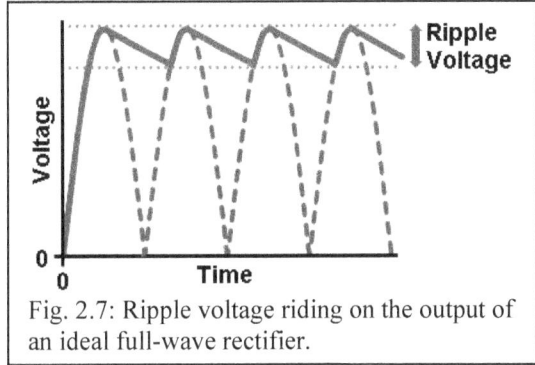

Fig. 2.7: Ripple voltage riding on the output of an ideal full-wave rectifier.

across it at any time during the discharge period, using the standard capacitor discharge formula:

$$v_c = V_0 \cdot e^{-T/CR} \qquad \text{volts} \qquad\qquad (2.1)$$

Where:
V_0 = initial voltage, i.e. peak transformer voltage;
t = discharge time;
C = reservoir capacitance;
R = load resistance.

If we make the above formula equal to $V_{peak} - V_{ripple}$ then we can rearrange to:

$$V_{ripple} = V_{peak} - V_{peak} e^{-T/RC}$$

This equation appears simple enough, except that one thing is unknown: the discharge time. From fig. 2.7 it is obviously slightly less than the time period of the rectified waveform, but exactly how much less is difficult to determine. However, we might observe that the smaller the ripple voltage, the closer this period gets to being equal to the rectified time period. We will therefore make a unilateral decision and say it *is* equal. This is true provided the ripple is a fairly small percentage of the total voltage, which is the case for most power supplies.

But we can go further. Having already made the concession that the ripple voltage will always be a small fraction of the total, we can therefore pretend the exponential discharge curve is actually linear, i.e. a straight line. Mathematicians would say the same thing by pointing out that from the Taylor series $e^x \approx 1+x$ when $x \ll 1$. This leads to:

$$e^{-t/RC} \approx 1 - \frac{t}{RC}$$

From which:

$$V_{ripple} = V_{peak} - V_{peak}.e^{-t/RC} \approx V_{peak} - V_{peak}\left(1 - \frac{t}{RC}\right)$$

$$V_{ripple} = V_{peak} \frac{t}{RC}$$

31

The time period of the pulses is $\frac{1}{2f}$ where f is the wall frequency. Substituting this in gives:

$$V_{ripple} = \frac{V_{peak}}{2fRC}$$

Or we can use Ohm's law to note that $V_{peak}/R = I_{dc}$, giving:

$$V_{ripple} = \frac{I_{dc}}{fC} \text{ (for half-wave rectification).}$$

$$V_{ripple} = \frac{I_{dc}}{2fC} \text{ (for full-wave rectification).}$$

Where:
V_{ripple} = Maximum peak-to-peak ripple voltage;
I_{dc} = DC load current in amps;
f = mains wall frequency in hertz;
C = reservoir capacitance in farads.

We can check this answer if we recall from our school physics lessons[*] that the charge drained off a capacitance is equal to Q = It, where I is the average load current and t is the time it flows for, which is the time between refills. The rectifier must supply an equal amount of charge during the period when the voltage is ramping back up from valley to peak –which is the ripple voltage– and we also remember from school that Q = CV. Equating the two we have CV = It, so the peak-to-peak ripple voltage is therefore:

$$V_{pp} = \frac{I_{dc}t}{C} \tag{2.2}$$

And since t = $\frac{1}{2f}$ we can substitute to get:

$$V_{ripple} = \frac{I_{dc}}{2fC} \quad \text{or} \quad C = \frac{I_{dc}}{2fV_{ripple}} \tag{2.3}$$

This confirms the earlier answer. With this formula we can select a reservoir capacitance large enough to achieve the desired ripple voltage, given an expected load current.

In the real world, ripple voltage will be smaller than predicted by this method (often about half), because the capacitor does not discharge linearly or for as long as we

---

[*] If you're anything like me you will remember nothing from school and will have to look it up.

pretended, and also because the source impedance of the transformer provides an additional filtering effect. Equation (2.3) is therefore a conservative estimate, which is a good thing when you consider the poor tolerance of capacitors.

A small point worth mentioning is that any mismatch between the two alternating current-paths in a full-wave rectifier will lead to some wall-frequency ripple on the output, rather than the strictly *twice* wall frequency (plus harmonics) predicted by theory alone. This is more noticeable on a two-phase supply since the two halves of the transformer winding normally have different DC resistances, even though they have the same number of turns (if you're wondering why, think about the length of wire needed on the bobbin as you wind it, turn upon turn). Mismatch between the two diodes inside a valve rectifier may further exacerbate this effect. This is not a problem as such, especially with modern smoothing capacitors, but it explains why vintage amps sometimes produce 50/60Hz hum even when other causes have been eliminated.

2.2.2: Ripple Current

From fig. 2.7 it is clear that the reservoir capacitor spends most of its time discharging, and has only a short time in which to recharge. But the average energy supplied to the capacitor must be equal to the average energy drained off by the load, if the average voltage across it is to remain constant (assuming nothing is lost along the way). In other words, whatever charge we draw from the capacitor while the rectifier is off, equal charge must be re-supplied when the rectifier is on, but there is less time available to provide it.

The rectifier therefore has no option but to dump charge into the capacitor in short but large pulses of current, known as ripple current. This is shown figuratively in fig. 2.8. This perhaps shows visually why the holy-rectifier-trinity should be kept in close physical formation, for ripple current does not make pleasant audio.

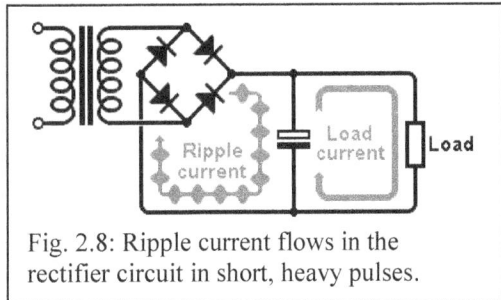

Fig. 2.8: Ripple current flows in the rectifier circuit in short, heavy pulses.

The concept of ripple current is so vital to the understanding of linear power supplies that it is worth saying twice: The current flowing around the transformer and rectifier, into the reservoir capacitor, flows in large, ugly, periodic pulses, while the current flowing out of the reservoir into the load is fairly smooth.

Fig. 2.9 shows the waveforms expected in an ideal full-wave rectifier. Note how the ripple current peaks are many times greater than the load current I_{dc}. The first pulse is greater still, since it must charge up the capacitor from zero volts immediately after switch on, which is called **inrush current**.

If we increase the reservoir capacitance then we increase the reservoir of energy. This means that for a given amount of current drained out of it, the voltage across it won't fall as much, meaning the rectifier has *even less* time in which to source the same amount of energy. So although more capacitance means less ripple *voltage*, i.e. cleaner DC, it also means more ripple *current*, and this puts greater strain on the rectifier and transformer, as well as generating a noisier magnetic field.

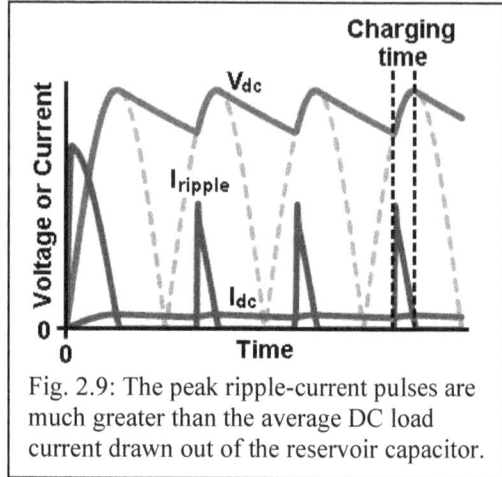

Fig. 2.9: The peak ripple-current pulses are much greater than the average DC load current drawn out of the reservoir capacitor.

A simple way to estimate the ripple current in an ideal rectifier is illustrated in fig. 2.10 (a half-wave rectifier is used here for simplicity). It plots the AC transformer voltage v_{ac}, and the DC voltage across the reservoir capacitor v_{dc}. The larger cosine-wave represents the current which would flow in the capacitor *if no rectifier were used*, that is, if the transformer were connected to nothing but the capacitor. The current leads the voltage by 90° as it would for any capacitor. In a rectifier circuit this phase difference will actually be somewhat less than 90°, but provided the reactance of the capacitor is small compared to the load resistance –which is true for any normal design– the error is very small.

The maximum possible value the current could reach, I_{max}, is equal to the peak voltage divided by the reactance of the capacitor, which comes out as:

$$I_{max} = 2\pi fCV_{peak}$$

Looking at the v_{ac} and v_{dc} plots we see that when they are equal at the 'valley' of the ripple voltage, the diode switches on and the current immediately reaches a peak value I_{pk}.

Fig. 2.10: Voltage and current waveforms for an idealised half-wave rectifier. Current flows in the reservoir during the shaded areas.

The diode then remains on until v_{ac} reaches its peak value. The charging pulse therefore occupies the shaded areas, each of which lasts for a period θ, which can alternatively be called the **conduction angle** in degrees or radians.

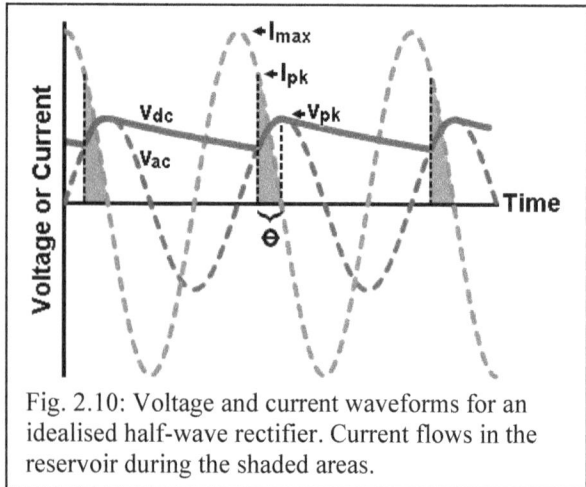

34

Rectification

From the figure it is clear that:

$$I_{peak} = I_{max} \sin \theta$$

And the valley occurs at:

$$v_c = V_{peak} \cos \theta \quad \rightarrow \quad \cos \theta = \frac{v_c}{V_{peak}}$$

But it was explained earlier that during the discharge period:

$$v_c = V_{peak}.e^{-t/RC} \approx V_{peak} \left(1 - \frac{t}{RC} \right)$$

Substituting this into the previous expression:

$$\cos \theta = 1 - \frac{t}{RC}$$

Now, if you look a trigonometry textbook you will find the following identity, $\sin \theta = \sqrt{1 - \cos^2 \theta}$, with which we can say:

$$I_{peak} = I_{max} \sqrt{1 - \left(1 - \frac{1}{fCR} \right)^2} \quad \text{(for half-wave rectification)}$$

$$I_{peak} = I_{max} \sqrt{1 - \left(1 - \frac{1}{2fCR} \right)^2} \quad \text{(for full-wave rectification)}$$

Where:
I_{peak} = peak ripple current in the reservoir capacitor;
$I_{max} = 2\pi fCV_{peak}$;
f = mains wall frequency in hertz;
C = reservoir capacitance in farads;
R = load resistance in ohms, which can be approximated as V_{peak}/I_{dc}.

The ripple current in the *rectifier and transformer* is the sum of the capacitor current, which we have just calculated, and the DC load current.

What these equations imply is that even with a mediocre 20% ripple voltage, the peak current ought to be about ten times greater than the DC load current, increasing still further if we aim for very low percentage ripple voltage. Ripple current is a powerful source of hum and buzz, as any nearby circuit loops will pick up its ugly magnetic field. Furthermore, any quiet circuits inadvertently connected to the 'dirty' wiring before the reservoir capacitor could end up with ripple current leaking into them, which is part of the reason for keeping the holy-rectifier-trinity in tight physical formation; it minimises loop area and discourages you from making wayward connections to the rest of the circuit. Nevertheless, if the equations above were a true reflection of reality we would have some real design difficulties, since we could have ripple currents of tens of amps pulsating noisily around even a modest little power supply. Fortunately, the real world is more forgiving (for a change), mainly due to the internal resistance of the transformer which we have so far ignored.

2.3: Estimating Practical Rectifier Performance

Having looked briefly at the performance of ideal –and therefore fictional– rectifiers we will move on to practical, real-world circuits. These have parasitic elements like diode drop and capacitor ESR, but the dominant hidden variable –which is universally ignored in standard electronics textbooks– is the winding resistance of the power transformer. This nudges the performance away from a pure capacitor-input rectifier and a little towards an RC filter. This causes both the ripple voltage and ripple current to be less than the ideal equations predict, which is lucky for us. On the other hand, it causes the average DC voltage to be lower too, and it will vary more with load current (power supply sag), which is not so lucky for us.

2.3.1: Average DC Output Voltage

One very important feature of the power supply is, of course, the average DC voltage produced. As designers we inevitably ask: what transformer voltage do we need, to obtain a certain DC voltage? Or conversely, given a transformer, what DC voltage will it provide? The first-principle answer is to say the reservoir capacitor will be charged up to the peak AC voltage of $\sqrt{2} \times V_{rms}$, as shown in fig. 2.7 earlier. For example, a $12V_{rms}$ transformer ought to give us $\sqrt{2} \times 12 = 17V_{dc}$. Most textbooks leave it at that. But for various reasons this simple approach is really quite inadequate for predicting real world behaviour.

First, voltage is lost across the rectifier diodes. For silicon diodes this is assumed to average around 0.7 to 1V for ordinary power diodes, or 0.3 to 0.5V for Schottky diodes. For valve diodes it will vary depending on the peak ripple current, which in turn depends on load current and reservoir capacitance.

Second, there is the discharging of the reservoir capacitor between conduction cycles. Using $\sqrt{2} \times V_{rms}$ only tells us the *maximum* DC voltage, not the *mean* average voltage as measured by an ordinary voltmeter –which is the figure we are normally interested in. This will vary depending on the reservoir capacitance and average load current.

Third, there is the voltage drop across the transformer's own source resistance (transformer regulation) which, again, varies depending on the peak ripple current. A so-called 12V transformer will only deliver a $12V_{rms}$ sine wave when *fully loaded* with a *resistive* load. But a reservoir capacitor is not resistive, and how often do we load a transformer to its very limit?

The only time we will get the $V_{dc} = \sqrt{2} \times V_{rms}$ figure quoted in the ordinary textbooks is under no-load conditions, but transformer manufacturers almost never say what the no-load voltages or source resistances are! And even if we knew them, this is not a simple potential divider problem since the rectifier/reservoir form a nonlinear circuit, which makes an analytical solution quite intractable (though not impossible[1]). A computer simulation can solve the problem, but let us explore a less abstracted route; it will lead to an intuitive understanding of how to estimate voltages without resorting to calculating machines.

What information do we really *need* to predict power supply performance with accuracy? Quite simply:
- Primary winding resistance;
- Secondary winding resistance;
- No-load primary voltage, i.e. the wall voltage;
- No-load secondary voltage.

If we already have the transformer in hand then we can easily measure the winding resistances, then connect it to the wall supply and measure the off-load voltages. To simplify things further we can refer the primary resistance to the secondary to find the total source resistance of the transformer, R_s, from the point of view of the rectifier. This is equal to the secondary resistance plus the resistance reflected across from the primary by the square of the voltage ratio (turns ratio):

$$R_s = R_{sec} + R_{pri} \times \left(\frac{V_{sec}}{V_{pri}} \right)^2 \qquad (2.4)$$

Here the voltages are the *no-load* voltages, remember.

If we don't have the transformer handy then the most we can hope for is that the manufacturer will provide the regulation percentage –usually vaguely. With this we can roughly estimate what the no-load voltage and secondary-referred resistance are. As covered in section 1.5.2, the percentage regulation is given by:

$$\%_{regulation} = 100 \times \frac{V_{no\,load} - V_{full\,load}}{V_{no\,load}}$$

Using this to find the transformer's secondary-referred resistance gives:

$$R_s = \frac{\%_{regulation} \times V_{full\,load}}{(100 - \%_{regulation}) \times I_{full\,load}}$$

And the no-load secondary voltage:

$$V_{no\,load} = \frac{100 \times V_{full\,load}}{100 - \%_{regulation}}$$

---

[1] Schade, O. H. (1943). Analysis of Rectifier Operation. *Proc. I.R.E*, (July), pp.341-61.

Once the effective source resistance of the transformer has been found, we can add any other resistance that happens to be in the charging path of the capacitor, whether before or after the rectifier diodes. This includes the internal resistance of a valve rectifier, for example. This total source resistance we will call R_s.

To predict the DC output voltage we also need to know how much current will be demanded by the load. With computer simulation we could go as far as to model the whole amplifier, but for more immediate gratification it is enough to represent the load as a simple resistance. In other words, simply estimate the DC voltage ($\sqrt{2} \times V_{rms}$ is close enough to start with) then divide by the expected load current to find the effective load resistance, which we will call R_l.

We now have enough information to create a very simple model of the power supply, as shown in fig. 2.11 (note that for a two phase rectifier R_s is the source resistance of one *half* of the secondary winding). The graph in fig. 2.12 can now be used to predict the output voltage with good accuracy. It plots the variation in DC output voltage relative to the no-load transformer voltage as a function of R_s/R_l. Two curves are plotted, corresponding to different values of $f \times C \times R_l$, where f is the mains wall frequency (the doubling of the ripple frequency is already accounted for in the graph). Nearly all practical circuits will fall within the range shown. Notice that increasing the reservoir capacitance a hundredfold gives only a minor increase in the average output voltage, and only at small values of R_s/R_l. Using a very large reservoir capacitance mainly just improves the ripple voltage (next section). With a little care it is easy to use this graph to get within a few percent of reality. Ordinary power supplies using solid-state rectifiers

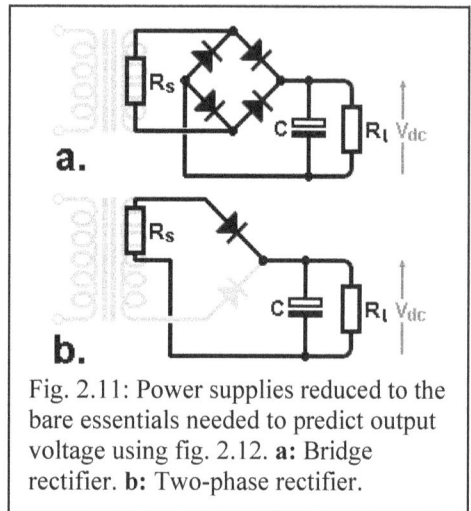

Fig. 2.11: Power supplies reduced to the bare essentials needed to predict output voltage using fig. 2.12. **a:** Bridge rectifier. **b:** Two-phase rectifier.

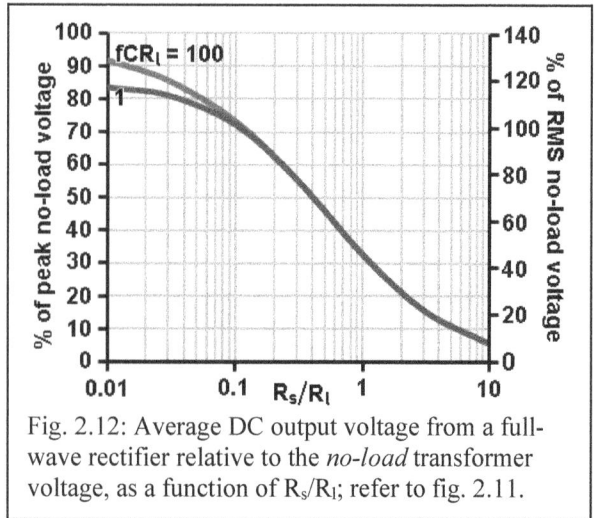

Fig. 2.12: Average DC output voltage from a full-wave rectifier relative to the *no-load* transformer voltage, as a function of R_s/R_l; refer to fig. 2.11.

38

will typically fall in the range where R_s/R_l is less than 0.1, while valve rectified circuits normally approach 0.1 or occasionally more.

When R_s/R_l is greater than about 0.1, the amount of reservoir capacitance no longer has much effect on the DC output voltage. This is because the circuit no longer behaves as a sample-and-hold but more like a simple RC filter. The DC output voltage is then the average value of the rectified sine wave, multiplied by the potential divider formed by R_s and R_l, that is:

$$V_{dc} = \frac{2V_{pk}}{\pi} \times \frac{R_l}{R_s+R_l} \qquad (2.5)$$

This is the case for bias supplies which use a large series dropping resistance, for example (chapter 8).

But if all this seems like too much work then here are the author's own personal rules of thumb for ordinary capacitor-input rectifiers with solid-state diodes and no additional series impedances (see fig. 2.17 for a summary):

- For >100VA transformers the *no-load* DC voltage will be around 1.4 times greater than the *rated* (i.e. full load) RMS secondary voltage.
- For >100VA transformers the *full-load* DC voltage will be about 1.35 times greater than the rated secondary voltage.
- For <12VA transformers the *no-load* DC voltage will be about 1.8 times greater than the *rated* secondary voltage
- For <12VA transformers the *full-load* DC voltage will be about 1.0 times greater than the rated secondary voltage.

Intermediate sized transformers should fall somewhere between these extremes. Also remember the DC voltage will vary in sympathy with the wall voltage; conservative designers will allow at least ±5% for this.

2.3.2: Ripple Voltage and Reservoir Capacitance

Unless we have infinite reservoir capacitance there will always be some residual ripple voltage riding on top of the average DC voltage. Fig 2.13 is a more realistic picture of the waveforms we might expect from a practical rectifier circuit. Notice that unlike fig. 2.7 earlier the output voltage never reaches the theoretical peak sine voltage, owing to losses across the diodes and transformer source impedance.

Ripple is usually expressed as a percentage of the maximum (peak)

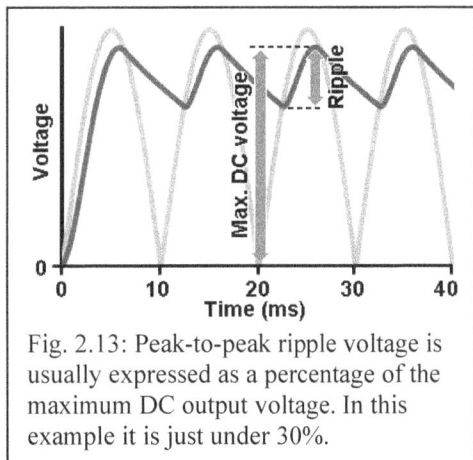

Fig. 2.13: Peak-to-peak ripple voltage is usually expressed as a percentage of the maximum DC output voltage. In this example it is just under 30%.

DC voltage, and a typical target figure might be 10% (though this is highly dependent on individual circuit requirements, of course). For example, if we were aiming for a $400V_{dc}$ supply with 10% ripple then this implies a ripple voltage of $400 \times 0.1 = 40V_{pp}$.

Formula (2.3) derived earlier is useful here because it overestimates ripple, compensating for the fact that electrolytic capacitors have poor tolerance and may have less than nominal capacitance. In practice, the ripple voltage will often turn out to be about half to two-thirds of what the formula predicts, and we are unlikely to need better accuracy than this:

$$V_{pp} = \frac{I_{dc}}{2fC} \quad \text{or} \quad C = \frac{I_{dc}}{2fV_{pp}}$$

Using more capacitance to get less ripple voltage means more ripple current (next section). More ripple current means more heating in the transformer, diodes, and capacitor ESR, and more radiated noise. This is not likely to be a problem in power supplies that deliver only tens of milliamps, but once we get into the hundreds of milliamps range it is usually a fool's errand to try to make really clean DC in one step. Instead, it is more prudent to deal with residual power supply noise using distributed smoothing, voltage regulation, or superior circuit PSRR.

2.3.3: Ripple Current

However much energy is drained off the reservoir capacitor by the load, the rectifier has to re-supply an equal amount to keep the average DC voltage constant. Unfortunately, while the load can drain current from the reservoir at a steady rate, the rectifier only has a very short time in which to refill it, leading to ripple current. Fig. 2.14 shows the shape of the waveforms in a realistic rectifier circuit. Compared with fig. 2.9 earlier we see that the pulses are not quite so sharp or 'peaky' as ideal theory imagines; they are slightly softened thanks to the transformer source resistance.

Recalling a basic rule of electronics, the *mean average* current in a capacitor is always zero. Hence the average incoming ripple current must be equal to the average outgoing load current. This might fool us into thinking the transformer current rating need only be equal to this figure. Unfortunately, we would be wrong, because what matters for heat dissipation is the *RMS* current, which depends on the *shape* of the waveform, and

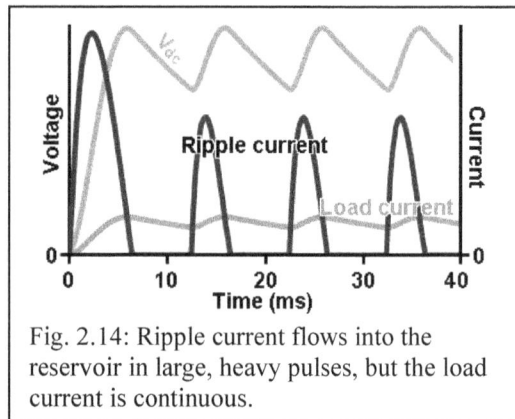

Fig. 2.14: Ripple current flows into the reservoir in large, heavy pulses, but the load current is continuous.

is always greater than the average. Fuses also work by resistive heating, so they have to tolerate the RMS ripple current too.

Trying to calculate the ripple current in a practical rectifier from first principles is tedious in the extreme, owing to the non-linearity of the circuit. But it is easy to simulate circuits on computer, and it can also be found graphically. Fig. 2.15 shows the ratio of peak and RMS ripple current to the average DC load current, as a function of R_s/R_l. This graph is the sister of fig. 2.12 earlier; using both together makes it easy to predict power supply performance on paper, with good accuracy.

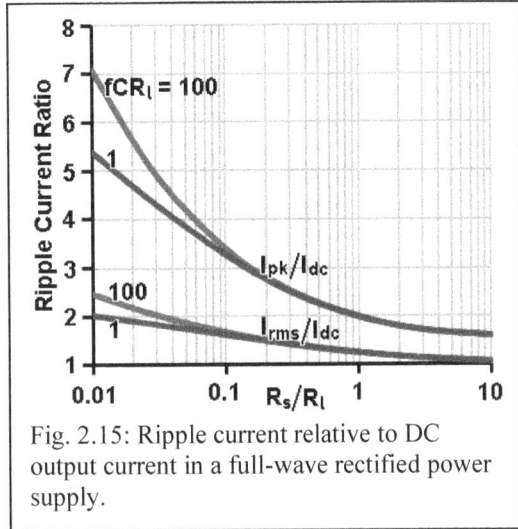

Fig. 2.15: Ripple current relative to DC output current in a full-wave rectified power supply.

The reservoir capacitor should have a ripple current rating at least equal to the total RMS ripple current. This is a conservative rule since the ripple current in the capacitor is actually less than in the transformer/rectifier, because a portion flows into the load instead. For ordinary power supplies the RMS ripple current can be expected to be in the range of 1.5 to 2.5 times the DC load current, while the ripple current in the capacitor will be perhaps 20% less than this. If this still sounds too complicated, simply make the capacitor ripple current rating at least equal to the transformer current rating and all will be well with the capacitor.

The *peak* ripple current could be up to seven times the DC current, but we rarely need to know the peak figure. Nevertheless, it is the peak current, multiplied by the transformer source resistance, that causes sudden voltage drop on the peaks of the AC waveform, which is why we never achieve the ideal $\sqrt{2} \times V_{rms}$ DC output voltage. It also causes the transformer AC voltage waveform to appear clipped rather than as a clean sine wave (section 1.4.4).

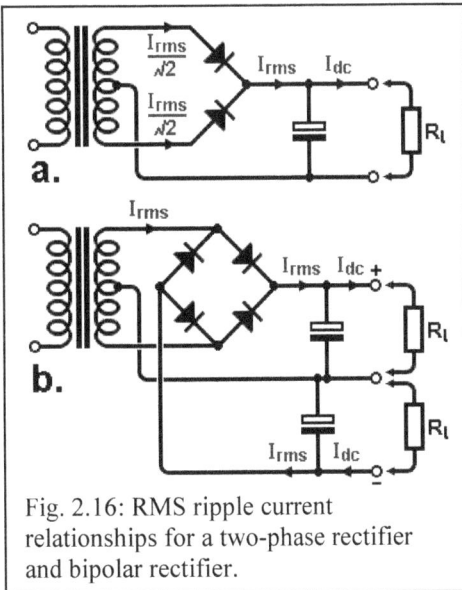

Fig. 2.16: RMS ripple current relationships for a two-phase rectifier and bipolar rectifier.

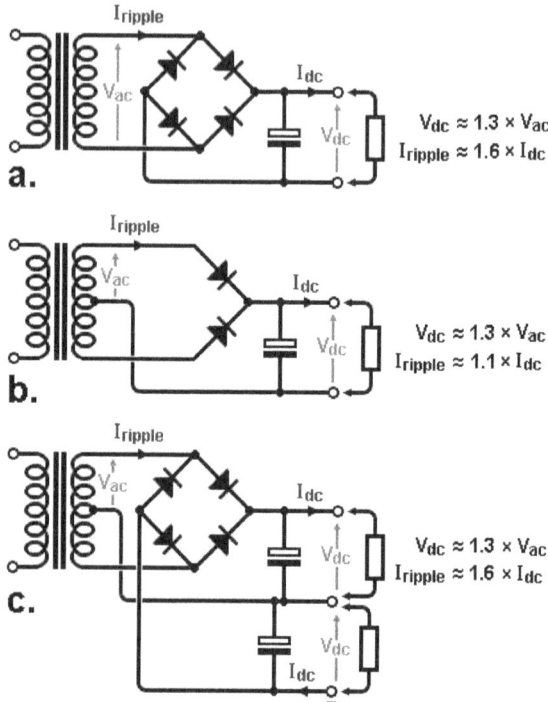

Fig. 2.17: Voltage and current relationships typically achieved in practical full-wave rectifiers (RMS quantities).

For a two-phase rectifier the RMS ripple current in each half of the secondary is not simply half the total indicated by fig. 2.15, but is closer to $1/\sqrt{2}$ times the total, as shown in fig. 2.16a (but the *peak* value is the same wherever we choose to measure it). A centre-tapped power transformer therefore needs a slightly higher VA rating than a single-winding transformer doing the same job. However, for a bipolar rectifier the RMS current in one half of the winding will be *equal* to the RMS ripple current indicated by fig. 2.15, *if* the two rails are equally loaded as in fig. 2.16b. If unequally loaded then it will approach $1/\sqrt{2}$ times the total, as before. Fig. 2.17 is a summary cheat-sheet showing the voltage and current relationships typically achieved with practical full-wave rectifiers.

2.3.4: Power Factor and VA Rating

The fact that the transformer must supply a larger RMS ripple current than is actually used by the load brings us to the concept of power factor. Of the total power delivered by the transformer, only some of it is dissipated in the load as heat or light, or sound, or whatever. The rest exists uselessly in the reservoir capacitor as *reactive* power.

Rectification

The *total* power handled by the transformer is called the **apparent power** and is equal to the *no-load* RMS secondary voltage multiplied by the RMS secondary current (i.e. ripple current), and has the units of volt-amps (VA). The *average* power delivered by the transformer is what does actual work in the load and is called the **real power**, in watts. The ratio of the two is called the power factor, PF:

$$PF = \frac{\text{Real power (watts)}}{\text{Apparent power (VA)}} \qquad (2.6)$$

If the transformer supplies nothing but a resistive load, such as AC valve heaters, then the voltage and current waveforms rise and fall together, in phase. All the power is dissipated in the load, which is a power factor of 1 or unity. This makes maximum use of the power transformed by the transformer. But as soon as we introduce a *reactive* component into the circuit –such as a reservoir capacitor– then our power factor suffers[*] because of the phase-shift between voltage and current. The maximum flow of current into the capacitive part of the load no longer corresponds to the maximum supplied voltage, leading to less efficient use of power.

For example, suppose we have a transformer which generates 100V at mains frequency of 50Hz, and we connect a 32µF capacitor across the secondary (no rectifier). The capacitor has a reactance of:

$$Xc = \frac{1}{2\pi fC} = \frac{1}{2\pi \times 50 \times 32 \times 10^{-6}} = 99.5\Omega$$

Applying Ohm's law we find that I = V/R = 100V / 99.5Ω ≈ 1A of current flows. But because of the natural operation of capacitors, this current leads the voltage by 90°, so the average power dissipated in the capacitor is zero. Now, although the capacitor dissipates no power, 1A of current is nonetheless flowing in the copper windings of the transformer, so we still require a transformer which is capable of supplying 100V at 1A, or in other words a 100VA transformer. That's a very large transformer considering no real work is actually done in the load; we have a power factor of zero.

A practical rectifier circuit with ripple current will fall between the extremes. The ripple current, being oddly shaped, comprises many harmonics of the switching frequency, but only the fundamental frequency delivers real power; the rest is simply an extra burden on the transformer and ultimately on the electricity supplier whose wires must be built to withstand the full RMS current. For this reason, appliances demanding more than about 500W may need to be fitted with power-factor correction circuitry in order to comply with national regulations.

---

[*] Note that the act of rectification does not affect the power factor; it is only the *reactive* components in the circuit that do. If the load is purely resistive then it does not matter if we supply it with AC or un-smoothed, full-wave rectified current, the power factor would be the same.

For an ordinary, linear power supply, the power factor can be determined from fig. 2.15. For example, if the RMS current in the transformer is found to be 1.6 times the DC load current as in fig. 2.17a, then the power factor is simply the inverse of this, or $1/1.6 = 0.625$, which is quite typical. We would need a transformer with a VA rating that is at least 1.6 times greater than the average load power. With low-voltage, high-current supplies like DC heaters, the ripple current often ends up around twice the DC load current, so the power factor will be worse at about 0.5. The transformer would then need a VA rating at least twice as great as the load power. For this reason, switching power supplies are much more attractive for low voltage appliances.

2.3.5: Capacitor Voltage Sharing

The reservoir capacitor must be rated for the maximum expected DC voltage, e.g. the peak transformer voltage plus allowance for mains variation. It is rare to find aluminium electrolytic capacitors rated for more than 450V, so it is advantageous to design circuits that use voltages less than this. Electrolytic capacitors boasting higher voltage ratings are occasionally encountered but should be regarded with some suspicion, since they push the limits of how thick an oxide layer can be grown or 'formed' inside the can.

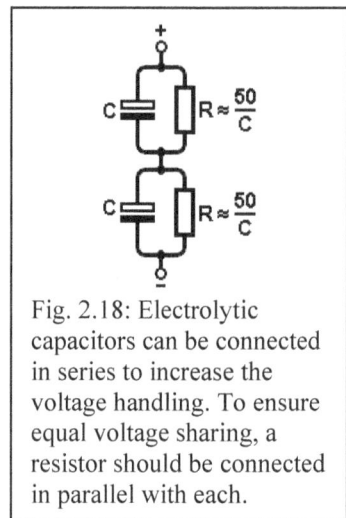

Fig. 2.18: Electrolytic capacitors can be connected in series to increase the voltage handling. To ensure equal voltage sharing, a resistor should be connected in parallel with each.

If a higher voltage rating is needed then capacitors can be connected in series to share the total drop. However, the leakage current is unlikely to be the same in multiple devices, which will cause unfair voltage sharing. To swamp this effect, a resistor should be connected in parallel with each capacitor as shown in fig. 2.18, and a rule of thumb is $R = 50/C$, where C is in Farads. The total useful capacitance is of course reduced since the capacitors are in series. It must also be pointed out that the metal body of a capacitor is usually connected to the negative terminal, so when two devices are connected in series the body of one of them will rest at half the total voltage and may present a shock hazard if exposed to the outside world.

2.4: Diode Ratings

When choosing rectifier diodes we must select devices that can safely handle the reverse voltages and forward currents expected in the circuit. When silicon rectifiers are subjected to excessive reverse voltage they have a tendency to fail short, with disastrous results for the rest of the circuit, so it is worth being conservative with rectifier voltage ratings.

Rectification

The popular 1N4007* and UF4007 are rated for a reverse voltage of 1000V and an average forward current of 1A (up to 75°C ambient), which is enough to accommodate many HT requirements. DC heater supplies are likely to need diodes with much larger forward current ratings. Discrete high-current diodes tend to be physically so large that it is often more convenient to use a bridge-rectifier package, even if you don't intend to use all four of the diodes inside.

2.4.1: Voltage Rating

The diodes used in a rectifier must be capable of withstanding the voltage imposed across them when they are reverse biased (off). The maximum reverse voltage a rectifier diode can safely block will be quoted on the datasheet as its **reverse-repetitive maximum** (V_{rrm}), or **peak inverse voltage** (PIV) on older datasheets. The minimum required rating depends on the circuit configuration used.

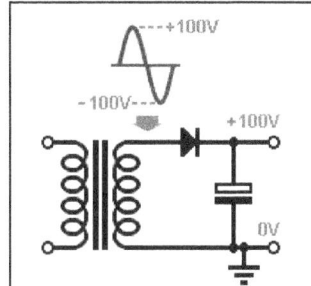

Fig. 2.19: A diode used as a half-wave rectifier must withstand the *peak-to-peak* AC voltage.

A half-wave rectifier diode must block one half of the AC cycle. Suppose we have a transformer which produces $100V_{pk}$ as shown in fig. 2.19. Whenever the AC voltage swings positive, the diode will pass current to the reservoir, and after a few cycles the reservoir capacitor will have charged up to the peak voltage (minus one diode drop which we can ignore). The cathode of the diode is therefore pinned at 100V. Now when the mains cycle reverses, the AC voltage swings negative to $-100V_{pk}$. If the anode is at $-100V$ and the cathode is at $+100V$ then the total voltage differential across the diode is 200V. Therefore, a half-wave rectifier diode requires a V_{rrm} rating in excess of the peak-to-peak AC voltage, or $2 \times \sqrt{2} \times V_{rms}$. However, we should add a further 10% to allow for mains voltage variation plus perhaps another 10% to allow for transformer regulation, bringing the total figure to at least $240V_{rrm}$. We can of course use a much higher-rated diode for extra robustness.

Since a two-phase rectifier is really just a pair of half-wave rectifiers working in anti-phase, they too must be rated to withstand

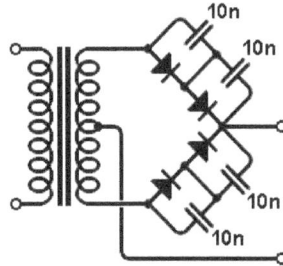

Fig. 2.20: The voltage-handling of a rectifier can be increased by connecting diodes in series. Capacitors should be connected in parallel with each diode to encourage equal voltage sharing. 1kV-rated ceramic capacitors are readily available.

---

* Manufacturers are now encouraging the adoption of the surface-mount S1x series in favour of the 1N400x through-hole series, but the through-hole parts are unlikely to disappear any time soon.

45

the peak-to-peak AC voltage across either half of the secondary. For example, a transformer delivering 350-0-350V would require diodes with a V_{rrm} rating of more than $\sqrt{2} \times 700 = 990V$. Adding another 20% or so for mains variation and transformer regulation brings the total to $1190V_{rrm}$. A 1N4007 would be inadequate.

Higher-voltage diodes are available (e.g. the RGP02 series rated for up to $2kV_{rrm}$, 500mA) but a common dodge is to use more than one diode in series so their V_{rrm} ratings add together. A 10nF to 100nF ceramic capacitor should be connected in parallel with each diode to encourage equal voltage sharing, as shown in fig. 2.21 (high-value resistors would be better, but it is a lot easier to find kilovolt-rated ceramic capacitors than kilovolt-rated resistors). It should perhaps be pointed out that these capacitors are sometimes omitted from cheap equipment, and premature rectifier death is the inevitable result.

When using a *pair* of two-phase rectifiers –to create a bipolar supply, for example– the diodes must *still* be rated for the peak-to-peak voltage across either half of the winding, even though it looks superficially like a bridge rectifier, as shown in fig. 2.21.

With an ordinary bridge rectifier the reservoir again charges up to the peak voltage, but whichever pair of diodes is 'on' there is always one diode connected between one end of the winding and the negative end of the reservoir, which prevents the AC voltage from swinging any more below zero than one diode drop. In other words, each end of the winding alternately swings from +100V peak to about −1V, as illustrated in fig. 2.22. The total voltage differential across any diode when 'off' is therefore about 100V. Hence a bridge rectifier only requires a V_{rrm} in excess of the peak AC voltage, or $\sqrt{2} \times V_{rms}$. For example, a transformer delivering $350V_{rms}$ would require diodes with a V_{rrm} rating of more than $\sqrt{2} \times 350 = 495V$, plus another 20% safety factor, bringing the total figure to $594V_{rrm}$. The 1kV-rated 1N4007 would therefore be more than adequate.

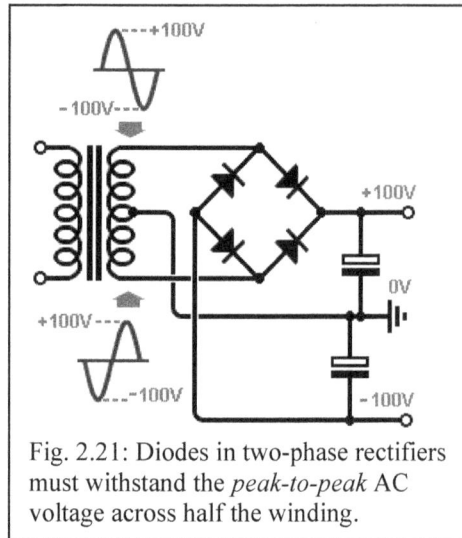

Fig. 2.21: Diodes in two-phase rectifiers must withstand the *peak-to-peak* AC voltage across half the winding.

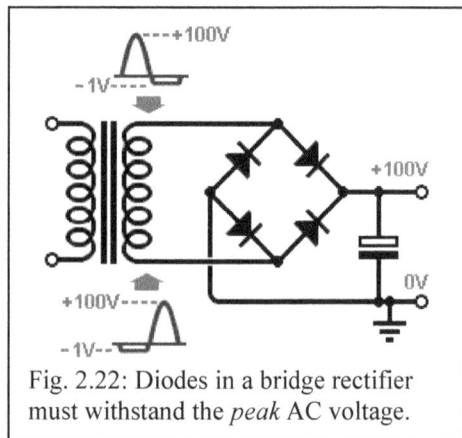

Fig. 2.22: Diodes in a bridge rectifier must withstand the *peak* AC voltage.

46

2.4.2: Current Rating

The maximum current rating for silicon diodes is normally quoted as a mean average value, $I_{F(av)}$. However, this figure applies to resistive loading and must be derated by at least 20% when the intended application is a standard capacitor-input rectifier [2], owing to the extra heating that arises from ripple current. For an ordinary rectifier (not a voltage multiplier –chapter 3) the total mean average ripple current is equal to the DC load current. In a full-wave rectifier this total is shared between more than one diode since they take turns, but the peak ripple current is the same everywhere, meaning the peak-to-average ratio may be quite severe for each individual diode. Therefore, a conservative rule of thumb is to use diodes with an $I_{F(av)}$ rating that is at least twice the expected DC load current. The popular 1N4007 is rated for $1A_{(av)}$ (up to 75°C ambient) and is therefore suitable for full-wave rectifiers delivering up to 500mA DC.

Rectifier diodes also have a rating for maximum forward surge current, I_{FSM}. This is the maximum non-repetitive current surge the diode can withstand. Such surges occur at switch-on when the empty reservoir has to charge up. The 1N400x series is rated for 30A in 8.3 milliseconds which is well beyond anything we will encounter in an ordinary valve amp. In other words, we don't normally need to consider surge current ratings for modern diodes.

2.4.3: Switching Noise and Fast Rectifiers

The audio press often cite diode switching noise as a malevolent spectre of power supplies, usually with only a vague explanation of what switching noise actually is. In fact, calling it 'diode switching noise' is a bit misleading as it makes it sound like the noise is coming from the diodes themselves, when in fact the problem has at least as much to do with the power transformer. 'Rectifier switching noise' is perhaps a better term as it encompasses the whole rectification process.

When a rectifier diode switches on and conducts, it allows current to flow from the transformer secondary into the reservoir capacitor. This also causes the transformer's leakage inductance –typically around 10mH to 100mH– to become energised or 'charged up' if you like. Eventually the diode comes to switch off again, but the leakage inductance would rather keep the current flowing just as it was, thank you very much, so it will generate a flyback voltage that keeps the diode turned on. In other words, the leakage inductance momentarily becomes the source of energy, releasing it into the reservoir as an additional little dollop of ripple current. In this way the peak of the AC voltage waveform, instead of descending in a smooth way, is held artificially high by the flyback voltage until the stored energy is finally spent. This leads to a 'hangover' or 'cliff face' effect on the AC waveform. This can be seen in the oscillogram of fig. 2.23 which shows the secondary voltage of a small 12V transformer when driving a rectifier circuit –it is not much of a sine wave! The sharp, steep edge contains high frequency Fourier components which can couple into the audio circuit via stray capacitance, leading to irritating switching spikes

---

[2] Gift, F. (1966). Rectifier Diodes, *Electronics World*, July, pp37-40.

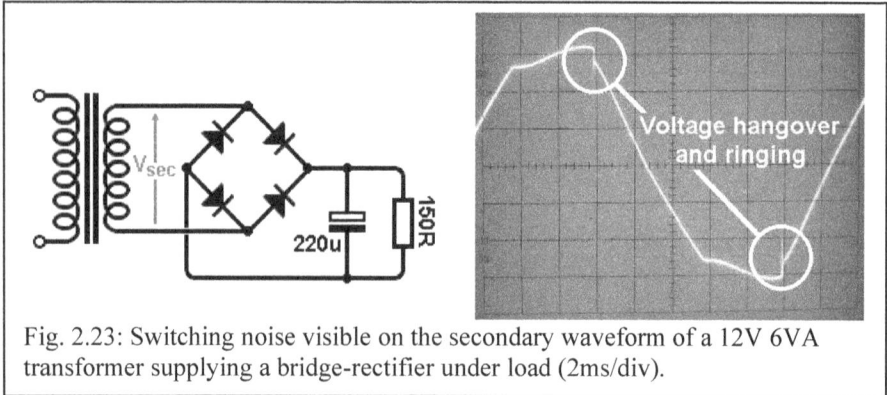

Fig. 2.23: Switching noise visible on the secondary waveform of a 12V 6VA transformer supplying a bridge-rectifier under load (2ms/div).

appearing in the signal path. It will also couple across to other windings on the same transformer, particularly from the HT to the heater supply, forging yet another path into the audio circuit.

But it gets worse. There is unavoidable stray capacitance across the transformer secondary, which includes the non-linear junction capacitance of the diodes themselves. We therefore have an RLC circuit formed by the transformer resistance, leakage inductance, and stray capacitance, as illustrated in fig. 2.24. The voltage hang-over is effectively a step input applied to this network, which will therefore resonate or ring. The effect is further worsened by the fact that a silicon diode does not switch off in an instantaneous way but will stay 'on' for a brief moment even after

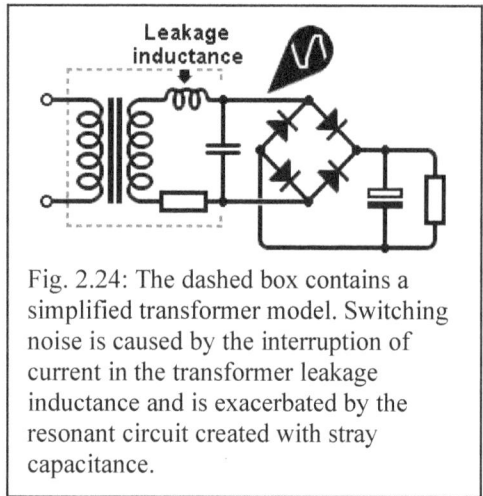

Fig. 2.24: The dashed box contains a simplified transformer model. Switching noise is caused by the interruption of current in the transformer leakage inductance and is exacerbated by the resonant circuit created with stray capacitance.

the voltage across it has become negative, which is called its **reverse recovery time**. During this period, current can flow in the opposite direction, as illustrated in fig. 2.25. For the 1N400x series the reverse-recovery time is about 2µs, while for their fast-recovery cousins, the UF400x, it is only 75ns.

The ringing will be high in frequency (visible only as spikes in fig. 2.23), so we have in effect a small RF transmitter, which can't be good news for nearby audio circuits. Fig. 2.26a shows a close-up of the ringing from fig. 2.23, revealing it to be around 125kHz. Rectifier switching noise is not a problem for all amplifiers, but if you wish to clean it up then there are three basic approaches: faster diodes, slugging, and snubbing.

Simply substituting the ordinary 1N4007s for the fast-recovery UF4007s produced the oscillogram in fig. 2.26b, which is already much better. Incidentally, when upgrading an existing power supply it is only necessary to add one fast diode in series with the existing rectifier, since this diode will impose its short recovery time on the whole circuit. This may be less invasive than replacing the whole rectifier. Schottky diodes also qualify as fast recovery diodes, though high-voltage devices are expensive.

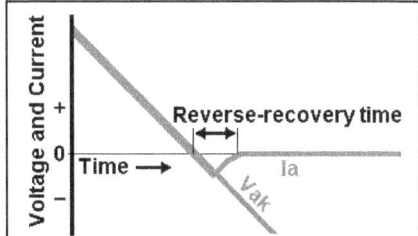

Fig. 2.25: The reverse-recovery time is the time taken for a diode to switch off after the anode-cathode voltage v_{ak} goes negative.

Slugging means adding more capacitance to reduce the resonant frequency and slow down the rate of change of voltage, so there is less energy likely to couple into other circuits. Often this takes the form of a capacitor connected in parallel with each rectifier diode. Confusingly, these are often referred to as snubbing capacitors when in fact they do no such thing. Snubbing implies burning energy off as heat, whereas slugging capacitors only spread the energy out, transferring it to lower frequencies. Also, the job can be done with a single capacitor in parallel with the transformer secondary. For example, four 100nF capacitors around a bridge rectifier are, as far as ringing is concerned, equivalent to one 100nF capacitor in parallel with the transformer. Such a capacitor added to the test circuit produced the oscillogram in fig. 2.26c. The ringing frequency has been lowered to around 20kHz (suggesting the transformer has about 25mH leakage inductance).

Snubbing is more sophisticated, consisting of an RC network across the transformer secondary, as indicated in fig. 2.27 (for a two phase rectifier, snubbing networks can be added to each half of the transformer secondary). The capacitor should be large enough that it swamps the effect of stray capacitance –about 10nF

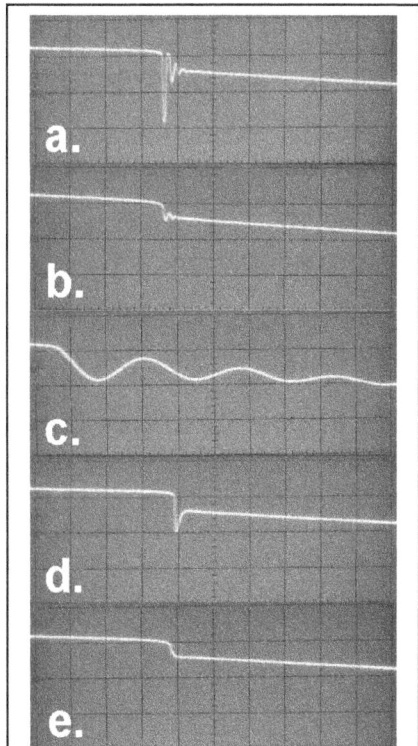

Fig. 2.26: Close up of the switching noise in fig. 2.23 (20µs/div) when using:
a: 1N4007 diodes;
b: UF4007 diodes;
c: 1N4007 diodes plus slugging capacitor;
d: 1N4007 diode plus snubber;
e: UF4007 diodes plus snubber.

49

should be ample– leaving only the resistor to choose. The aim is to burn off the oscillatory energy as quickly as possible in the resistor, and there will be an optimum value that provides the most damping. Too little resistance and we will have slugging rather than snubbing; too much and the snubber will have little effect. In some cases it is possible to monitor the circuit on an oscilloscope and tune the snubbing resistance to the optimum value, but if not,

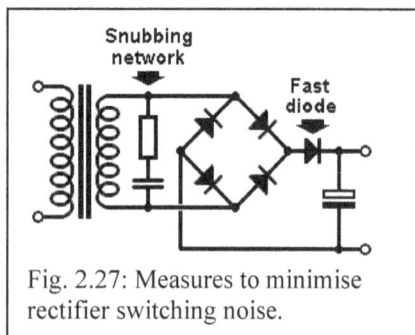

Fig. 2.27: Measures to minimise rectifier switching noise.

a value between 1kΩ and 4.7kΩ ½W is usually a good compromise. Fig. 2.26d shows the effect on the test circuit of adding an optimised 10nF+1.5kΩ snubbing network with 1N4007s. The picture is quite good, but a small negative spike remains due to the diodes' lazy reverse-recovery time. Fig. 2.26e shows the result when using UF4007s with the snubber; the switching is now about as clean as we could hope for.

Simply adding resistance in series with the transformer can also be regarded as a sort of heavy-handed snubbing since it damps the resonant network and adds compliance. Valve rectifiers have a great deal of internal resistance and are often lauded as being 'quiet' for this reason. But a silicon rectifier with the same resistance deliberately added has much the same effect, as well as being a lot cheaper and more reliable. Either way, the extra series resistance wastes power, reduces the DC output voltage and degrades the 'regulation' of the whole power supply, so it is an inefficient route to take.

2.5: Some Rectifier Circuit Arrangements

Most of the power supplies found in audio amplifiers follow conventional design practices, and the common arrangements or topologies will become familiar after looking at various amp designs. But it is worth noting that the standard-practice designs are not compulsory; there are many ways to arrange the same transformer and rectifier depending on the output voltage/s we need.[3] Remember, the transformer, rectifier and reservoir together represent a dirty battery, which is entirely floating until we connect some part of it to a reference voltage of our choosing, e.g. ground or zero volts.

For example, the circuit in fig. 2.28 is an ideal half-wave rectifier that charges a reservoir up to 140V. If we choose to ground the negative end of the capacitor then the other end must be +140V as in fig. 2.28a, whereas grounding the positive end must force the other end to −140V instead. Either way there is 140V across the capacitor and both circuits function identically; all we have done is move our point of reference. Alternatively, we could reverse the polarity of the diode and capacitor to produce the variations in c. and d.

---

[3] Brandon, D. A. (1962). Power Rectifier Circuits, *Practical Wireless*, July, pp245-6, 249-50, 258.

Although fig. 2.28a and c. are the textbook arrangements, there is no particular reason why b or d could not be used to achieve the same results. This ability to shift our frame of reference and see how the circuit works rather than how familiar it is, is essential to being a good designer and frees us from dogmatic

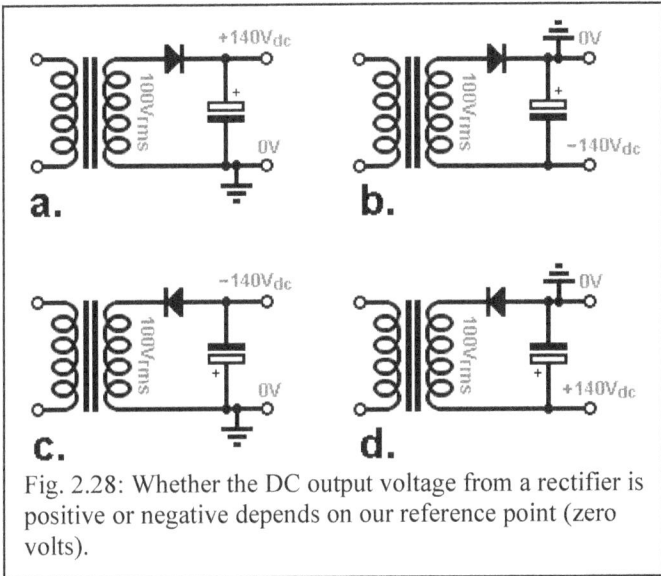

a. **b.**

c. **d.**

Fig. 2.28: Whether the DC output voltage from a rectifier is positive or negative depends on our reference point (zero volts).

adherence to convention. In particular, some arrangements may be more conducive to physical layout.

Fig. 2.29 shows some full-wave rectifier arrangements. The voltages shown are just examples to illustrate the basic relationship between AC input and DC output. Here the circuits are shown with the components in fixed positions while the reference node (ground) changes places, but by convention circuit diagrams are normally drawn with positive conductors towards the top of the page, above the zero-volt conductors, with negative conductors below and towards the bottom, when possible. If in doubt, remember that the rectifier diodes always 'point towards' the positive output terminal, and work from there.

By combining two two-phase rectifiers we can obtain positive and negative voltages as in fig. 2.29g. Alternatively, by moving the ground reference we could obtain two different positive or negative voltages as in h. and i. However, this comes with restrictions as we will see in chapter 3.

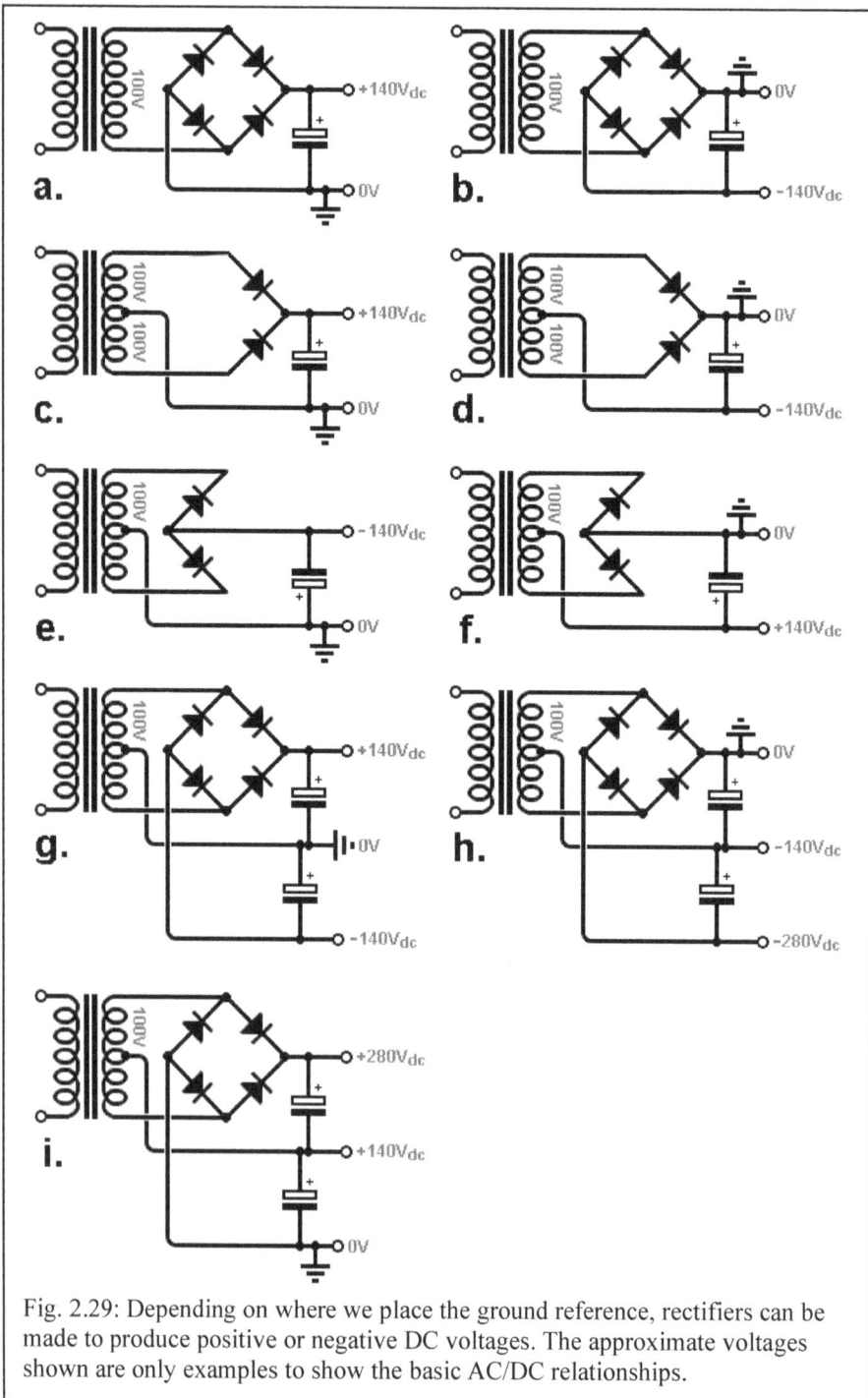

Fig. 2.29: Depending on where we place the ground reference, rectifiers can be made to produce positive or negative DC voltages. The approximate voltages shown are only examples to show the basic AC/DC relationships.

2.6: Valve Rectifiers

The traditional tool of rectification is of course the *valve* rectifier. Valve rectifiers normally contain one diode, or two diodes with a shared cathode, conventionally intended for half-wave or two-phase rectification. However, you can achieve bridge rectification at virtually no extra cost by adding a couple of silicon diodes (section 2.6.5). These do not need to be fast rectifiers since the valve imposes its superior switching characteristics on the whole circuit. These days there is no *technical* justification for using a valve rectifier at all; ordinary silicon diodes are far more robust, efficient, and inexpensive. Power resistors can be deliberately added if we want to simulate the voltage losses and reduced switching noise of a valve, and auxiliary circuits can provide a power-on delay if necessary, without adding great complexity. On the other hand, valves clearly have superior aesthetics and marketing appeal.

The high cost of the most popular rectifier valves has driven more enthusiasts to use 'junk box' alternatives, such as television damper/efficiency diodes like the PY500A. These were originally used as snubbing devices in televisions, but their high voltage and current ratings make them suitable for power supply duty too. In the author's opinion, however, valve rectifiers are suitable for low-current, light service only, where their inefficiency is of little consequence to the rest of the design. When used very conservatively, their service life can be extremely long, removing worries about dwindling supplies of old-stock devices. Conversely, for high- or wildly-varying current applications like power amplifiers, valve rectifiers are too lossy, too expensive, and too unreliable to be justifiable. There is something quite perverse about paying the same money for a GZ34 as for an EL34, when the rectifier performs such a mundane job compared to an amplifier valve, and so badly compared to a bit of silicon.

2.6.1: Electrical Ratings

Capacitor-input rectification is a brutal task for a valve, owing to the high peak ripple current. There are severe limitations on what valve rectifiers can handle, and table 2.1 provides figures for some popular rectifier types. The datasheets may be quite extensive, including multiple graphs showing permissible operating conditions, and there is not space here to go into the various ways different manufacturers presented their information. Neither is there really any need to, because the true limitations are quite straightforward. Just like modern diodes, they have ratings for:

- Maximum repetitive reverse voltage (usually quoted as an RMS rather than peak value);
- Maximum average current, i.e. DC load current;
- Maximum surge current, e.g. inrush.

Max. Rating	GZ34 5AR4	5Y3GT	EZ81 6CA4	EZ80	EZ90 6X4*	GZ32 5V4G	5U4GB
V_{heater}	5V	5V	6.3V	6.3V	6.3V	5V	5V
I_{heater}	1.9A	2A	1A	600mA	600mA	2A	3A
$V_{hk(max)}$	-	-	500V	500V	450V	-	-
$r_{a (beam)}$	50Ω	350Ω	120Ω	300Ω	300Ω	115Ω	150Ω
$V_{rrm(rms)}$	450V	450V	450V	350V	325V	375V	450V
$I_{dc(max)}$	250mA	100mA	150mA	90mA	70mA	125mA	275mA
$I_{ripple(peak)}$	750mA	440mA	500mA	270mA	210mA	525mA	1A
$I_{surge(200ms)}$	3.7A	2.5A	1.8A	?	?	3.5A	4.6A
C_{max}	60μF	20μF	50μF	50μF	50μF	40μF	40μF

Table 2.1: Maximum ratings of some rectifier valves in a full-wave capacitor-input circuit; see also fig. 2.30.
*The popular Chinese 6Z4 has the same characteristics as the EZ90/6X4 but with a different pinout; but it is completely different from the old American 6Z4!

But there is another, crucial rating: the maximum peak repetitive current, i.e. peak ripple current. This is taken for granted with silicon rectifiers, but valve cathodes can only handle so much current before they are at risk of saturation, degradation, and eventually flashover. This figure is not always stated explicitly on the datasheet but is bound up within the *minimum* **limiting resistance** and *maximum* reservoir capacitance ratings. These two things work together to keep both the peak ripple current and inrush current at safe levels.

The minimum limiting resistance is the total resistance which must be in series with each anode, for a given reservoir capacitance. Most of this resistance will come from the source resistance of the transformer itself:

$$R_s = R_{sec} + R_{pri} \times \left(\frac{V_{sec}}{V_{pri}}\right)^2$$

However, if the transformer does not provide enough source resistance by itself then more must be added to make up the deficit. Fig. 2.30 illustrates how these two things together form the total limiting resistance (per anode). Be aware that any added resistors must withstand the RMS ripple current, which for a two-phase valve rectifier is normally about 1.1 times the load current.

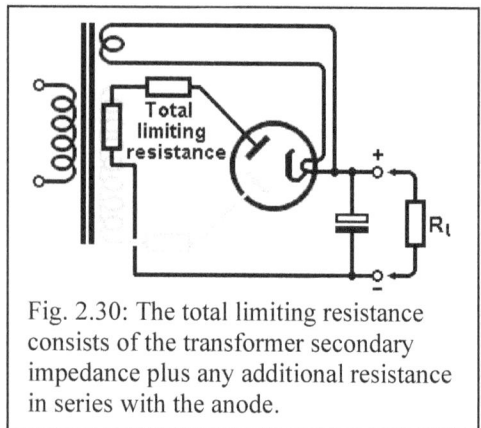

Fig. 2.30: The total limiting resistance consists of the transformer secondary impedance plus any additional resistance in series with the anode.

Alternatively, a single limiting resistor can be added between the cathode and reservoir capacitor, or in the centre tap of the transformer, but it will have to handle about 1.5 times the load current.

Rectification

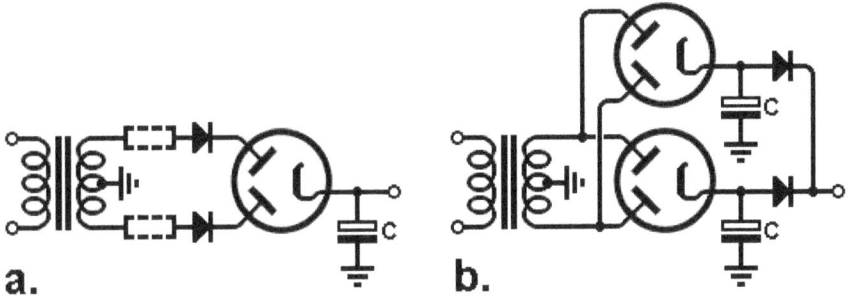

Fig. 2.31: Silicon diodes can help to protect a valve-rectified power supply from certain failure modes.

Furthermore, a common protective measure against valve rectifier failure is to add silicon rectifier diodes in series with each anode, as in fig. 2.31a. These diodes provide a fail-safe action if the valve happens to fail short, and they also reduce the reverse-voltage seen by each valve anode. The latter is very useful for ensuring reliable operation of a valve rectifier working close to its maximum voltage rating, and for protecting it against transformer flyback at switch off. Similarly, when valve rectifiers are used in parallel it is tempting to double the reservoir capacitance, but this of course means that if one valve fails it will leave its partner overloaded, inducing it to fail too. This can be avoided by giving each valve its own reservoir capacitor, with silicon diodes to ensure each valve only sees its fair share, as in fig. 2.31b.

Among designers more familiar with silicon, one of the most common abuses of valve rectifiers is to use too much reservoir capacitance in an attempt to make a nice, stiff power supply with low ripple. This inevitably leads to a short and painful life for the valve, ended by bright blue flashes inside the bulb as the cathode material evaporates and arcing occurs. However, it should be appreciated that it is not the reservoir capacitance _per se_ that kills the valve, but the ripple/inrush current. The maximum value quoted on the datasheet is only a derived limitation based on the manufacturer's assumption that the device will be used in an ordinary, traditional power supply design. In theory, we can use any reservoir capacitance we like, provided we also take steps to keep the inrush and peak ripple currents within the recommended limits. As a rough rule of thumb, however much the reservoir capacitance is increased above the datasheet recommendation, the total series limiting resistance should be increased by at least the same factor. This of course incurs further voltage and power loss which is why manufacturers assume no one would want to do it. It is also why the author considers valve rectifiers to be suitable only for mild, unvarying current applications where extra smoothing or regulation can easily be added, removing the need for a big reservoir capacitor in the first place.

55

2.6.2: Voltage Drop

The voltage lost across a valve rectifier is equal to the peak ripple current multiplied by the internal (beam) resistance of the diode, which is relatively high. What's more, the voltage drop will vary in sympathy with changes in load current, so the HT voltage will rise when the current demands are low, and sag when the demands are high. This can be a problem in vintage amplifiers where the power supply capacitors are only rated to withstand the HT voltage under *full load* conditions. If the amplifying valves are removed, so there is little or no load current, the supply voltage will climb to its maximum peak value, possibly leading to bursting capacitors. A similar problem occurs if a valve rectifier is substituted with a modern silicon rectifier, in which case it may be necessary to add some resistance to simulate the previous rectifier's internal anode resistance. Even then, a silicon rectifier has no warm up time, so the maximum voltage will be present for many seconds before the amplifier valves warm up and conduct enough to pull the voltages down to normal working levels.

Fig. 2.32 shows the anode characteristics of some popular rectifier valves. The internal beam resistance (per diode) is found by dividing the anode voltage by the anode current, at the desired point on the curve; table 2.1 summarises some typical values. This resistance adds to the source resistance of the transformer, from which circuit simulation or fig. 2.12 will yield the expected DC output voltage. Typically, R_s/R_l will end up close to 0.1, so the

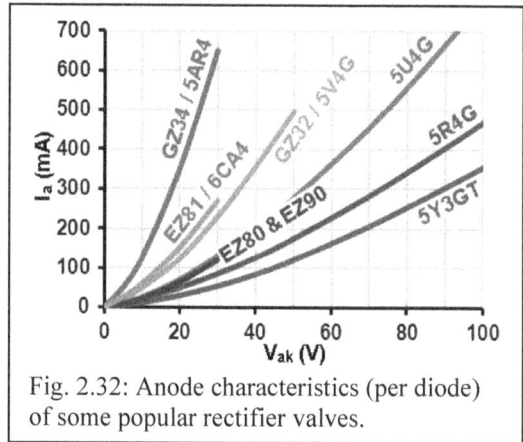

Fig. 2.32: Anode characteristics (per diode) of some popular rectifier valves.

DC voltage at full load is likely to be numerically similar to the *off load* RMS transformer voltage. In other words, a transformer that measures 300-0-300V_{rms} *off load* is likely to yield about 300V_{dc} at *full load* with a valve rectifier, roughly speaking. With negligible loading, however, the same circuit would settle closer to 420V_{dc}. If we further allow for 10% mains variation then we would need capacitors rated for more than 460V. This may require the use of multiple capacitors connected in series, with equalising resistors –section 2.3.5.

2.6.3: Heater Supply

Valve rectifiers handle heavy cathode currents and therefore have quite hungry heaters to match, which is a serious disadvantage compared to silicon diodes. Some small rectifiers such as the EZ81/6CA4 have completely separate heater and cathode. They also have a deliberately high $V_{hk(max)}$ rating so they can be operated from the same heater supply as other valves in the amplifier, without needing heater elevation. Most rectifiers, however, are either directly heated or else have the heater connected

56

to the cathode internally (in which case they are sometimes referred to as 'directly heated' as a shorthand). This forces us to use a separate heater supply dedicated to the rectifier, which will float on top of the HT, thereby eliminating any possibility of straining the heater-cathode insulation. Many valve power transformers include a 5V winding for just this purpose.

Directly heated rectifiers are normally powered from a heater winding with a centre tap, as in fig. 2.33a. Taking the DC output from this centre-tap forces the DC ripple current to distribute more evenly through the filament, avoid a hotspot as well as cancelling residual heater hum on the HT. For indirectly-heated rectifiers where the cathode is internally connected to the heater, the DC output should be connected to the actual cathode pin as in fig. 2.33b.[4] Connecting it as in c. is wrong since it forces ripple current to flow through the heater and transformer winding.

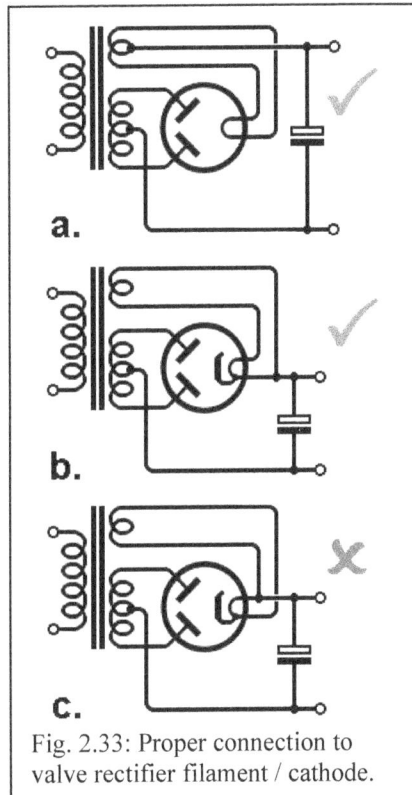

Fig. 2.33: Proper connection to valve rectifier filament / cathode.

2.6.4: Hot Switching and Inrush Current

Valve rectifiers should always be allowed to charge the reservoir naturally from cold, as the heater warms up (see section 5.3.4). 'Hot switching' refers to pre-heating the heater/cathode before switching on the anode current. This is the principle of most guitar-amplifier standby switches, for example. Depending on how such a switch is implemented this can mean the rectifier is fully warmed up, but the reservoir empty, when the switch is finally thrown. The valve will then suffer the massive inrush current into the reservoir capacitance. This is likely to cause momentary saturation, which can be very damaging to the cathode, and valve rectifier failure (arcing) in guitar amps is uncomfortably common for this reason. Hot switching of rectifier valves was expressly discouraged by valve manufacturers, as the RCA 5U4GB datasheet notes: *"Even occasional hot-switching with capacitor-input circuits permits the flow of plate current having magnitudes which can aversely affect tube life and reliability."*

A related problem may also present when *opening* a standby switch, by inducing a ghastly flyback voltage across the transformer secondary, large enough to cause

---

[4] Delaney, W. J. (1949). Power Pack Problems, *Practical Wireless*, September, pp.353-4.

arcing in the valve. A precaution against this is again to add robust silicon diodes in series with each anode of the valve rectifier, as in fig. 2.31, to take the strain of reverse voltage.

2.6.5: Hybrid Rectifiers

If for some reason we want to use a valve rectifier –whether to add sag, delayed warm up, charm or marketing hype– but the power transformer has no centre tap, then a hybrid bridge[5] rectifier can be built as in fig. 2.34a. This yields effectively the same end result as a standard valve rectifier configuration.

Alternatively, we could connect the valve rectifier in series with the output of a standard bridge rectifier. With the valve sections in parallel, each vacuum diode carries half the total ripple current, rather than taking turns as in a. This mean the voltage lost across the valve will be approximately $\sqrt{2}$ times less than in a., thereby producing a slightly higher DC voltage.

Moreover, we could place the valve after the reservoir capacitor as in c. It now behaves like a dropping resistor and handles only the relatively steady load current. Voltage loss/sag, and stress on the valve, are smallest with this arrangement. Indeed, since the valve no longer handles the bulk of the ripple current we could use a diode-connected amplifier valve instead, as in fig. 2.34d. This is a possible use for junk-box TV valves.

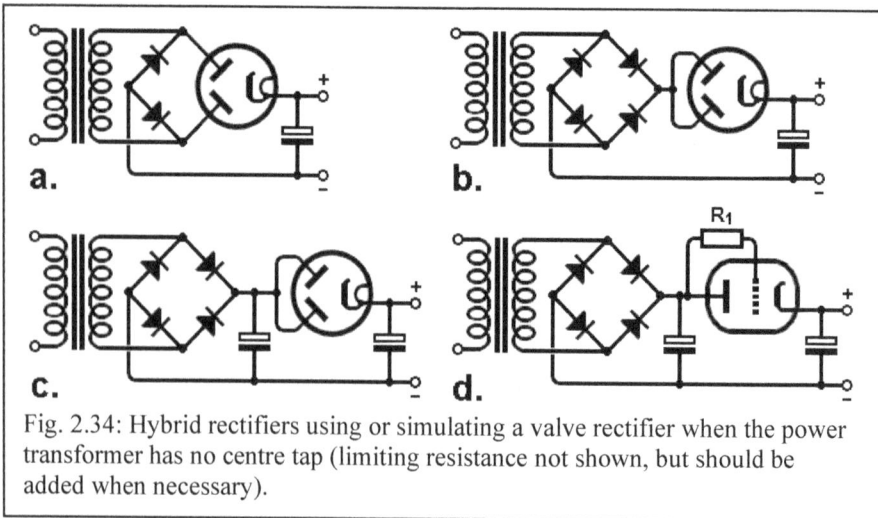

Fig. 2.34: Hybrid rectifiers using or simulating a valve rectifier when the power transformer has no centre tap (limiting resistance not shown, but should be added when necessary).

2.6.6: Load Regulation and Sag

Since valve rectifiers have significant internal resistance compared with silicon diodes, valve rectification suffers poorer load regulation, i.e. the DC voltage sags more under load. This does not matter if the load is constant, but many valve power

---

[5] Marshall, J (1963). Hybrid Bridge Power Supplies, *Electronics World*, June, p82-3.

amps are class-AB, which means their average current will increase at higher audio levels as operation moves away from class-A. This will limit the available audio output power, and reduces the headroom, potentially overdriving the output valves harder, or shifting their bias point colder.

The effect of a sagging HT is undesirable for hi-fi but is often deliberately exploited in guitar amps; at least one manufacturer patented a circuit for switching between silicon or valve rectification, as in fig. 2.35.[6] Of course, the same audible effect can be replicated with a sag resistor that mimics the anode resistance of a real valve rectifier, as in fig. 2.36 (see also section 3.2.3). A real valve rectifier is non-linear, but the nonlinearity is so minor that it makes no difference to the resulting sound or playability of the amp when compared with a simple resistor.

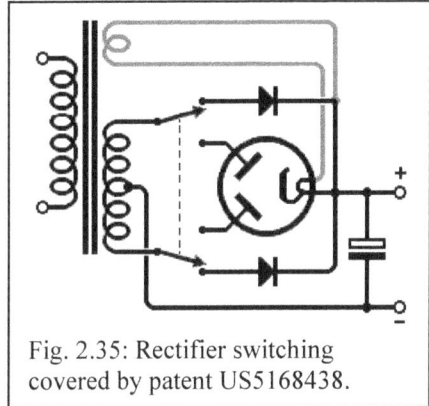

Fig. 2.35: Rectifier switching covered by patent US5168438.

The resistor can easily be made switchable as shown, and some enthusiasts have gone as far as to include different selectable values to simulate different types of rectifier valve, though the utility of this is debatable. Note that in fig. 2.36d the resistor is placed in the ground side of the circuit which reduces the voltage stress on the resistor and switch. However, if a bias supply is obtained from the same transformer winding then the negative voltage will increase (go more negative) if a resistor is added here (see chapter 8).

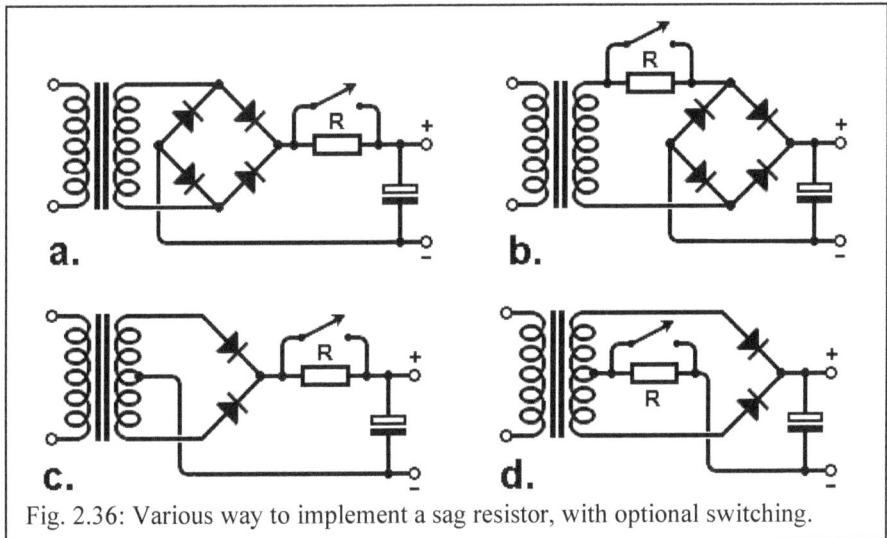

Fig. 2.36: Various way to implement a sag resistor, with optional switching.

[6] Smith, R. C. (1992). Selectable Dual Rectifier Power Supply for Musical Amplifier. US patent 5168438A

In a traditional valve-rectified power supply R_s/R_l will often be about 0.1, so a sag resistor will need to approximate this. Using fig. 2.15 we see the total RMS ripple current will be about 1.6 times the DC load current, and since this current flows in the sag resistor it may need a surprisingly large power rating. For example, a typical 50-watt amplifier might have a maximum average load current of 180mA, so the ripple current will be around 288mA. If a 150Ω sag resistor is installed, $P = I^2R$ shows it will dissipate 12.5W at full output! Even a 15W aluminium-clad resistor would bake.

2.6.7: Gas Rectifier Tubes

Gas rectifier tubes contain a gas of one sort or another, which enables them to achieve a lower voltage drop than vacuum diodes. The best known are mercury-arc rectifiers –also called mercury-vapour rectifiers– which contain a drop of mercury that vapourises after being warmed up by the filament. When anode current flows, the gas ionises and emits a much-admired soft blue glow, with the average voltage drop being typically about 14V. They were designed for high-voltage, relatively high-current applications such as radio transmitters. For example, a pair of 872As could deliver a terrifying 2.5A at $5000V_{dc}$, and that is small by mercury rectifier standards! Still rarer are xenon rectifier tubes which achieve even lower voltage drop. Today, gas rectifiers are mainly the preserve of DHT transmitter tube enthusiasts and 1930s revivalists.

Importantly, gas rectifiers cannot tolerate significant ripple current, so are normally used in a choke-input rectifier configuration (next section). For the mercury type, a specific mounting orientation will be specified to prevent liquid mercury from contaminating the wrong parts of the device. They also require the (extremely hungry) filament to be preheated before the high-voltage is switched on, and extra time should be allowed if the tube has been moved or left unused for a long time. Be sure to check the datasheet for all these quirks.

2.7: Choke-Input Rectifiers

All the previous discussion has been about capacitor-input rectifiers, which are by far the most common type. However, an alternative arrangement is the choke-input rectifier, which has slightly different characteristics. Whereas a reservoir capacitor tries to maintain a constant voltage across itself by taking heavy pulses of ripple current, a 'reservoir choke' tries to maintain a constant current through itself by varying the voltage across itself. Choke-input rectifiers were more popular in the days of valve rectifiers when it was expensive to regulate the HT by electronic means. These days the reverse is true, and it has become more difficult to obtain suitably-rated chokes for this purpose. Nevertheless, a few enthusiasts still embark on this path.

Fig. 2.37 shows the basic choke-input rectifier configuration (a two-phase rectifier could also be used, of course). Since rectified half-cycles are unidirectional they contain a DC component, but also a whole series of harmonics which form an AC

component. It is these harmonics
which the choke should 'block'
while allowing the DC component to
pass as freely as possible. From
basic electronics we know that the
DC component of any signal is equal
to its mean average, and for a full-
wave rectified sine wave this is
$2 \times V_{pk}/\pi$. If the choke blocks all the

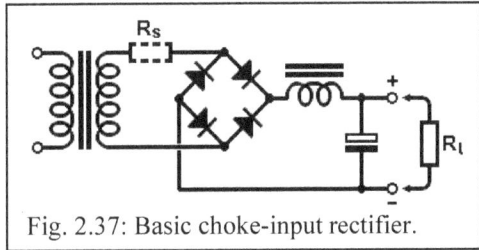

Fig. 2.37: Basic choke-input rectifier.

AC component then the output voltage will be pure DC equal to 64% of the peak AC
voltage, or 90% of the RMS voltage. This contrasts with a perfect capacitor-input
rectifier which produces DC equal to the *peak* AC voltage.

With infinite inductance, the DC output voltage would remain constant, regardless of
how much load current was drawn, but in reality we should expect some practical
limitations. The main one is that for a given inductance there is a critical value of
load current which must be exceeded before the choke will properly regulate the
output voltage. If the load current is less than this value, the DC voltage will climb
towards the peak voltage, just as it does with a capacitor-input rectifier. It might be
supposed that a bleeder resistor could be used to keep the current above the critical
value at all times, but at switch-on the load voltage will oscillate at the resonant
frequency of the inductor-capacitor combination, so it may still reach or even exceed
the peak AC voltage. Therefore, with a choke-input filter, the power supply
capacitors (and anything else) must be generously rated to withstand *more* than the
peak transformer voltage, unless an automatic soft-start or over-voltage protection
circuit is added. This is best modelled on computer.

2.7.1: Design Calculations

For a given resistance in series with the choke, the minimum value of inductance
required for regulation is:[7]

$$L_{min} = \frac{R_s + R_l}{6\pi f} \qquad (2.7)$$

Where:
R_s = the effective resistance in series with the choke, including transformer
impedance, valve rectifier anode resistance (if used), and the DC resistance of the
choke itself;
R_l = load resistance;
f = mains frequency.

[7] Schade, O. H. (1943). Analysis of Rectifier Operation. *Proceedings of the I.R.E,*
July, pp341-61.

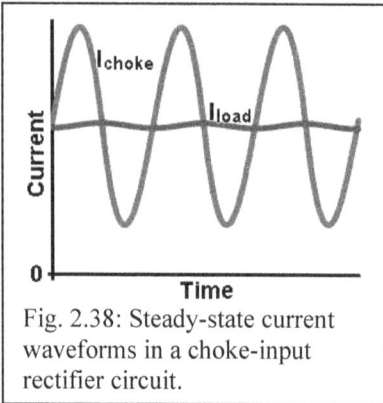

Fig. 2.38: Steady-state current waveforms in a choke-input rectifier circuit.

L_{min} is the effective inductance of the choke *with the direct load current also flowing through it.* This is likely to be much less than the inductance we would measure without any direct current, so it is important to buy a choke that is designed for this sort of duty, which quotes the inductance with DC current (they are sometimes called 'reactors'). But that is not all. The current in the choke is the sum of the DC load current plus an AC current component which gets stored as a magnetic field in the choke's core, as shown in fig. 2.38. If the choke is not sufficiently rated to handle this current it will saturate, quickly leading to failure, sometimes even by catching fire. The current rating of the choke must be greater than the *peak* current if we are to avoid saturation, and it is this requirement above all that makes suitable chokes hard to find –an ordinary smoothing choke will not do.

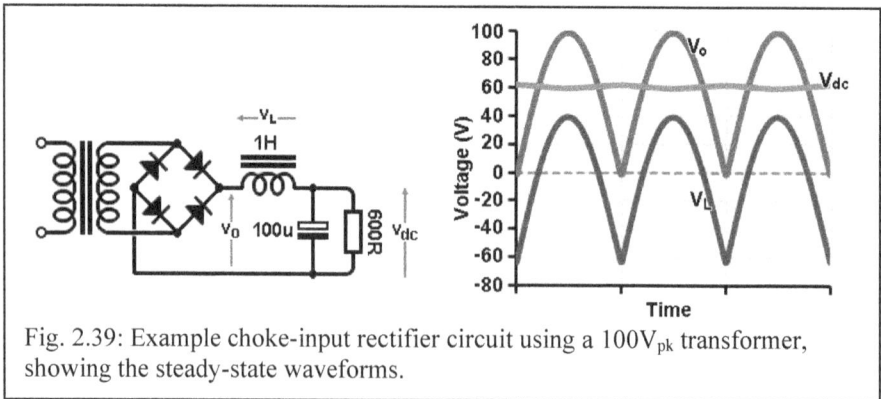

Fig. 2.39: Example choke-input rectifier circuit using a $100V_{pk}$ transformer, showing the steady-state waveforms.

To find the peak current, let us assume there is zero ripple voltage on the capacitor, so the whole AC component of the rectified sine wave is dropped across the choke. Fig. 2.39 shows an example circuit using a $100V_{pk}$ transformer (chosen because the voltages nicely translate into percentages) along with the resulting steady-state waveforms. From Fourier's theorem the frequency content of V_o can be written:

$$V_o = V_{pk}\left(\frac{2}{\pi} + \frac{4}{3\pi}\cos(2\omega t) - \frac{4}{15\pi}\cos(3\omega t) + \frac{4}{35\pi}\cos(4\omega t)...\right) \quad (2.8)$$

From the above we can see that the second harmonic of mains frequency dominates the rest, and since the reactance of the choke rises with frequency, all the higher harmonics can safely be ignored.

Rectification

Since we are really interested in the *peak* AC voltage across the choke we will put $\cos(2\omega t) = 1$.

$$V_{choke(peak)} \approx \frac{4\sqrt{2}}{3\pi}.V_{rms}$$

So the peak *alternating* current in the choke is:

$$i_{choke(peak)} \approx \frac{\dfrac{4\sqrt{2}}{3\pi}.V_{rms}}{2\pi fL} = \frac{4\sqrt{2}}{6\pi^2 fL}.V_{rms}$$

Where f is *twice* the mains frequency.

To this we must add the DC current to find the *total* peak current in the choke:

$$I_{choke(peak)} = i_{choke(peak)} + I_{load} \approx \frac{4\sqrt{2}V_{rms}}{6\pi^2 fL} + \frac{\dfrac{2\sqrt{2}V_{rms}}{\pi}.\dfrac{Rl}{Rl+Rs}}{Rl}$$

$$I_{choke(peak)} = \frac{4\sqrt{2}V_{rms}}{6\pi^2 fL} + \frac{2\sqrt{2}V_{rms}}{\pi(Rl+Rs)}$$

Simplifying:

$$I_{choke(peak)} \approx \frac{0.1V_{rms}}{fL} + \frac{0.9V_{rms}}{Rl+Rs}$$

Where f is again twice the mains frequency.

$$I_{L(pk)} = \frac{0.6V_{rms}}{2\pi fL} + \frac{0.9V_{rms}}{R_1} \qquad (2.9)$$

Where f is *twice* mains frequency.

This formula is a slight overestimation because it does not take into account R_s or the effect of higher harmonics, but it is better to err on the side of caution where the choke current rating is concerned.

The average current in the choke is relatively constant so we can easily predict the DC output voltage by treating R_s and R_1 as a potential divider (which we could not do with a capacitor input rectifier). Provided the load current is greater than the critical value, and ignoring diode drop, the DC output voltage will be:

$$V_{dc} = 0.9V_{rms} \cdot \frac{R_1}{R_1 + R_s} \qquad (2.10)$$

Fig. 2.40 shows a comparison of the load regulation of choke-input and capacitor-input rectifiers. The capacitor-input filter was set so that $fCR_l = 100$ when $R_s/R_l = 0.1$. Two choke-input filters are represented, one set so the critical load resistance is $100R_s$, and another at $10R_s$. Notice that with the choke circuit the same degree of regulation is always achieved; increasing the inductance simply means the regulation is maintained down to smaller load currents. The amount of smoothing capacitance used after the choke does not improve the regulation either, it only reduces the ripple voltage.

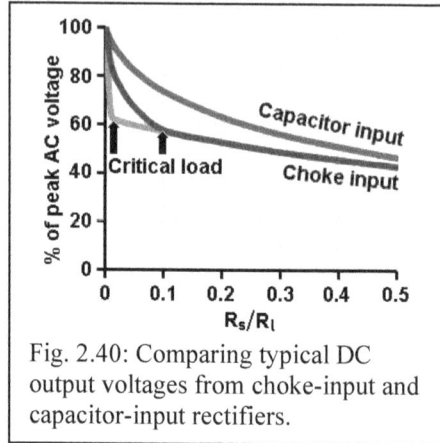

Fig. 2.40: Comparing typical DC output voltages from choke-input and capacitor-input rectifiers.

Thanks to the constant-current action of the choke, the RMS current in the transformer is also the same as the DC load current, so the required transformer VA rating is simply $V_{rms} \times I_{dc}$. In other words, a choke-input rectifier provides us with a power factor of approximately unity (they are still used in kilowatt industrial applications for this reason).

One further point worth mentioning is that when the rectifier diodes switch off, the choke will generate a considerable flyback voltage. It is therefore advisable to connect a 220nF to 470nF capacitor immediately prior to the choke –as shown in fig. 2.41– to suppress transients, and to use diodes with a very generous V_{rrm} rating. It is also worth pointing out that because the

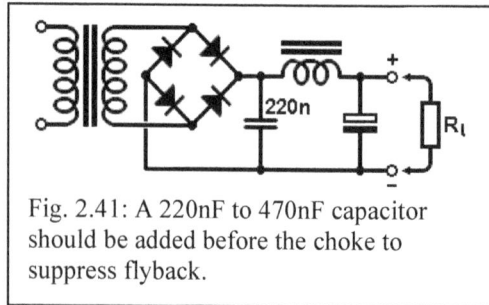

Fig. 2.41: A 220nF to 470nF capacitor should be added before the choke to suppress flyback.

voltage across the choke occurs in rectified pulses, the electric field around it is very noisy, so it should not be placed too close to any sensitive circuitry unless it is electrostatically shielded. Any loose laminations will vibrate and buzz, so everything must be firmly clamped.

To sum up, the choke-input rectifier gives good voltage regulation and, because the current drawn from the transformer is roughly constant, the power factor is also very good. The disadvantages are the low output voltage and high cost of the choke, and the possibly wild oscillations and flyback voltages that can occur at start up, switch off, or if the load current drops below the critical value.

Chapter 3: Voltage Multipliers

This chapter will explore rectifier configurations that not only convert AC into DC, but which also multiply the available DC voltage. Probably the main attraction of voltage multipliers for the valve circuit designer is that they allow high voltages to be generated from low-voltage transformers –the kind that can be bought from almost any electronics supplier or scrounged from discarded appliances. This can be a major cost saving compared with 'proper' valve power transformers. This chapter will therefore end with a collection of practical circuits for obtaining useful HT voltages from common, modern transformers.

Rather than assault readers immediately with the generalised voltage multiplying principle –which involves infinitely repeating 'ladder' circuits that disappear off the page– we will begin with the simplest sort: voltage doublers. These are the sort we are most likely to use for a valve circuit in any case. Once the operation of these simple doublers is grasped, the general ladder-multiplier principle should appear much less intimidating. Voltage triplers, quadruplers, and indeed all other multiplier circuits then need no further explanation, they are merely extensions of the general ladder-multiplier principle.

3.1: Half-Wave Voltage Doubler

Fig. 3.1 shows a simple voltage doubling rectifier. On the first half-cycle, current flows through D_1, charging up C_1 to the peak AC voltage. This much is a simple half-wave rectifier. But on the reverse half-cycle, current is able to flow through D_2 which charges up C_2, but the voltage already held across C_1 is effectively added to the transformer voltage too, since C_1 is in series with the charging loop. C_2 therefore gets charged up to the peak AC voltage *plus* the voltage across C_1, making $2 \times V_{pk}$ in total.

Fig. 3.2 shows this functioning in more detail with voltage waveforms (note that the negative terminal of C_2 has been chosen as ground or zero volts). The waveform at node ① is simply the AC transformer voltage, relative to ground. D_1

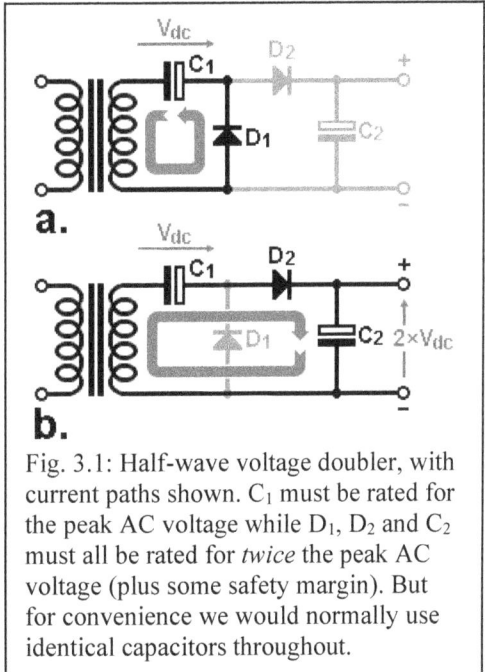

a.

b.

Fig. 3.1: Half-wave voltage doubler, with current paths shown. C_1 must be rated for the peak AC voltage while D_1, D_2 and C_2 must all be rated for *twice* the peak AC voltage (plus some safety margin). But for convenience we would normally use identical capacitors throughout.

allows C_1 to charge up to the peak voltage on negative half cycles, and the voltage ② *across* this capacitor (i.e. not relative to ground), is shown by the thin line. The voltage at ③ is the sum of these two waveforms. In other words, C_1 is effectively bobbing up and down on top of the AC waveform to create a new sinusoidal waveform at the C_1-D_1 junction, which is 'jacked up' by the voltage stored across C_1. The positive-going half cycles of this waveform allow C_2 to be charged through D_2, producing the final DC voltage ④, which is twice the original peak voltage. All the voltage multipliers in this chapter work on this 'ratchet' or 'bucket brigade' principle.

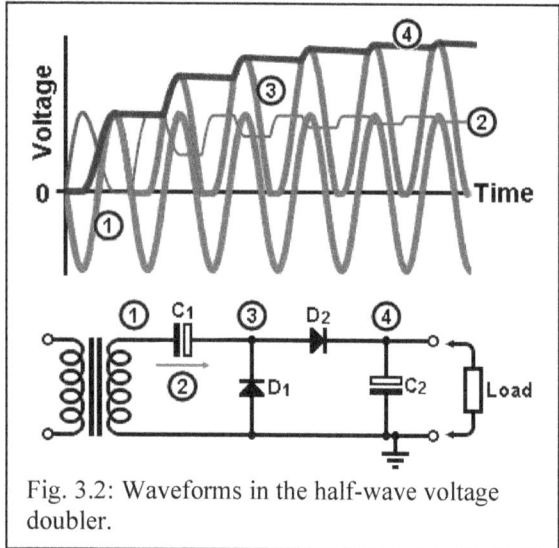

Fig. 3.2: Waveforms in the half-wave voltage doubler.

From the point of view of the transformer the circuit is full-wave since both halves of the AC waveform are used equally, which avoids transformer saturation. But C_2 is only charged up once per mains cycle, and since the load is connected across this capacitor, the circuit is only half-wave from the point of view of the load. In other words, the ripple voltage is the same as for a simple half-wave rectifier:

$$V_{ripple} = \frac{I_{dc}}{fC_2} \qquad (3.1)$$

(The ripple voltage increases disproportionately if the circuit is extended to higher multiplication factors as described in section 3.3.)

When considering ripple current, we must look at the circuit from the point of view of the transformer, where it appears to be full-wave. In fact, the ripple current for this voltage doubler is identical to an ordinary bridge rectifier supplying the same DC load power and reservoir capacitor C_2. Notice that we stipulate the same load *power*, i.e. twice the load current or one *quarter* the load resistance, since our fictitious bridge rectifier would not double the voltage.

Thus, one way to design a voltage-doubler of this kind is first to imagine we need half the voltage and twice the load current that we are aiming for, but the *same* ripple voltage. We then design an ordinary bridge-rectifier that would satisfy these figures, using any and all of the methods already familiar to us from the previous chapter. Finally, we convert it into a voltage doubler configuration, leaving the transformer unchanged. The voltage doubler will then meet our actual voltage and current requirements. The main differences are that the ripple frequency will be 50/60Hz rather than 100/120Hz, and it will take longer to charge up after switch on.

In practice, this type of doubler is seldom found in linear power supplies, since the full-wave voltage doubler delivers better performance with the same parts (next section). However, the basic principal is useful for multiplying the output from a pre-existing half-wave rectified supply with minimal surgery. For example, the author has used this method to multiply the output voltage of off-the-shelf DC-DC boost converters, as shown in fig. 3.3a. Similarly, it can be used to obtain a negative rail voltage, creating an approximately bipolar supply as in b. (the negative voltage will be slightly smaller in magnitude than the positive one). Remember, Schottky diodes are required for switching power supplies.

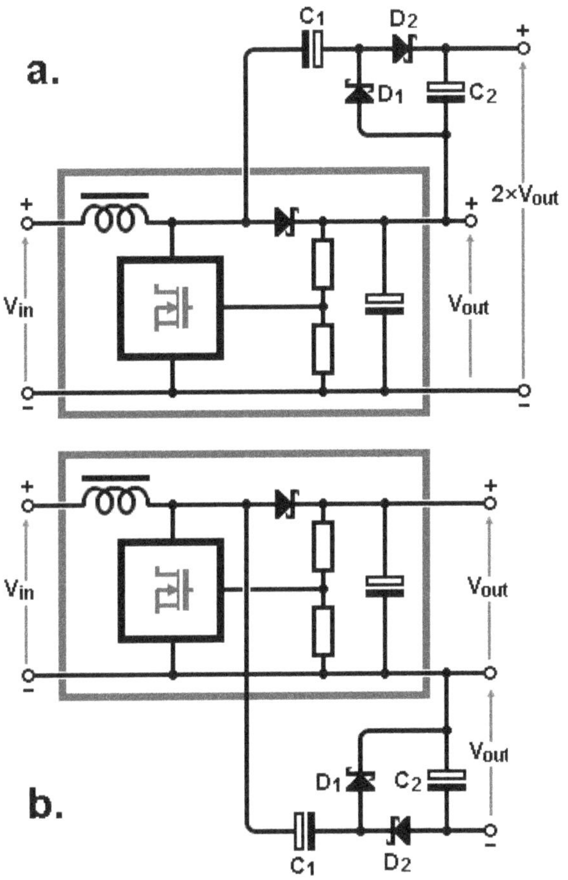

Fig. 3.3: A half-wave voltage doubler can be used with an existing boost converter to **a:** obtain a voltage doubled output; **b:** obtain an approximately bipolar supply.

3.2: Full-Wave Voltage Doubler

A different way to achieve voltage doubling is to build two opposing half-wave rectifiers and take the output from between them, as in fig. 3.4. This is sometimes called a Delon circuit in old textbooks. It may alternatively be drawn like a bridge rectifier but with capacitors replacing two of the diodes, as shown in c.

One capacitor is charged up to the peak AC voltage on the positive half cycle, and the other is similarly charged on the negative half cycle. When the load is connected across these two capacitors it sees the difference between them, or $2 \times V_{dc}$. Because the two capacitors together comprise the reservoir capacitance, and each is charged

67

on alternate half-cycles, the total reservoir
receives a charging pulse *twice* every mains
cycle, so it is considered to be a full-wave
rectifier. In other words, the ripple frequency
will be twice mains frequency, unlike with
the voltage doubler in the previous section.
What's more, with this circuit both
capacitors need only be rated for the peak
AC voltage (the diodes must be rated for
twice the peak AC voltage) plus some safety
margin. However, it is important that the
two capacitors are reasonably well matched,
otherwise the larger one will demand more
ripple current than the smaller, leading to net
DC in the transformer. Use capacitors with
±20% tolerance or better. Another possible
trap for designers is that the transformer
secondary now floats at half the DC output
voltage. The transformer might therefore
need special insulation between primary and
secondary if the difference is unusually
large.

The total ripple current and ripple voltage for
this voltage doubler are identical to an
ordinary bridge rectifier supplying the same
DC load power, but with a reservoir
capacitance equal to *one* of the capacitors in
the doubler:

Fig. 3.4: Full-wave voltage
doubler drawn in different ways.

$$V_{ripple} \approx \frac{I_{dc}}{2fC}$$

Where C is the capacitance of *one* of the capacitors in the voltage doubler (assumed
identical). This makes sense since the total reservoir comprises the two capacitors in
series, so we would expect twice the ripple voltage for the *same* load current when
compared with a bridge rectifier.

The process of voltage doubling means the transformer supplies half the voltage but
twice the current compared with an ordinary bridge rectifier (but the diodes still only
need an I_F(av) rating equal to the DC load current, and this includes a healthy safety
margin). For example, in an ordinary bridge rectifier the RMS ripple current is
commonly about 1.6 times the DC load current, whereas for a voltage doubler it is
twice as great, about 3.2 times the load current.

3.2.1: Bipolar Supply

By grounding the junction between the two capacitors we split the doubled voltage
into two, creating a bipolar supply as shown in fig. 3.5. This is a useful trick when a
68

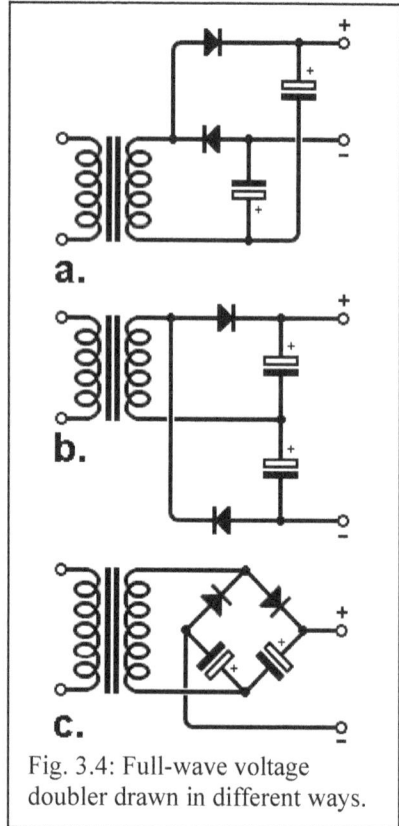

centre-tapped transformer is unavailable.
However, since each side of the circuit
is effectively a half-wave rectifier, it is
essential that both rails are equally
loaded, otherwise a net DC current will
exist in the transformer, leading to core
saturation. It is also worth remembering
that in this configuration the ripple
voltage on each rail will be at mains
frequency and 90° out of phase with one
another, i.e. in quadrature. This might
spoil the PSRR of some bipolar circuits
that rely on symmetrical ripple for hum
cancellation. Section 3.4 describes an
alternative bipolar rectifier which does
not have these shortcomings.

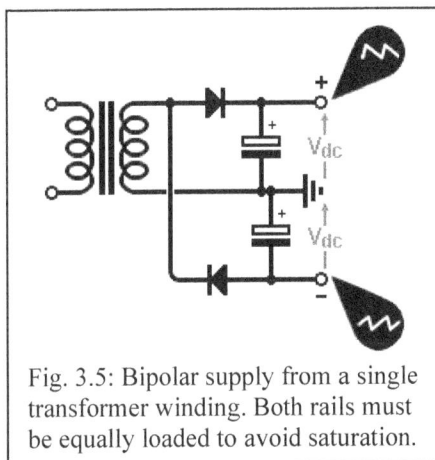

Fig. 3.5: Bipolar supply from a single
transformer winding. Both rails must
be equally loaded to avoid saturation.

3.2.2: Switchable Doubling

The full-wave voltage doubler presents a
useful way to switch between two DC
voltages, as shown in fig. 3.6. With the
switch open as shown, the circuit is an
ordinary bridge rectifier. The two
reservoir capacitors appear in series, so
voltage equalising resistors (section
2.3.5) may be required, as shown faint.
Closing the switch converts the circuit
into a full-wave voltage doubler,
producing twice the DC output voltage.
D_3 and D_4 now become reverse biased
and play no part in rectification. The

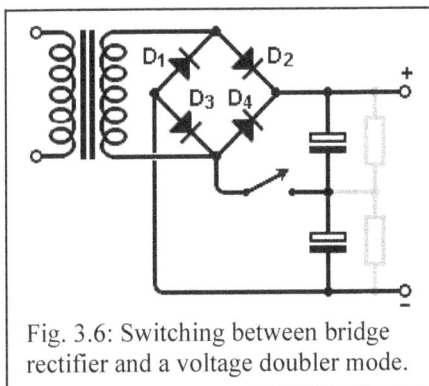

Fig. 3.6: Switching between bridge
rectifier and a voltage doubler mode.

author has used this principle many times to build versatile power supplies for
prototyping. The same trick is used in many off-line SMPSs to select between 115V
or 230V wall voltages.

3.2.3: Tweakable Output Voltage

For certain applications, the full-wave voltage doubler offers an efficient way to
fine-tune the average DC output voltage, independently of the ripple voltage. This
can be done by adding an extra capacitor across the whole output, as shown in fig.
3.7. For a given load, we can make the DC voltage sag down to any value between
full voltage-doubling and practically zero, simply by choosing the right value for C_1
and C_2. The smaller we make this pair of capacitors, the more the average voltage
will sag under load. Of course, by itself this would make the ripple voltage
progressively worse, so the job of C_3 is to clean up the output and give us whatever
final ripple voltage we desire.

69

What we are really doing here is controlling the effective output resistance of the doubler. Indeed, the end result is similar to using a dropping resistor, but without burning the voltage off as heat. With no load the voltage will always climb back up to the full voltage-doubled value, so this trick is mainly useful for situations where the load current is constant, such as a DC heater supply, but it could conceivably be used to emulate the sag effect of a valve rectifier in a guitar amp. The actual values suitable for

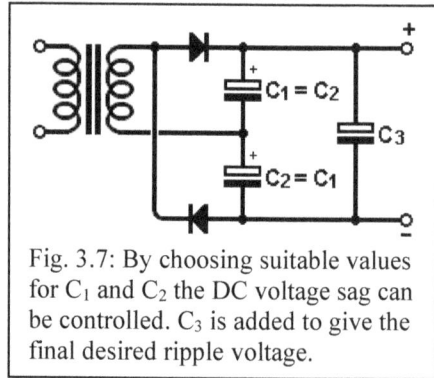

Fig. 3.7: By choosing suitable values for C_1 and C_2 the DC voltage sag can be controlled. C_3 is added to give the final desired ripple voltage.

a given application are best found by experiment or computer simulation.

3.3: Ladder / Greinacher Multiplier

The voltage doublers described previously are really just special cases of a more general voltage multiplying concept described by Swiss physicist Heinrich Greinacher over a century ago.[1] The simplest form is reproduced in fig. 3.8. In English-language textbooks this is often called a Cockcroft-Walton multiplier, after the physicists who applied the same technique to high-voltage experiments.[2]

The half-wave voltage doubler is just the first two of stages in this ladder (Cockroft called it a stairway[3]) whose every 'rung' multiplies the voltage one more time. An advantage of this method is that each diode and capacitor need only be rated to withstand twice the peak AC voltage, no

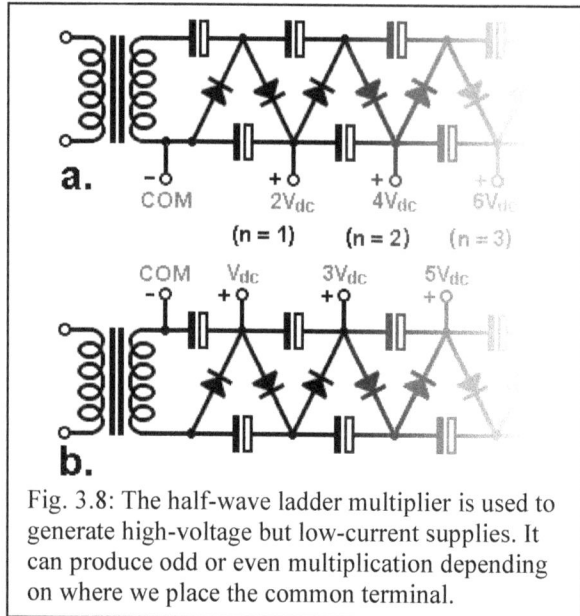

Fig. 3.8: The half-wave ladder multiplier is used to generate high-voltage but low-current supplies. It can produce odd or even multiplication depending on where we place the common terminal.

[1] Greinacher, H. (1914). Das Ionometer und seine Verwendung zur Messung von Radium- und Röntgenstrahlen, *Physikalische Zeitschrift*, 15, pp410–15.
[2] Cockcroft, J. D. & Walton E. T. (1935). System for the Voltage Transformation of Direct Current Electrical Energy. British Patent 652750.
[3] Cockcroft, J. D. (1932). Improvements in High Voltage Direct Current Systems. British Patent 367785.

matter how large the multiplication factor. Odd or even multiplication is possible depending on the choice of common reference point or ground, as shown in the figure.

In theory the ladder can be extended indefinitely. However, losses such as capacitor leakage rapidly make the ladder multiplier impractical for anything other than low-current power supplies. The greater the multiplication factor, the worse the efficiency and load regulation.[4] Large multiplication factors are generally only practical for ionisers and electrostatic equipment that needs very-high polarising voltage but little or no current. Nevertheless, this circuit is the gateway to some other circuit snippets that *are* useful to us, so it is worth exploring it a little further.

3.3.1: Full-Wave Ladder Multiplier

A logical next step is to observe that the previous circuit is only half wave. By connecting a duplicate circuit to the transformer, with the diodes orientated in the other direction, and taking the output between them, we obtain full-wave rectification. This is shown in fig. 3.9. For example, taking the output across A+ and A− gives full-wave voltage doubling. Indeed, it is exactly the same circuit described earlier in section 3.2, but we now appreciate it to be part of the ladder-multiplier principle. Taking the output between B+ and B− gives full-wave voltage

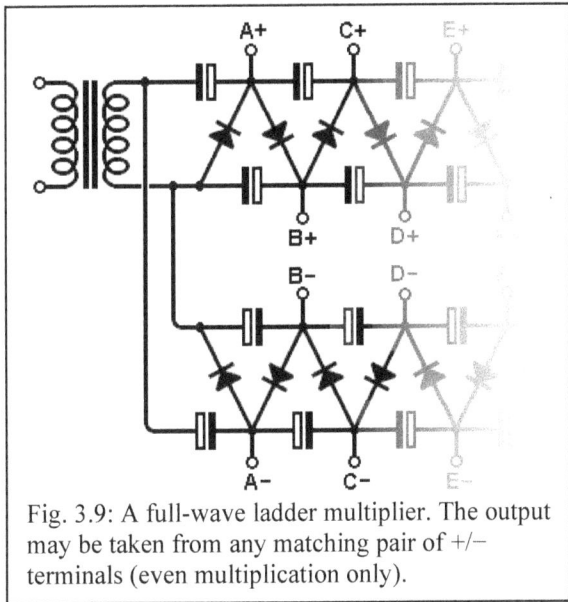

Fig. 3.9: A full-wave ladder multiplier. The output may be taken from any matching pair of +/− terminals (even multiplication only).

quadrupling; between C+ and C− is voltage sextupling, and so on. Odd multiplication is not possible with this architecture.

Each capacitor in the ladder so far described is stacked upon a previous capacitor, so losses accumulate along the chain, in series. We can sidestep this issue by returning each capacitor, on either side of the ladder, directly to its respective transformer terminal, as shown in fig. 3.10. This alternative topology results in slightly better load regulation (less sag), and less ripple voltage, since the total reservoir capacitance is now always comprised of the last matching pair of capacitors. The

---

[4] Ruzbehani, M. (2017). A Comparative Study of Symmetrical Cockcroft-Walton Voltage Multipliers, *Journal of Electrical and Computer Engineering*, **2017**, pp1-10.

disadvantage of doing this is that each capacitor is charged to progressively higher voltage, i.e. the first capacitor in each ladder must be rated for the peak AC voltage, the second capacitor for twice as much, the third for three times as much, and so on. In practice we would probably use identical high-voltage-rated capacitors throughout. See section 3.6 for some practical multiplier circuits based on this prototypical ladder, redrawn more simply and going as far as sextupling.

3.3.2: Two-Phase Ladder Multiplier

From chapter 2 we know that an alternative method of full-wave rectification is to use a centre-tapped transformer with a pair of half-wave rectifiers feeding the same reservoir, producing the familiar two-phase rectifier. We can also do this with a pair of half-wave ladder multipliers, as shown in fig. 3.11. Like fig. 3.8 earlier, both odd and even multiplication is possible depending on where we place the common node or,

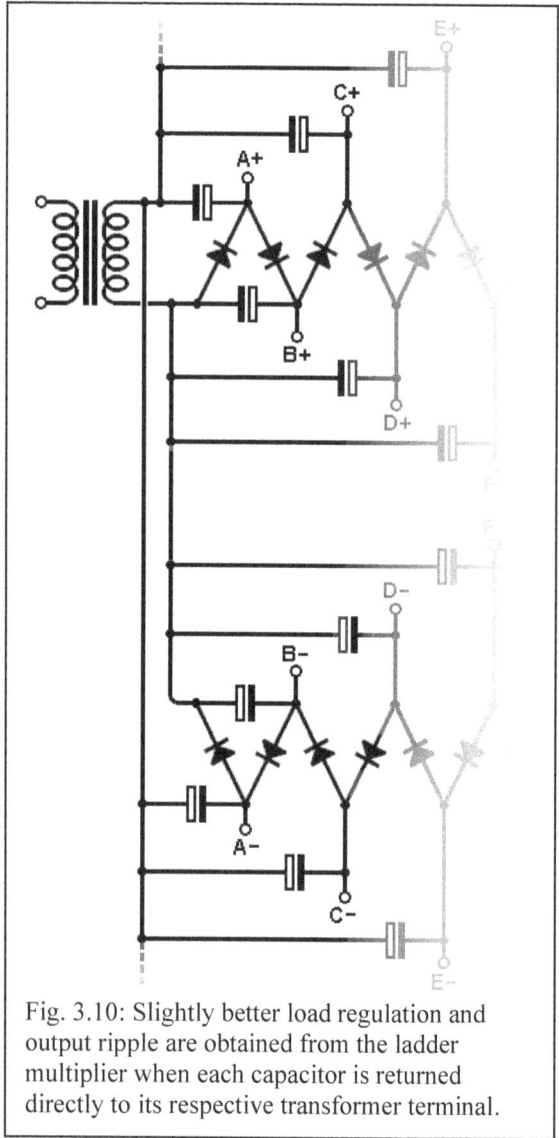

Fig. 3.10: Slightly better load regulation and output ripple are obtained from the ladder multiplier when each capacitor is returned directly to its respective transformer terminal.

since it is often convenient to maintain the centre tap as the common terminal, where we begin the ladder. And like fig. 3.12, slightly improved load regulation and ripple voltage is possible if the capacitors in the two outer 'legs' are returned directly to their respective transformer terminals rather than being stacked on top of one another, with the same caveats about voltage rating.

Keen-eyed readers may notice that in fig. 3.11b the first pair of diodes are in fact an ordinary two-phase rectifier, and the next four diodes look like a bridge rectifier stacked upon it. This presents an alternative way of looking at the circuit which is perhaps more intuitive, and is shown in fig. 3.12. Here we see a familiar two-phase rectifier and, above it, a capacitor-coupled bridge rectifier. The lower reservoir capacitor is charged up to the peak AC voltage as usual, while the upper reservoir is charged to twice this value (the peak voltage across the *whole* of the transformer secondary). However, we have a choice about where to reference the upper reservoir. If we connect it to the transformer centre tap (typically ground) then we have two entirely separate rectifiers: an ordinary two-phase rectifier, and a voltage doubler which is the same as the first section of fig. 3.11a. On the other hand, if we stack the voltage doubler on top of the lower reservoir then the

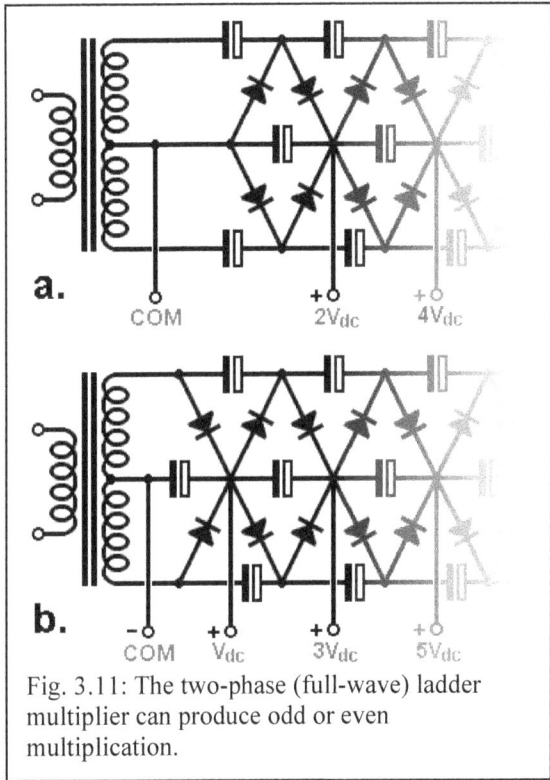

Fig. 3.11: The two-phase (full-wave) ladder multiplier can produce odd or even multiplication.

two voltages add together to produce three times the peak voltage, just like the first two stages of fig. 3.11b.

These circuits are enormously useful for generating disparate rail voltages for elaborate amplifier designs, but be aware that loading on either rail will cause the

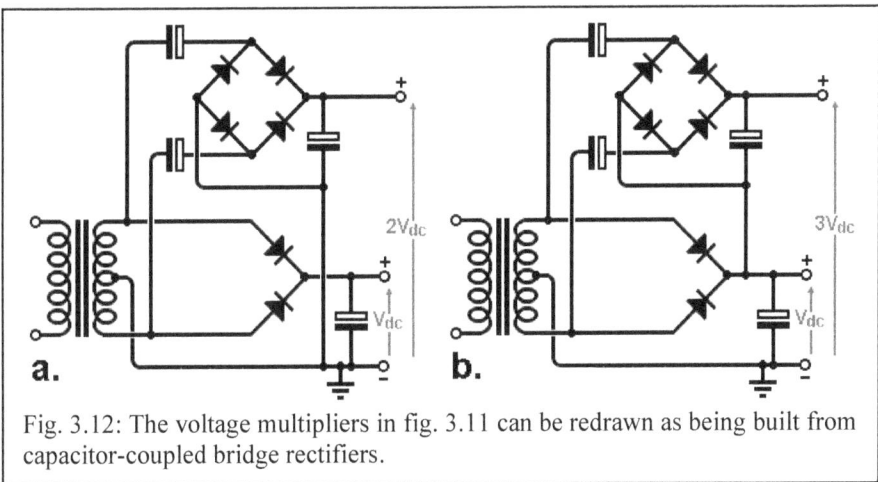

Fig. 3.12: The voltage multipliers in fig. 3.11 can be redrawn as being built from capacitor-coupled bridge rectifiers.

voltage on the other one to sag too, since they share the same transformer source impedance. The same treatment can be given to negative voltages of course, i.e. the circuit can be made bipolar.

A subtlety of the circuit in fig. 3.12b is that, depending on the moment of switch on, either of the coupling capacitors may be subjected to a brief but large reverse voltage. This is best avoided by adding protection diodes in parallel with each of these capacitors. Once the capacitors have charged up the diodes will be reverse biased and play no further part in circuit operation.

Predicting ripple current in these circuits gets complicated. The ripple current in the bridge rectifier is exactly what we would expect from the previous chapter –typically 1.5 to 2 times the DC load current. The same current flows in the transformer of course, and the capacitors must have ripple current rating in excess of this this figure. The ripple current in the two-phase rectifier is similar but will be $1/\sqrt{2}$ times smaller in the transformer. The voltage tripler is more complicated since load current is drained from both reservoir capacitors. Suffice it to say that the ripple current in the transformer will typically be 3 to 3.3 times the DC load current. Drawing current from two outputs simultaneously makes things really intractable. The RMS ripple currents for each rectifier do not simply sum –it is not as bad as that. But neither can we use the root of the sum of the squares (RSS addition) –it is worse than that! It will be something in between, so if an accurate figure is needed then it is best found by computer simulation.

3.4: Capacitor Coupled Bridge Rectifier

By followed the evolutionary process of voltage multipliers we have seen a new species emerge: the capacitor-coupled bridge rectifier. Intriguingly, this animal is not limited only to centre-tapped transformers but can perform the remarkable trick of generating bipolar power rails from a single-winding transformer, as shown in fig. 3.13. This method uses more parts than the simple voltage doubler in section 3.2.1, but it has the advantage that the ripple voltage will be at twice mains frequency and can be made symmetrical on each rail, rather than in quadrature. Many bipolar circuits rely on this fact to achieve good PSRR. Another advantage is that there is no risk of transformer saturation even if the two rails are not equally loaded.

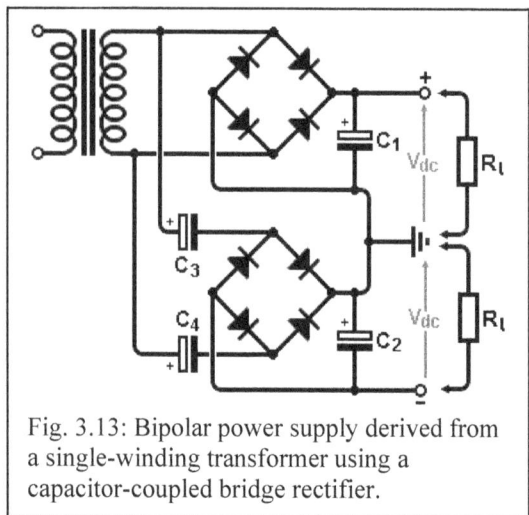

Fig. 3.13: Bipolar power supply derived from a single-winding transformer using a capacitor-coupled bridge rectifier.

Voltage Multipliers

Referring to fig. 3.13, the reservoir capacitors C_1 and C_2 should be identical to ensure equal-but-opposite ripple voltages. C_3 and C_4 should also be identical to one another and be larger than C_2, otherwise the ripple voltages will not be exactly in phase and the capacitor-coupled rail will suffer more sag under load than the opposing rail. Moreover, the reactance of C_3 and C_4 at mains frequency must be small enough that large AC voltages cannot appear across them, since electrolytic capacitors cannot tolerate reverse voltages larger than about one volt. An easy rule of thumb is to make them larger than $V_{pp}/(2\pi f R_1)$, or $I_{dc}/314$. They should be rated to withstand the peak transformer voltage and RMS ripple current, which will typically be about 1.6 times the load current, as for any bridge rectifier. Note that in this configuration the transformer winding must be truly floating.

The capacitor-coupled bridge rectifier is also useful for obtaining a low-voltage DC supply from the AC heater supply of an existing amplifier. Care must be taken with the ground referencing as this will determine the polarity of the coupling capacitors. If either the AC transformer secondary, or the DC side of the rectifier, is truly floating, then the polarity does not matter, but the coupling capacitors must be larger than $I_{dc}/314$.

Typically, however, the AC heater supply will have a ground reference or be elevated to a large positive voltage, while the DC supply we have created will probably end up with a ground reference too. The anodes (positive terminal) of the coupling capacitors must therefore point towards whichever side of the

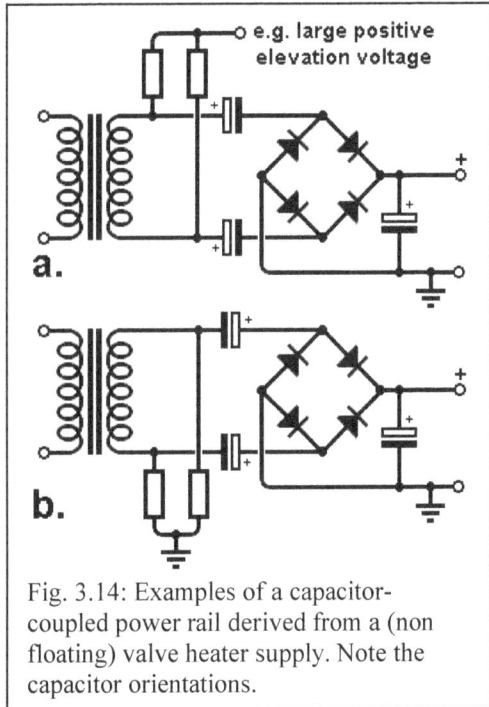

Fig. 3.14: Examples of a capacitor-coupled power rail derived from a (non floating) valve heater supply. Note the capacitor orientations.

circuit is more positive. For example, if the negative terminal of the rectifier is grounded to create a positive DC power rail, but the transformer secondary is elevated to an even larger positive voltage, then clearly the AC side of things is the more positive and the coupling capacitors must be orientated to suit, as shown in fig. 3.14a. They must also have sufficient voltage rating to withstand the elevation voltage, but since they now do not suffer reverse bias, they need not necessarily be as large as $I_{dc}/314$. This can get confusing so, if in doubt, simulate the circuit on a computer. See also section 8.4.4 for a bias supply using this sort of rectifier.

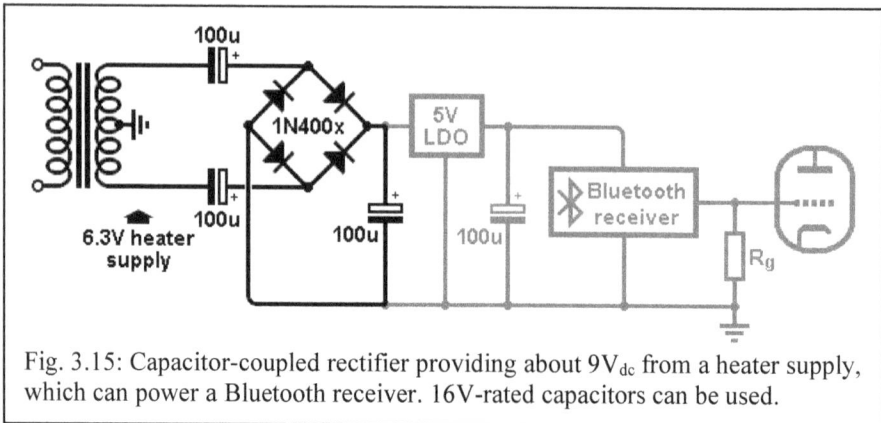

Fig. 3.15: Capacitor-coupled rectifier providing about 9V$_{dc}$ from a heater supply, which can power a Bluetooth receiver. 16V-rated capacitors can be used.

Fig 3.15 shows a practical example which will produce about 9V$_{dc}$ up to 50mA, from an ordinary 6.3V heater supply. The author has used this (with an additional low-dropout regulator shown faint) to add a Bluetooth receiver module to a vintage amplifier, for example. This approach suppresses the annoying buzz that usually results from running a Bluetooth receiver from a shared DC power source.

3.5: Disparate HT Voltages

Many valve amplifier designs call for disparate HT voltages. For example, high-power amplifiers using KT88s often need a high voltage for the output stage anodes but a lower voltage for the screen grids. Conversely, an amplifier with a cathode follower output stage will typically run the output valves from a lower-voltage HT rail, and the preamp/driver from a higher-voltage HT rail.

Supplying the entire amplifier with one, main HT rail, and dropping any excess voltage with resistors is extremely wasteful for anything more than a few milliamps, and introduces heat management problems. It is more efficient to generate the right HT voltages in the first place, which is easy to do if one rail needs to be twice the voltage of the other. Fig. 3.16 shows some ways to do this. A simplified output stage, with the anode supplied from the high-voltage rail and screen grid from the low-voltage rail, is shown faint as an example.

It may be tempting to use a full-wave voltage-doubler as in fig. 3.16a, since we know the total voltage is split equally across the two reservoir capacitors. However, this is not always a good idea. In this case the valve anode current is drawn from both capacitors, which is fine, but the lower capacitor also supplies the screen grid. This capacitor will therefore demand more ripple current than its partner, leading to net DC in the transformer. Even a few milliamps of DC will cause core saturation – loud buzzing and overheating. The safe option, if we cannot change the power transformer, is to utilise a capacitor-coupled rectifier as in c. A centre-tapped transformer could alternatively be used, as in b.

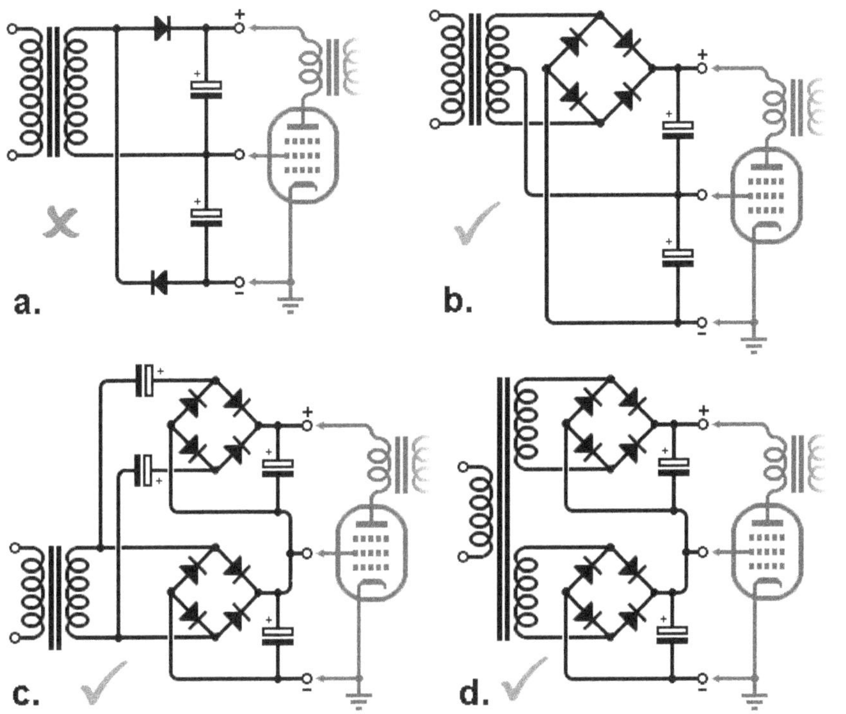

Fig. 3.16: Right and wrong ways to generate disparate HT voltages. The arrangement in **a.** should not be used because it will cause net DC in the power transformer and probable saturation.

The stacked-rectifier arrangement in fig. 3.16d is the most flexible since the two transformer secondary windings need not be of the same voltage, so the high-voltage rail could be something other than twice the lower one. Two separate transformers could be used in the same way, of course.

3.6: Some Practical Multiplier Circuits

Voltage multiplier circuits provide a simple way to get a useful HT supply from a modern off-the-shelf power transformer not normally intended for valve applications. If the transformer can be scrounged from a discarded appliance, along with a second transformer to provide heater voltage, then we have the main ingredients for a complete valve-amplifier power supply, for very little money. A few circuits are suggested here. Remember, however, that with a full-wave multiplier the transformer secondary floats at half the DC output voltage. It is therefore unwise to use a full-wave multiplier to generate voltages significantly higher than $300V_{dc}$, say, unless the transformer has special insulation between primary and secondary.

3.6.1: Voltage Doubler

Fig. 3.17 shows an HT supply built using 50VA, 0–55, 0–55V transformer. The secondaries are connected in series to obtain 110V_{ac} which is then voltage doubled and will realistically produce about 290V_{dc} at full load. Using 220µF capacitors as shown, the ripple voltage will be about 7V_{pp}. Notice that the

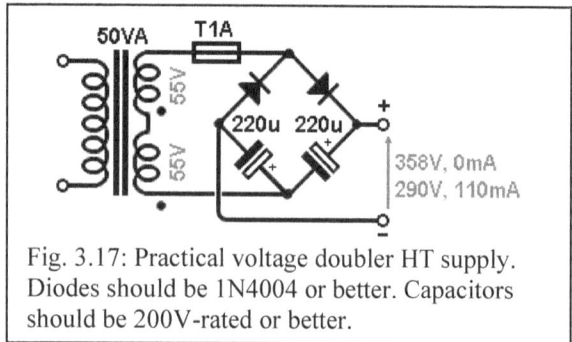

Fig. 3.17: Practical voltage doubler HT supply. Diodes should be 1N4004 or better. Capacitors should be 200V-rated or better.

maximum load power is 32W but a 50VA transformer is required, i.e. the power factor is 0.64, which is no different from an ordinary full-wave power supply.

3.6.2: Voltage Quadrupler

Fig. 3.18 shows an example of a voltage quadrupler using a 20VA, 0–24, 0–24V transformer from which 240V_{dc} is obtained at 60mA maximum current. The regulation is poor with such a low-power transformer, so this technique is best reserved for a constant-current application such as a class-A amplifier or line stage.

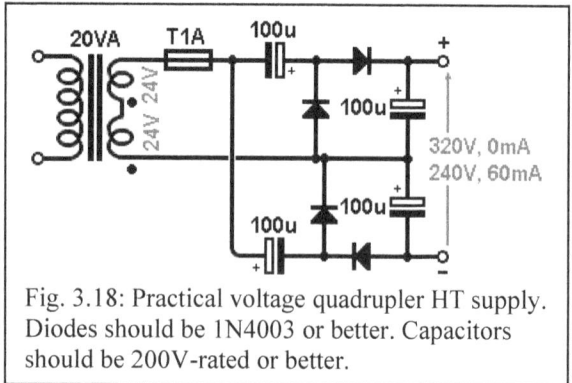

Fig. 3.18: Practical voltage quadrupler HT supply. Diodes should be 1N4003 or better. Capacitors should be 200V-rated or better.

Using 100µF capacitors the ripple will be about 7V_{pp} (pay close attention to their orientation). At full load they handle about half the total RMS ripple current, or about 230mA in this case.

3.6.3: Voltage Sextupler

Fig. 3.19 shows an example of a voltage sextupler using a 50VA, 0–15, 0–15V transformer from which 185V_{dc} is obtained at 180mA maximum current. With increased multiplication comes ever declining regulation, even with this improved variant of the multiplier. Nevertheless, this supply might be useful for a small class-A project like a radio, stereo headphone amp, or guitar practice amp. Using 220µF capacitors the ripple voltage will be about 10V_{pp}, and the capacitors should be rated for at least 700mA ripple current.

3.6.4: Voltage Octupler

Fig. 3.20 shows a low-current power supply suitable for a standalone valve preamp or guitar effects box. This uses a very cheap 12V, 12VA transformer – often available in the form of a mains plug-pack adapter or wall wart, making it ideal for those who want to avoid live mains wiring or hair-raising voltages.

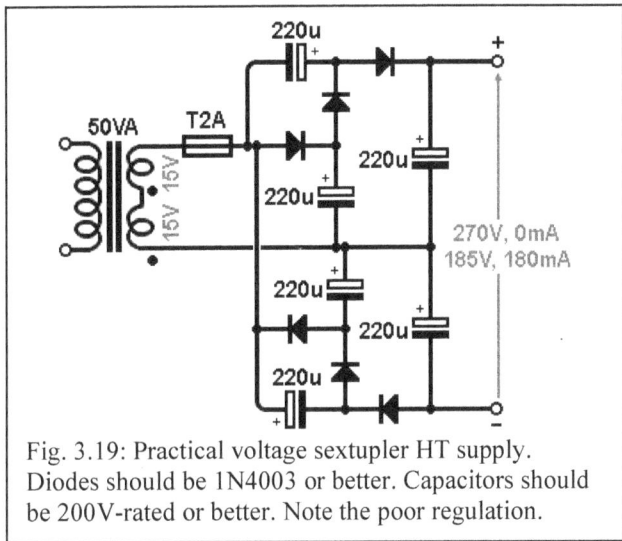

Fig. 3.19: Practical voltage sextupler HT supply. Diodes should be 1N4003 or better. Capacitors should be 200V-rated or better. Note the poor regulation.

Up to 600mA of heater current could be supplied directly with $12V_{ac}$ from the transformer. However, an alternative is to use the voltage doubled output, in which case $32V_{dc}$ is available at up to 150mA as noted in the figure (notice that the first two capacitors are ten times larger than the rest). This leaves enough headroom to supply a couple of 12.6V heaters in series from a 24V linear regulator, eliminating heater hum. But remember, whether AC or DC, the heater supply will be effectively floating at half the HT voltage so should not be grounded. The two 1000µF capacitors should be rated for at least 500mA ripple current but need only be 25V rated. The 100µF capacitors should be 100V-rated, but their ripple current is less than 100mA which is negligible as far as capacitor ratings go.

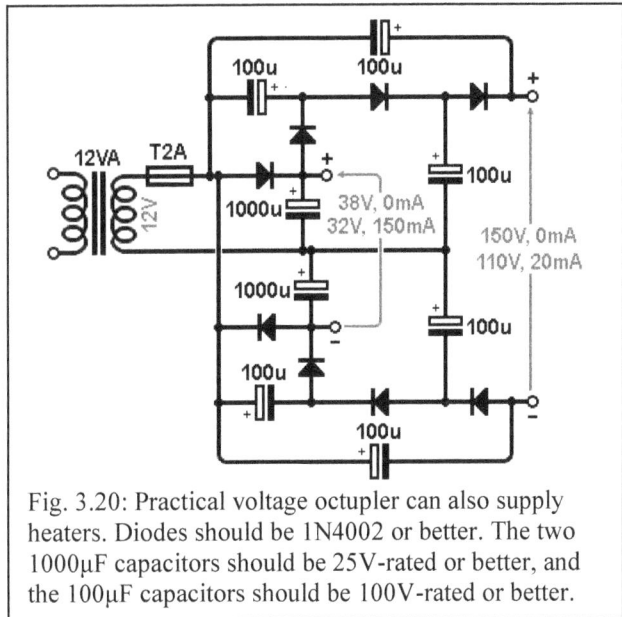

Fig. 3.20: Practical voltage octupler can also supply heaters. Diodes should be 1N4002 or better. The two 1000µF capacitors should be 25V-rated or better, and the 100µF capacitors should be 100V-rated or better.

3.6.5: Power Supply for a Circlotron

Fig. 3.21 shows an example of an HT supply for a circlotron amplifier. This uses a capacitor-coupled rectifier to create a second independent supply from a single, 120V isolation transformer. Using 220µF reservoir capacitors the ripple voltage will be about 7V$_{pp}$, and the

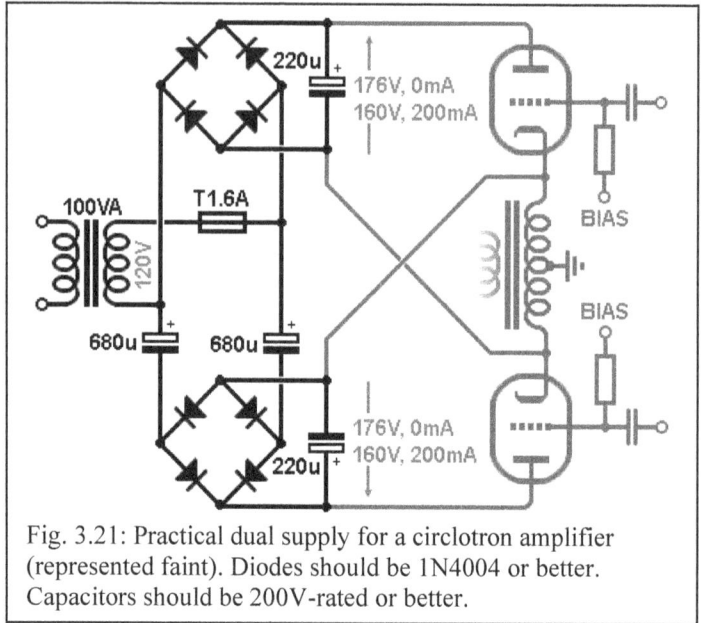

Fig. 3.21: Practical dual supply for a circlotron amplifier (represented faint). Diodes should be 1N4004 or better. Capacitors should be 200V-rated or better.

capacitors should be rated for at least 500mA ripple current.

3.7: Multiplier Cheat Sheets

The following quick-reference formulas can be used as a guide when designing a new multiplier circuit. These formulas assume a typical power transformer will be used, achieving a typical power factor of about 0.65. Figures will be less accurate for a transformer smaller than about 20VA or with a secondary voltage lower than about 20V$_{ac}$, owing to diode drop and poorer regulation.

3.7.1: Full-Wave Voltage Doubler

Refer to fig. 3.22.
No-load output;
$V_{dc} = 2.8\ V_{ac(rms)} = 2\ V_{ac(peak)}$

Full-load output (typical):
$V_{dc} = 2.65\ V_{ac(rms)}$
$I_{dc} = 0.25\ I_{ac(rms)}$

Minimum diode reverse voltage rating:
$V_{rrm} = 2.8\ V_{ac(rms)} = 2\ V_{ac(peak)}$

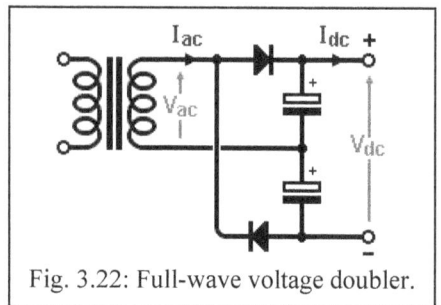

Fig. 3.22: Full-wave voltage doubler.

Minimum diode average forward current rating:
$I_{F(av)} = I_{dc}$

Minimum capacitor voltage rating (all capacitors identical):
$V = V_{dc} / 2$.

Minimum capacitor ripple current rating (all capacitors identical):
$I = I_{ac(rms)} / 1.4$

3.7.2: Full-Wave Voltage Quadrupler

Refer to fig. 3.23.
No-load output;
$V_{dc} = 5.6 \ V_{ac(rms)} = 4 \ V_{ac(peak)}$

Full-load output (typical):
$V_{dc} = 5.35 \ V_{ac(rms)}$
$I_{dc} = 0.12 \ I_{ac(rms)}$

Minimum diode reverse voltage rating:
$V_{rrm} = 2.8 \ V_{ac(rms)} = 2 \ V_{ac(peak)}$

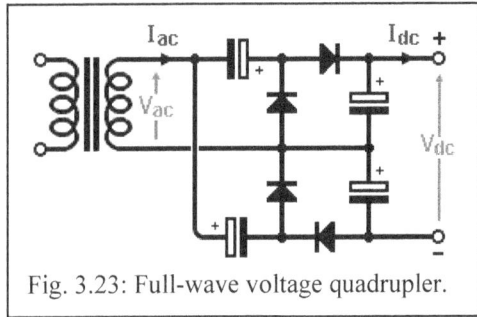

Fig. 3.23: Full-wave voltage quadrupler.

Minimum diode average forward current rating:
$I_{F(av)} = I_{dc}$

Minimum capacitor voltage rating (all capacitors identical):
$V = V_{dc} / 2$

Minimum capacitor ripple current rating (all capacitors identical):
$I = I_{ac(rms)} / 2$

3.7.3: Full-Wave Voltage Sextupler

Refer to fig. 3.24.
No-load output;
$V_{dc} = 8.4 \ V_{ac(rms)} = 6 \ V_{ac(peak)}$

Full-load output (typical):
$V_{dc} = 8.0 \ V_{ac(rms)}$
$I_{dc} = 0.08 \ I_{ac(rms)}$

Minimum diode reverse voltage rating:
$V_{rrm} = 2.8 \ V_{ac(rms)} = 2 \ V_{ac(peak)}$

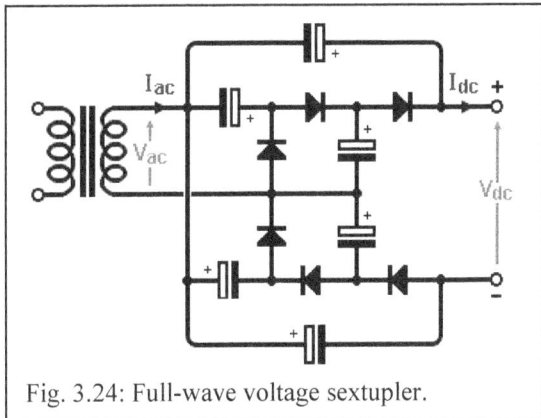

Fig. 3.24: Full-wave voltage sextupler.

Minimum diode average forward current rating:
$I_{F(av)} = I_{dc}$

Minimum capacitor voltage rating (all capacitors identical):
$V = V_{dc} / 2.$

Minimum capacitor ripple current rating (all capacitors identical):
$I = I_{ac} / 2.4$

Chapter 4: Smoothing Filters

Despite wading through three whole chapters, all we have done so far is to turn AC voltage into dirty DC voltage. This may be good enough for the relatively insensitive audio output stage, but the rest of the amplifier or preamplifier will usually need something cleaner. Admittedly, the highest expression of design is to make an amplifier that doesn't need a good power supply in the first place, i.e. to improve the circuit's PSRR to the point where it happily ignores the shortcomings of the supply. Unfortunately, this is seldom a trivial exercise.[*]

4.1: Smoothing, Bypassing, and Decoupling

Ripple voltage is not the only source of noise on the power supply. A second sort of noise appears due to the amplifier circuit itself, because the source impedance of the power supply is not zero. When the audio circuit draws more current, the supply voltage will sag, and when the demand subsides the supply voltage will recover. In this way the supply rail is modulated by the audio signal. We must therefore fight a battle on two fronts; one against the rectifier and one against the amplifier itself. The second is perhaps more important because, if the audio-induced signal couples from one amplifier stage back to a previous stage, it can result in positive feedback. This may cause distortion, peaks in the amplifier's frequency response, ringing (unsustained oscillation), or even full oscillation.

These noisy signals on the power supply can be suppressed with the use of RC or LC filters. In this context they are variously referred to as smoothing, bypass, or decoupling filters. These alternative names derive from the fact that there are really three interrelated jobs being performed:
1. Smooth/filter out residual ripple voltage;
2. Bypass/provide a local energy storage for sudden current demands;
3. Decouple/isolate each amplifier stage from others.

All three are ultimately attempts to make the power supply low-impedance or 'stiff', that is, not easily perturbed by varying rectifier voltage, or varying current demands. We want the power supply to behave like an ideal voltage source as far as possible (although guitar amps may deliberately violate this principle for tonal effect). The filters may also be designed to produce specific supply voltages at different points in the amplifier, although this is not always important. Later in this book we will explore active stabilising circuits and regulators that perform similar jobs with more sophistication.

---

[*]See Chapter 12, Section 10 in: Langford-Smith, F. (1957) *Radio Designer's Handbook* (4th ed.), Iliffe and Sons ltd., London.

83

4.1.1: Smoothing

A simple, passive smoothing circuit is normally built from one or more RC or LC filter stages. A passive filter is really a kind of potential divider or attenuator, as illustrated in fig. 4.1. The ripple attenuation offered by the circuit is just another name for its voltage gain, which in this context is a loss, that is, a gain of less than 1. For a given AC voltage (e.g. ripple) entering the filter, we can easily predict how much will come out by applying its transfer function –a fancy name for a gain equation. The general formula for the gain (loss) of a general attenuator as in fig. 4.1a is:[*]

$$B = \frac{Z_2}{Z_1 + Z_2} \qquad (4.1)$$

The gain is the ratio of the impedance across which the output is taken, to the total impedance $Z_1 + Z_2$.

If we replace Z_1 with a resistor, and Z_2 with a capacitor as in fig. 4.1b, then the gain must be the ratio of the *reactance* of the capacitor X_C, to the total *impedance*. However, we cannot simply add the reactance of the capacitor to the resistance of the resistor (R + X_C) because reactance introduces a 90° phase shift between voltage and current. Instead we must take the vector sum, by squaring the terms, then adding them, then taking the square root: $\sqrt{R^2 + X_C^2}$. This leads to:

$$B = \frac{X_C}{\sqrt{R^2 + X_C^2}} \qquad (4.2)$$

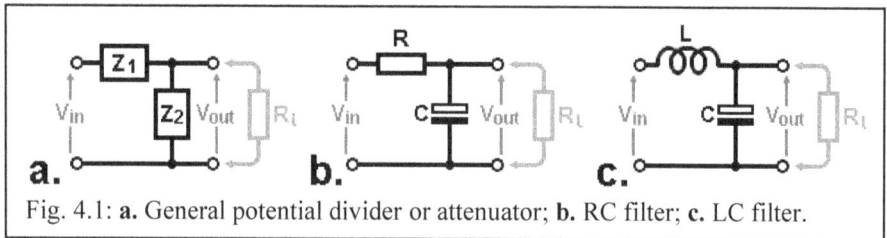

Fig. 4.1: **a.** General potential divider or attenuator; **b.** RC filter; **c.** LC filter.

The reactance of the capacitor falls with frequency, effectively shunting the unwanted ripple to ground. When the reactance is exactly equal to the resistance the gain will be ×0.71, or −3dB. This is called the cut-off, roll-off, or corner frequency. Below this point, signals (or in our case ripple) pass with very little attenuation, so this region is called the passband, as illustrated in fig. 4.2. Above the cut-off frequency the rate of attenuation quickly approaches −20dB per decade (−6dB per octave), which is called the stopband. With a little derivation that we need not reproduce here, we arrive at the familiar formula:

---

[*]It is common to use the letter A for the gain of an active circuit, but the letter B for a passive circuit, though this is by no means universal.

$$f = \frac{1}{2\pi RC} \quad or, \quad R = \frac{1}{2\pi fC} \quad or, \quad C = \frac{1}{2\pi fR} \qquad (4.3)$$

Where f is the cut-off frequency. In a power supply we really only want to pass DC (0Hz), so the cut-off frequency can hardly be set low enough. Often it is convenient to enter a value of 1Hz since this makes f effectively vanish from the formulae.

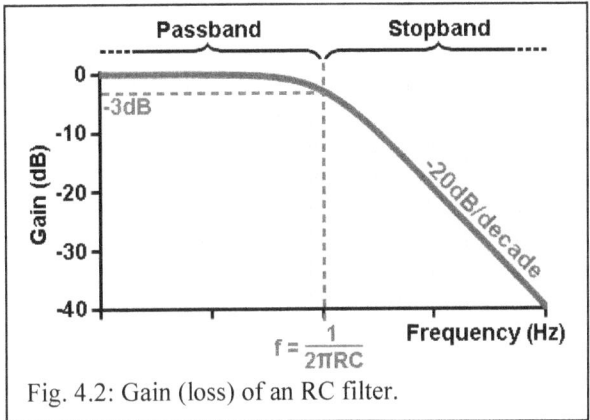

Fig. 4.2: Gain (loss) of an RC filter.

A single RC filter provides up to −20dB/decade attenuation and is called first order. Two RC sections in cascade provide up to −40dB/decade attenuation, which is called second order, and so on. Fig. 4.3 shows this for up to five identical RC sections. It is an essential bit of electronics knowledge that filter slopes *always* come in multiples of −20dB/decade. Note, however, that the attenuation at the cut-off

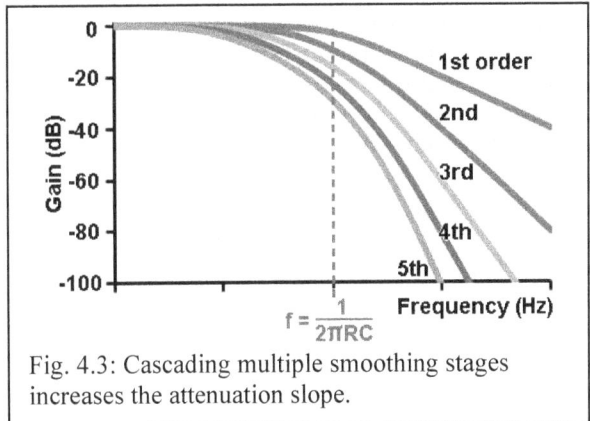

Fig. 4.3: Cascading multiple smoothing stages increases the attenuation slope.

frequency increases by more than −3dB each time; this is because each passive filter section has an additional loading effect on the previous one.* The use of multiple cascaded smoothing sections brings other advantages too, as explained later.

If we go back to fig. 4.1 and this time replace Z_1 with an inductor and Z_2 with a capacitor as in c., then equation (4.1) leads to:

$$B = \frac{X_C}{\sqrt{X_L^2 + X_C^2}} \qquad (4.4)$$

---

* For this reason it is unwise to call the cut-off frequency the '−3dB frequency', since it is not always −3dB.

The reactance of the inductor rises with frequency so will tend to block AC, and the reactance of the capacitor falls with frequency so it will tend to shunt AC to ground. We therefore have *two* reactive components working simultaneously to reduce ripple. An LC filter is therefore a second-order filter and hence delivers a higher degree of ripple

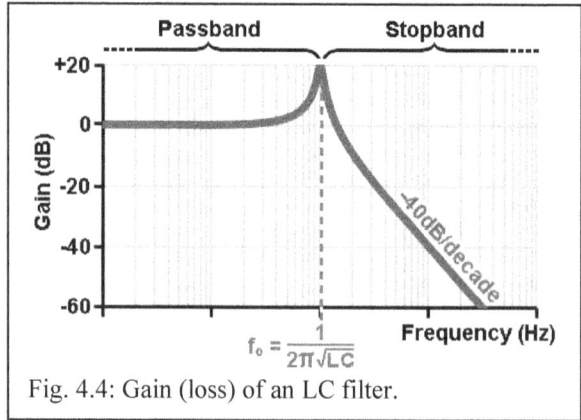

Fig. 4.4: Gain (loss) of an LC filter.

rejection than a single RC filter stage, as illustrated in fig. 4.4.

However, the behaviour of an LC filter at the point where the magnitude of the two reactances are equal is a little more exciting this time. This is because inductive and capacitive reactances introduce phase shift in opposite directions, leading to resonance or 'magnification'. Although this critical frequency could again be called the cut-off frequency, it is more usually called the resonant frequency, f_0. From the previous equation the following can be derived:

$$f_0 = \frac{1}{2\pi\sqrt{LC}} \quad \text{or,} \quad L = \frac{1}{(2\pi f_0)^2 C} \quad \text{or,} \quad C = \frac{1}{(2\pi f_0)^2 L} \qquad (4.5)$$

Resonance in LC smoothing filters introduces problems that do not exist for RC filters, and this is dealt with in more detail in section 4.3.1.

4.1.2: Bypassing

A smoothing capacitor can also serve as a power-supply bypass capacitor. To do so it should be physically close to the amplifier stage, not at the wrong end of a long feed wire which has unwanted resistance and inductance. When it is physically close to the amplifier stage, the capacitor provides a low-impedance, local reservoir of energy, just for that stage. Signal currents will then be preferentially drained from this local bypass capacitor, as illustrated in fig. 4.5. This keeps the precious signal current flowing around the amplifier stage as actually designed, rather than being influenced by the less-well-controlled impedances in the raw power supply feed and ground wiring. You can imagine each bypass capacitor acting a bit like a rechargeable battery,

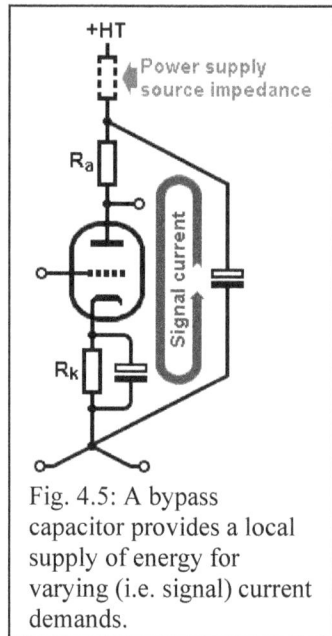

Fig. 4.5: A bypass capacitor provides a local supply of energy for varying (i.e. signal) current demands.

dedicated to each amplifier stage. The raw power supply serves only as the battery charger, fulfilling the long-term average current demands rather than the instantaneous ones.

For this to be effective, the capacitive reactance must be small enough that it can be considered a short circuit to ground at the lowest frequency of interest. That is, it must be small compared to the *intended* impedance in series with the valve (e.g. the anode resistor). Too little capacitance and the reactance will rise too high at low frequencies, so the total impedance seen by the valve will increase, leading to a frequency response anomaly or some other unintentional aberration, depending on circuit details. Note that this is an entirely separate issue from whether the capacitor provides enough smoothing of ripple voltage.

To drive this point home, suppose we had a hair-raising 1kV raw power supply but we only needed 300V/1mA for an amplifying stage. This implies we could use a 700kΩ dropping resistor. This is so large we might think we can get away with a 1μF smoothing capacitor; after all, this would create a cut-off frequency of just 0.2Hz and provide about 52dB ripple reduction at 100Hz. But looking at it from the bypassing perspective, a 1μF capacitor has a reactance of 8Ω at 20kHz but 7958Ω at 20Hz, so the amplifier stage will see an ever increasing load impedance at low frequencies. For a typical valve stage this could well be enough to cause an uneven frequency response. In other words, the output impedance of the smoothing filter grows too large at low frequencies. For this reason, valve stages will nearly always need at least 10μF of bypass capacitance, irrespective of smoothing and decoupling requirements, in order to keep the output impedance of the filter small enough.

Ideally, every valve would have its own bypass capacitor (e.g. fig. 4.6b), mounted as close to it as possible to minimise the impedance in series with the capacitor. However, this is often impractical with bulky components, so a typical compromise is to provide one decoupling capacitor for every two gain stages, which often means one per bottle (dual triode). The two valves should be contiguous in the signal path – it would be a bad idea to allow, say, the first and last valves in a multi-valve audio chain to share the same bypass capacitor, as this would result in the worst possible decoupling –see next.

4.1.3: Decoupling

When several amplifying stages share the same power supply node then there will be some degree of unwanted feedback (and indeed feedforward) between them. This is because the power supply does not have zero source impedance, so audio currents being drained from it will develop a corresponding voltage drop across it, which is shared by all the amplifying stages, as illustrated in fig. 4.6a. Another way to describe it is to observe that in the figure the load resistors form potential dividers with the power supply impedance, so the output signal voltages from each stage will appear (attenuated) on the shared power supply rail, readily mixing and feeding back into each of them. Since smoothing capacitors have rising reactance at low frequencies this tends to be a low frequency problem. A mild symptom might be increased LF distortion or crosstalk in a stereo amplifier, while more severe positive feedback can cause sustained LF oscillation called **motorboating**, or sub-1Hz oscillation called **breathing**.

This problem is reduced by feeding each stage from its own RC or LC filter so that each stage is decoupled from the rest, and the lower the cut-off frequency the better. Any signals trying to creep along the power supply rail from other stages will then be attenuated by the local decoupling, thus reducing the loop gain around the amplifier through the power supply. Yes, this is similar in principle to smoothing, except we're now talking about audio-induced ripple rather than rectifier ripple. One massive capacitor in the attic could be thoroughly effective at smoothing rectifier ripple, yet be useless for decoupling since the long wire leading from it to the

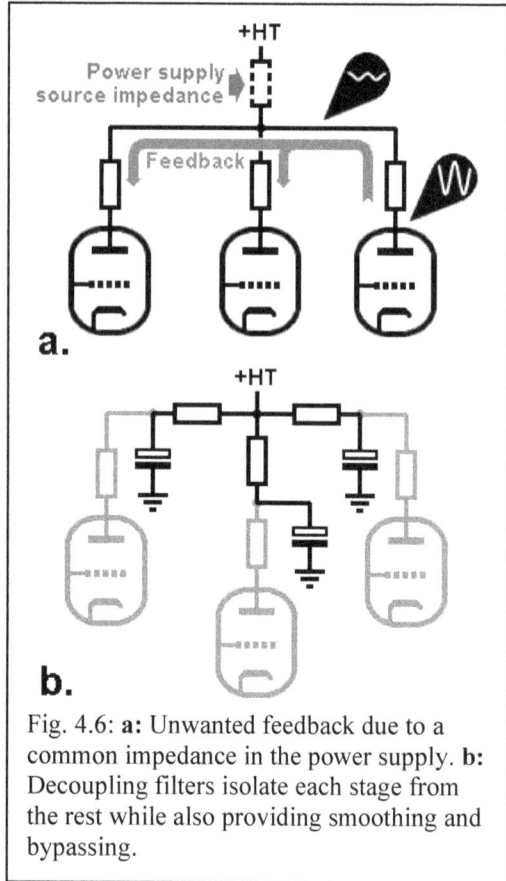

Fig. 4.6: **a:** Unwanted feedback due to a common impedance in the power supply. **b:** Decoupling filters isolate each stage from the rest while also providing smoothing and bypassing.

amplifier would represent a shared (common) impedance across which a program-induced voltage would appear.

Ideally, we would provide one filter for every amplifier stage, but the usual compromise of one capacitor for every two stages usually suffices for decoupling. If stereo channels share the same power supply then similar valves from opposite channels can share a filter, as in fig. 4.7, since positive feedback is not an issue with this arrangement (the problem is instead crosstalk, but as it should only exist at sub-audio frequencies it is less important). However, wide-bandwidth or digital circuits should also have ceramic or plastic bypass capacitors mounted very close to each amplifying stage or IC, in addition to general electrolytic bypassing, as even a little inductance can form a common impedance that leads to HF (rather than LF) instability. This is much less likely in an audio circuit, especially the valve kind –see the next section.

Whether to use a system of cascaded filters or individual decoupling filters all connected to a common star point (see fig. 4.9), is at the designer's discretion. For similar component values a cascaded arrangement provides better ripple smoothing

88

whereas the star system gives better decoupling and greater freedom to choose DC voltage drop. Of course, a combination of the two (a branching network) may also be used.

A further point of note is that when amplifying stages share the same power supply filter, increasing the series dropping resistance (or inductance) will improve the smoothing but *worsen* the decoupling between the two valves, since it raises the output impedance of the filter. By contrast, increasing the capacitance will improve *both* things. Therefore, if you

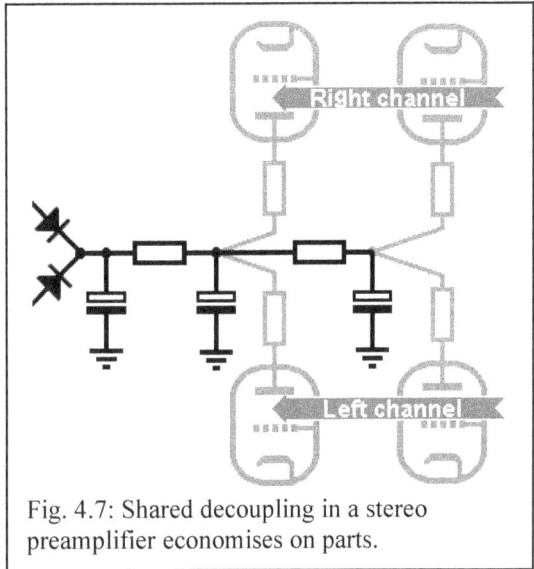

Fig. 4.7: Shared decoupling in a stereo preamplifier economises on parts.

have the option, use smaller dropping resistances and larger smoothing capacitors.[*]
On the other hand, farads are more expensive than ohms.

4.1.4: High-Frequency Performance

All capacitors have some unavoidable self-inductance which will resonate with the capacitance at a particular frequency, and for electrolytics this lies in the region of tens of kilohertz. Above the resonant frequency, the device stops behaving like a capacitor and starts behaving like an inductor, i.e. it becomes less and less effective at shunting high frequencies. If the electrolytic had nothing but capacitance and inductance then, at resonance, its impedance would drop to nothing, but in practice there is also the equivalent series resistance (ESR) to damp the circuit. In

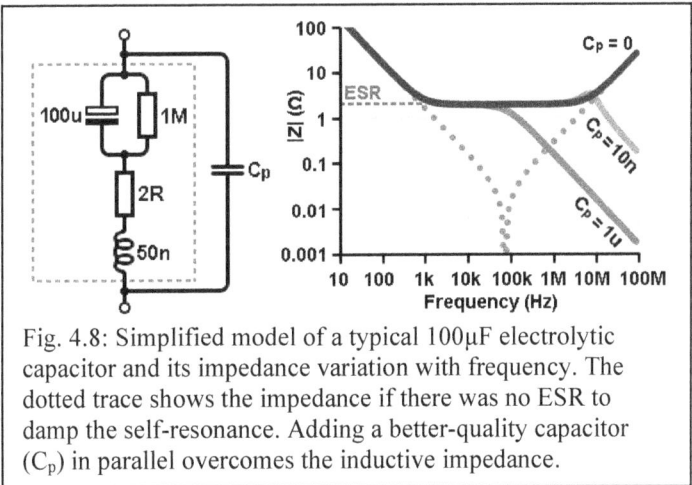

Fig. 4.8: Simplified model of a typical 100µF electrolytic capacitor and its impedance variation with frequency. The dotted trace shows the impedance if there was no ESR to damp the self-resonance. Adding a better-quality capacitor (C_p) in parallel overcomes the inductive impedance.

[*]This can be an effective way to reduce 'ticking' in any guitar amp which happens to have a smoothing filter shared between an audio stage and a tremolo oscillator.

most electrolytic capacitors this causes the impedance to bottom out somewhere in the audio range.

To counteract this effect, a plastic or ceramic capacitor can be added in parallel with the electrolytic so that, at high frequencies where the electrolytic starts to give up, the smaller capacitor takes over and maintains a low impedance path. The effect on total impedance is illustrated in fig. 4.8. There is no particular rule governing the size of the second capacitor –a few tens of nanofarads or more will do. However, this technique does rather rely on the second capacitor itself having low inductance, which might not be the case in a hand-wired amp where the leads are not cut very short.

Although much is written in the audio press about paralleling capacitors, it is usually a moot point for audio circuits, especially valve amps. A competent audio circuit will usually have its bandwidth curtailed not far above 100kHz, or even lower in a valve amp, so it usually does not matter if electrolytic capacitor impedance starts to creep up beyond 10MHz. There is almost never any need to add parallel ceramic capacitors in a valve amp, though neither will it do any harm. Having said that, in some designs you may encounter several parallel capacitors of ever-diminishing value. This is usually the mark of a designer who has read a little *too* much and potentially made things worse. It is better to use identical values in parallel to avoid multiple resonant effects, especially in circuits operating over 50MHz.

4.2: RC Smoothing

An RC smoothing stage is a low-pass filter whose attenuation increases with frequency. The only frequency we really want to pass is 0Hz or DC, but since infinite capacitors are only available for government work, we must compromise. For a given capacitance, a larger resistance lowers the cut-off frequency and therefore improves the smoothing. However, there will also be some DC drop across the resistor due to the load current flowing in it (fig. 4.10), so the choice of this **dropping resistor** is usually restricted by how much voltage we can afford to throw away. The attenuation of the filter also depends on the load impedance, but we can ignore this since it is normally much larger than the reactance of the capacitor.

Note that the dropping resistor must be capable of withstanding the full supply voltage and charging current at start up. In the HT supply this usually demands ½W resistors or better, even though their steady-state dissipation might be much less.

If smoothing rectifier ripple was the *only* task at hand, then it could be done in one go, with a series resistor or inductor plus one big capacitor. Unfortunately, the more subtle requirements of decoupling and bypassing make such an approach impractical and uneconomical, so a network of more modest filters is normally used instead.

4.2.1: RC Filter Chains

In a multi-stage amplifier, it is normal for every one or two valve stages to be supplied from its own smoothing filter. Typically, the smoothing filters will be arranged in one of two ways as illustrated in fig. 4.9. Either they will be supplied from a common feed point (usually the reservoir capacitor) to form a 'star' network is in a., or else they will be cascaded as in b. In practice there is little to choose between the two approaches, but the cascade seems to be the most popular. Branching combinations of both are also possible.

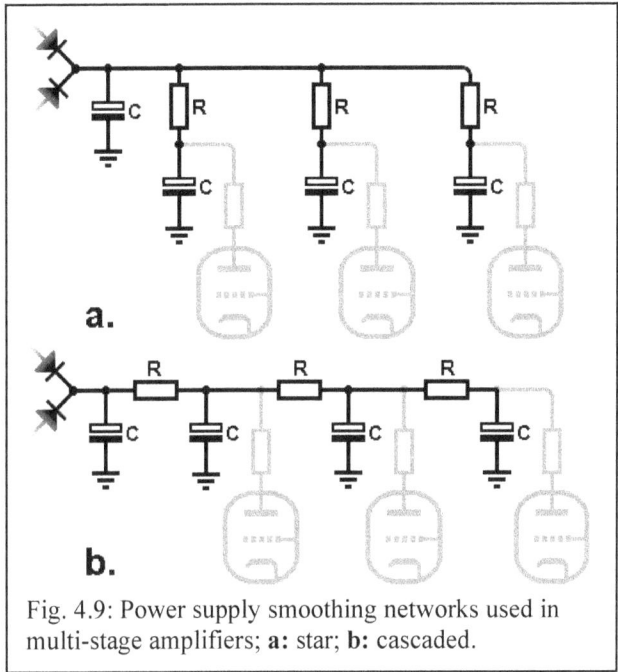

Fig. 4.9: Power supply smoothing networks used in multi-stage amplifiers; **a:** star; **b:** cascaded.

For a given set of resistors and capacitors, the star network offers better decoupling between neighbouring amplifier stages, and complete freedom to set the DC voltage at each stage. On the other hand, the cascade offers better ripple filtering as you progress down the line. The order of valve stages should flow in the opposite direction, i.e. the input valve should be fed from the final (best smoothed) filter section, and the output stage should be feed from the first (least well smoothed) filter section.

The anode currents of each stage accumulate towards the start of the filter chain, as illustrated in fig. 4.10. This makes the DC voltage drops at each node somewhat interactive, but in most amplifiers the exact HT voltage at any point is not critical. The drop across each resistor

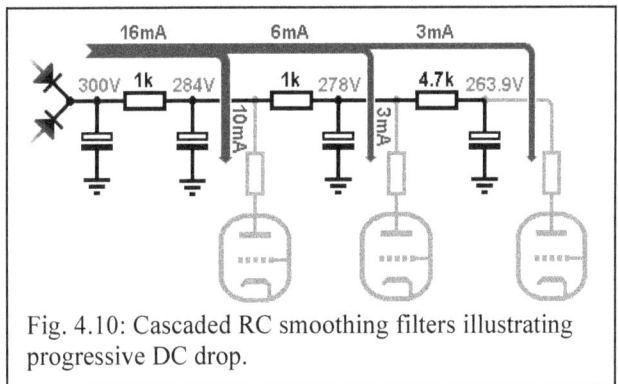

Fig. 4.10: Cascaded RC smoothing filters illustrating progressive DC drop.

91

(and its consequent power dissipation) is easily calculated if we know how much anode current each stage draws.

Of course, the cascade arrangement also means that if one amplifier stage draws a surge of current, slightly draining its smoothing capacitor, then the voltages at all succeeding supply nodes will sag too. This is just another way of saying the decoupling between neighbouring stages is not as good as with a star network. This 'successive sagging' effect can be minimised, or at least slowed down, by inserting a diode (e.g. 1N4007) into the

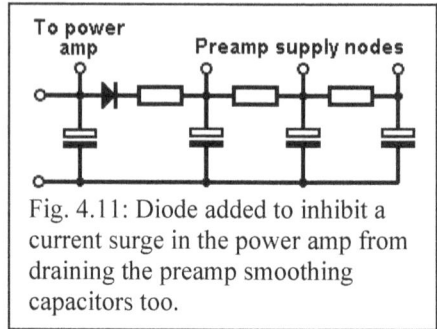

Fig. 4.11: Diode added to inhibit a current surge in the power amp from draining the preamp smoothing capacitors too.

power supply as in fig. 4.11. This prevents the offending stage –usually a class-AB or -B power output stage– from instantaneously draining current from the succeeding filters. This technique can also be used to help reduce the interaction of stereo channels operating from a shared reservoir. Of course, if the current surge is sustained for a *long* time, the preamp voltages *will* eventually sag too.

4.2.2: Tuned RC Filters

Power supply smoothing need not be restricted to simple RC filters. The twin-T filter is a familiar way to create a notch, providing deep attenuation at one particular frequency. In a power supply it is usually twice mains frequency (100Hz or 120Hz) that is targeted for elimination. There are various component combinations that could be used,[1] but the version in fig. 4.12 utilising three identical capacitors is probably the most convenient. In this case R must be equal to:

$$R = \frac{1}{\pi f C \sqrt{2}} \qquad (4.6)$$

Where f is the notch frequency.

Capacitor tolerance and use of nearest-standard values will inevitably throw off the notch slightly, but it will usually be close enough for rock and roll. If not, it can be manually tuned by varying the shunt resistance. Fig. 4.13 shows a practical example tuned to 100Hz, together with its frequency response and output impedance. All capacitors must be rated to withstand the supply voltage. Further decoupling can be added after the notch filter, as required.

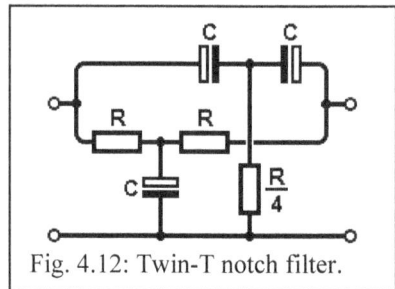

Fig. 4.12: Twin-T notch filter.

---

[1] Langford-Smith, F. (1957) *Radio Designer's Handbook* (4th ed.), p1196. Iliffe and Sons ltd., London.

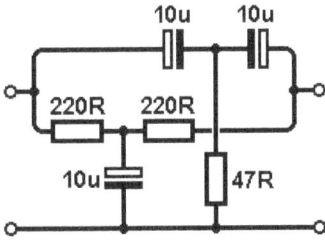

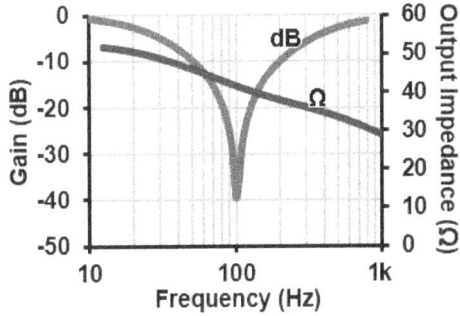

Fig. 4.13: Practical twin-T notch smoothing filter. For a 120Hz notch use 180Ω and 39Ω.

The twin-T filter can produce an impressive notch, but it needs six components which makes it rather expensive, both in money and space. Perhaps we can work smarter, not harder? A typical amplifier design may already include a few cascaded RC smoothing filters, and by adding one capacitor we can create a bridged-T notch filter. Fig 4.14 shows an example where a bridging capacitor C_1 has been added across a pair of ordinary smoothing stages. The accompanying graph shows the frequency response of the power supply filtering. At node ① it remains unchanged, but at node ② a 100Hz notch is created. The corresponding valve stage –which might be the amplifier input stage– therefore gets better smoothing at this dominant ripple frequency, at the expense of higher frequencies. Whether this gives a better audible result can only be judged by ear. Since filter sections vary enormously between different amplifier designs, determining the exact value for C_1 is best done by computer modelling initially, then cut-and-try on test.

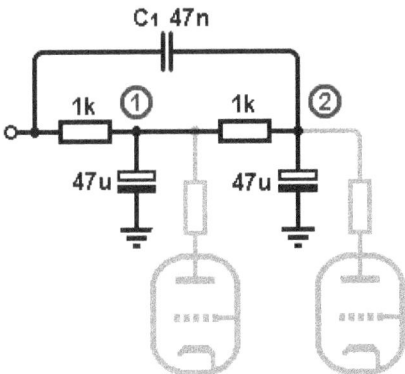

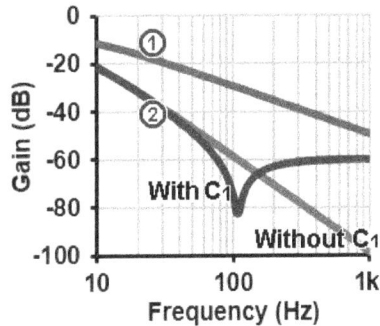

Fig. 4.14: Bridging capacitor C_1 converts two existing RC smoothing filters into a bridged-T notch filter. For 120Hz use 39nF.

A further variation on this theme is possible when there are three or more cascaded RC filters to play with.[2] With at least three filters, phase shift in the final stage will reach and exceed 180°. If some of the original signal is fed forward in the right proportion, phase cancellation will occur at one particular frequency, again creating a type of notch filter. Fig. 4.15 shows an example. Since the bridging component is resistor, rather than a capacitor, it is easier to tune.

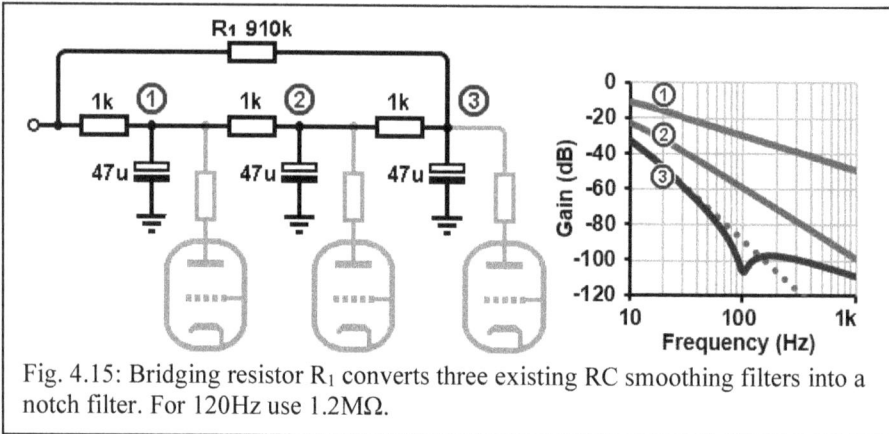

Fig. 4.15: Bridging resistor R_1 converts three existing RC smoothing filters into a notch filter. For 120Hz use 1.2MΩ.

4.3: LC Smoothing

The previous RC filters could in principle be replaced by LC filters, where the dropping resistor is instead a large inductor or choke, as in fig. 4.16. This would double the rate of attenuation compared with RC smoothing, and a further advantage is that a choke normally has low DC resistance, which minimises DC voltage drop. For this reason, a common arrangement is to use an LC filter between the

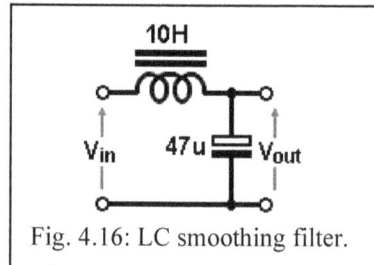

Fig. 4.16: LC smoothing filter.

anode and screen supplies of the power valves, since this maximises the screen voltage (a cheap way to maximise output power) while giving the screens the improved smoothing they need. But as always, these advantages come at a price.

The disadvantages of LC filtering in an HT supply are that suitable chokes are bulky and expensive, both because they must be constructed with an air gap and because they must be suitably rated to withstand the full supply voltage between conductor and core. This makes them a specialist item. A more disconcerting disadvantage is that the choke will resonate with the capacitor at a frequency of:

---

[2] Scott, H. H. (1939) Simple Improvements in R-C Power Supply Filters, *Electronics*, August, pp42+44+46+48.

$$f_o = \frac{1}{2\pi\sqrt{LC}}$$

To illustrate this problem, fig. 4.17 shows the response of the LC smoothing filter from fig. 4.16. The conspicuous feature is the resonant peak at 7.3Hz. At this frequency the filter magnifies rather than attenuates any noise on the power supply. Of course, the rectifier ripple frequency will be much higher than this, so we might think there is no problem.[*] Unfortunately, the output impedance of the LC filter also shoots up around the resonant frequency, so in this region we get little or no effective bypassing or decoupling either. This may induce unpleasant audio behaviour when triggered by the envelope of the audio signal, or it may simply motorboat. If the resonance happened to occur within the audible spectrum –which is quite possible with a small choke– then the situation would be even worse.

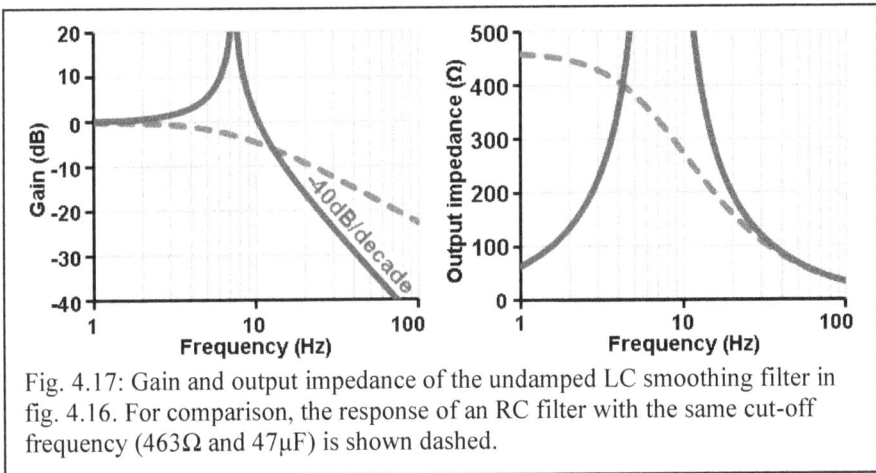

Fig. 4.17: Gain and output impedance of the undamped LC smoothing filter in fig. 4.16. For comparison, the response of an RC filter with the same cut-off frequency (463Ω and 47μF) is shown dashed.

4.3.1: Avoiding Resonance in LC Filters

One way to avoid the resonance problem might be to use such large values of L or C that the resonant peak is pushed down to ultra-low frequencies, which we imagine can't possibly exist in an audio amplifier. Unfortunately, this is naïve, because the envelope of an audio signal can contain energy at any and all frequencies, almost down to DC. For example, the sound of a kick drum being thumped at 120bpm will drink pulses of current from the power supply at a rate of two per second, or 2Hz, and that is a trivial example. The wall voltage also fluctuates and wanders around at

---

[*]Actually, it *was* a problem in a circuit I encountered that used an LC filter consisting of a 2.2μH inductor and 2200μF smoothing capacitor, which together resonate at 2.3kHz. The power source was an SMPS which contained enough hash at this frequency to excite the filter and introduce an annoying low-level whine into the audio circuit. Adding a 0.2Ω damping resistance cured it.

a slow rate, which will be passed on to our smoothing filter and may be magnified by LC resonance.

The inductance of an iron choke varies with the current flowing in it, particularly when it is a mixture of DC and AC. Smoothing chokes are built with an air gap in the core to suppress DC flux, but it is not completely effective. The inductance may be more-or-less constant when the DC current is close to the published maximum, but it will fall if the DC current or ripple voltage are too large (causing the core to saturate), or when the DC current is very small (because the permeability of iron is smaller at low flux levels). At switch-on there is an inrush of current to charge the smoothing capacitors and this may cause the HT voltage to oscillate at the resonant frequency, so the DC current and the ripple voltage across the choke will both be abnormally large for a short time. This may push the choke momentarily into saturation, causing its inductance to fall and the resonant frequency to rise into the audio range. For this reason, it is not uncommon to hear strange throbbing or squealing sounds during the switch-on period, and possibly at switch-off too. This may be worse if a standby switch is used since the inrush will be heavier, and the valves will be ready and primed to amplify any signal.

A better way to avoid resonance is to damp it by adding resistance in series with the choke, as in fig. 4.18. Some of this is already provided by the choke's own winding resistance, but we can always add more (an 'audiophile' choke made with low-resistance wire may therefore give more trouble with ringing and oscillation than a less expensive choke!). The resonant peak is completely suppressed when the resistance is equal to $R = X\sqrt{2}$, where X is the reactance of the inductor *or* capacitor at the resonant frequency. This is called a Butterworth response. However, this will still

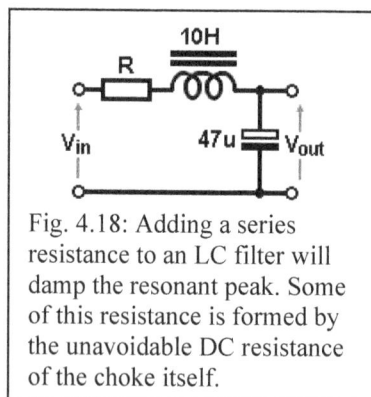

Fig. 4.18: Adding a series resistance to an LC filter will damp the resonant peak. Some of this resistance is formed by the unavoidable DC resistance of the choke itself.

allow a small overshoot or 'bounce' in the supply voltage if the circuit is triggered with a step change in load current. It takes a little more resistance to completely suppress this overshoot, and this occurs when the resistance is equal to

$$R = 2\sqrt{L/C} = 2X$$, which is called critical damping.

Fig. 4.19 shows some examples of the gain and output impedance of the filter in fig. 4.18 with different resistance values. A value of 652Ω produces a Butterworth response while 922Ω gives critical damping. In practice the load resistance also adds some damping, so we might settle on a convenient value of 470Ω. It is clear from the figure that the damping resistance determines the output impedance at low frequencies. In other words, bypassing/decoupling close to the resonant frequency is improved, but at lower frequencies it is degraded (to add insult to injury, the output impedance is slightly worse than a simple RC filter using the same RC combination).

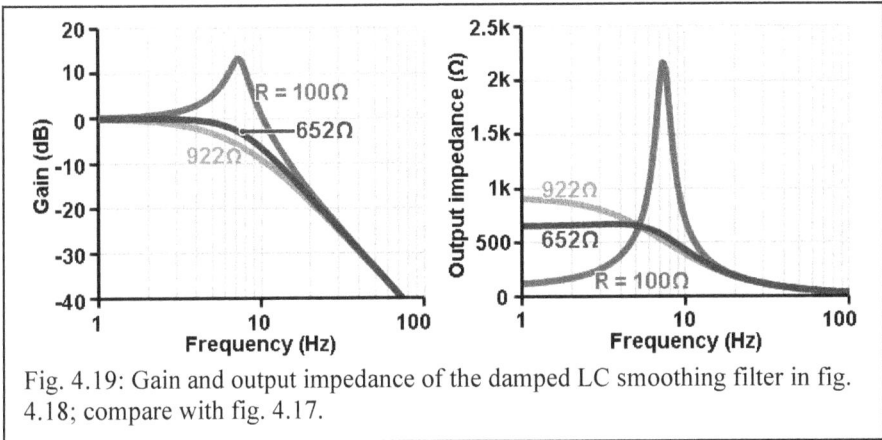

Fig. 4.19: Gain and output impedance of the damped LC smoothing filter in fig. 4.18; compare with fig. 4.17.

Moreover, by adding resistance we also increase the DC drop, which is unfortunate since minimal drop was one of the advantages of using a choke in the first place. When using an LC filter there is, therefore, a balance to be struck between acceptable DC drop and effective decoupling/bypassing. Most amplifier circuits employ one smoothing choke at most, and it is wise to avoid cascades of (undamped) LC filters since this invites multiple resonances.

4.3.2: Tuned LC Filters

By adding a capacitor in parallel with a choke, as in fig. 4.20, an LC filter can be tuned to reject one frequency more effectively than all others. This technique is not often used nowadays since we can usually buy as much smoothing as we need with inexpensive capacitors, but if power-supply hum does give trouble then it may be possible to suppress it using this method. The resonant frequency of the paralleled

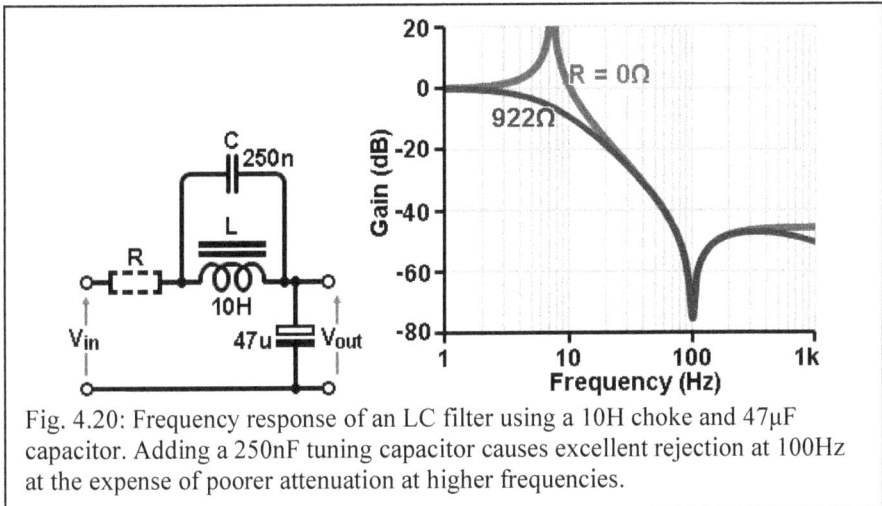

Fig. 4.20: Frequency response of an LC filter using a 10H choke and 47μF capacitor. Adding a 250nF tuning capacitor causes excellent rejection at 100Hz at the expense of poorer attenuation at higher frequencies.

choke and capacitor is:

$$f = \frac{1}{2\pi\sqrt{LC}}$$

Solving for C gives:

$$C = \frac{1}{(2\pi f)^2 L}$$

In practice some adjustment-on-test is inevitable since both components usually have wide tolerances. It may be necessary to use a few different capacitor values in parallel to hit the critical number. Any additional resistance used to damp the low frequency resonant peak should be added in series with the whole LC combination (shown dashed in fig. 4.20) so it does not affect the depth of the notch. For reliability, the capacitor ought to be rated for the HT voltage, to cope with switch-on and switch-off transients and flyback.

The filter in fig. 4.20 is tuned to 100Hz with a 250nF capacitance, producing a deep notch, but at higher frequencies the response flattens off since the filter now looks like a capacitive divider. This is the price which is paid for tuning to a particular frequency; energy is mined out to create the notch, but it must be piled up somewhere else on the spectrum.

Another application of tuned LC filters arises when dealing with switch-mode power supplies. The switching frequency may be above the audio band, but this does not stop it intruding into wideband distortion and noise measurements. It may also modulate with other frequencies to produce products that *are* within the audio band. An LC filter at the output of the power supply, tuned to the switching frequency, can be very effective in suppressing these problems. Since the switching frequency is high, the inductor can usually be in the microhenry range, meaning it will be small and readily available. Moreover, many AC-DC ('offline') power supplies actually

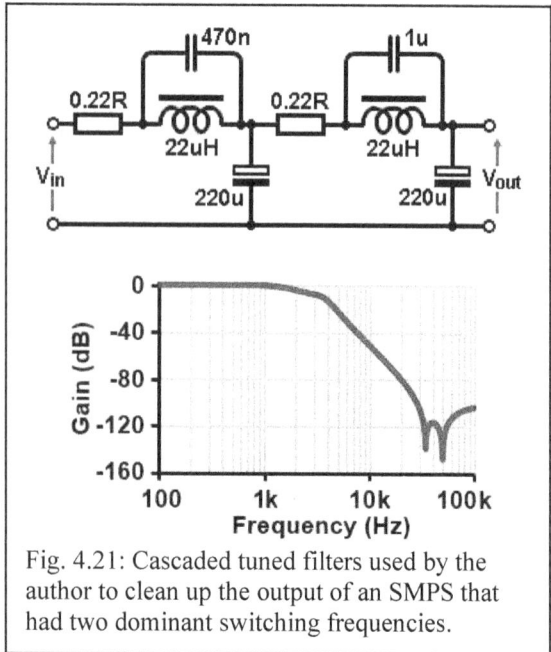

Fig. 4.21: Cascaded tuned filters used by the author to clean up the output of an SMPS that had two dominant switching frequencies.

98

employ two switching frequencies; the main one used for conversion which may be impressively high, and a lower-frequency they don't tell you about, which is apparently used for power-factor correction. If you can identify the offending frequencies with a spectrum analyser you can use cascaded LC filters to tune each of them out. Fig. 4.21 shows an example the author used in a commercial product (the chokes themselves had about 50mΩ of coil resistance which added to the damping).

4.3.3: Measuring Smoothing Chokes

A smoothing choke must be rated for the maximum average DC current that is expected in the circuit. Exceeding this value for too long may cause the core to saturate, which will cause it to heat up. In extreme cases, overloaded chokes have been known to catch fire. A possible failure mode exists if the reservoir capacitor dies or disconnects, leaving any subsequent choke and capacitor to handle the heavy ripple current, effectively creating a choke-input rectifier.

If you have an unknown HT smoothing choke then a conservative estimate for the DC current rating is:

$$I_{dc(max)} = \frac{10}{R} \qquad (4.7)$$

Where R is the resistance of the coil.

Commercial chokes will usually advertise a DC voltage rating too, which is the maximum that can be sustained between the winding and the frame. For an unknown choke there is no way to be sure of this rating, but if you know what circuit it was removed from then this will serve as a guide. If the voltage rating is unknown or not high enough for the application then a possible solution is to place the choke in the ground path of the power supply instead, as in fig. 4.22b. This is only possible in situations where the reservoir capacitor itself does not need to be directly connected to audio ground, obviously. Many early radios also used this approach for back-biasing – section 8.4.6.

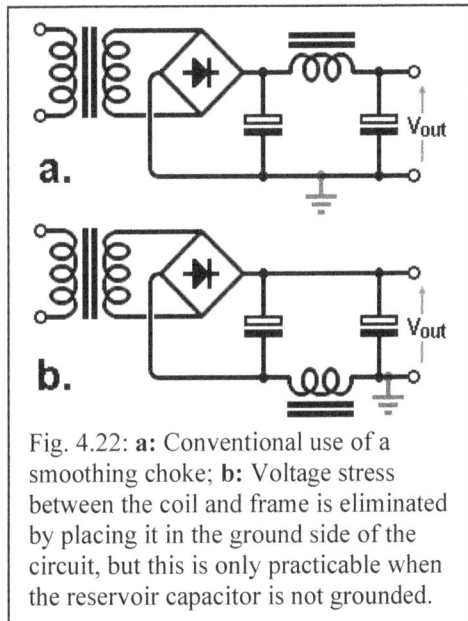

Fig. 4.22: **a**: Conventional use of a smoothing choke; **b**: Voltage stress between the coil and frame is eliminated by placing it in the ground side of the circuit, but this is only practicable when the reservoir capacitor is not grounded.

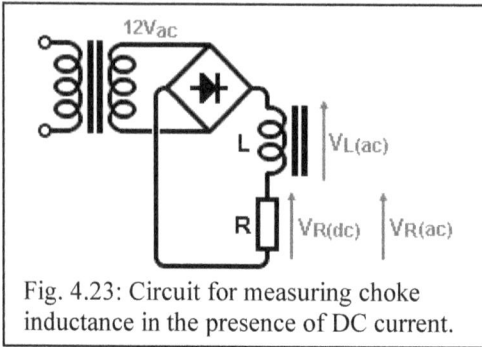

Fig. 4.23: Circuit for measuring choke inductance in the presence of DC current.

The effective inductance of a smoothing choke can vary significantly depending on both the DC current through it, and the amplitude of the AC ripple voltage imposed across it. Trying to measure the inductance with an ordinary digital LCR meter is hopeless, as such meters apply only a small, high-frequency AC test voltage, and they are confounded by large coil resistance and stray capacitance. Smoothing chokes are best measured under conditions that resemble how they will actually be used in practice. Fig. 4.23 shows a method that will work for most HT smoothing chokes. A small 12V transformer provides a convenient voltage which is rectified. The rectified signal contains a 17V_{pp} AC component that is mainly 100/120Hz, and a 10.8V mean average DC component. The voltage is applied to the choke under test, L, and a known resistor, R, in series. Use the following method:

Set your digital multimeter to measure DC volts. Measure the voltage across R; call this $V_{R(dc)}$. From Ohm's law the DC current in the circuit is: $I_{dc} = V_{Rdc}/R$ Adjust the resistance to get the desired DC test current. R will usually fall between about 30Ω and 1kΩ (use power resistors, 2W or better); a value smaller than 10Ω will make it difficult to get an accurate reading in the next step. The resistance of the choke coil itself will limit the maximum available current, but it should be possible to get close to the DC rating of most commercial chokes with this method.

Now set your multimeter to measure AC volts. Measure the voltage across R and across L; call these $V_{R(ac)}$ and $V_{L(ac)}$. The effective inductance can now be calculated:

$$L = \frac{R}{2\pi f} \times \frac{V_{L(ac)}}{V_{R(ac)}}$$

Where f is twice the wall frequency. This ignores coil resistance, but for chokes of standard design this has negligible effect on the answer.

Fig. 4.24 shows the author's results when measuring two different chokes across a range of DC currents. The upper set were for a device advertised as 10H 100mA$_{dc}$ –a specification which it appears to meet with reasonable accuracy. The lower set were for an unknown choke removed

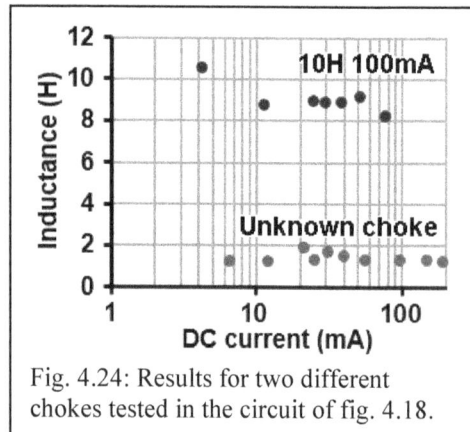

Fig. 4.24: Results for two different chokes tested in the circuit of fig. 4.18.

100

from old equipment –it appears to be a 1.5H device, probably rated for at least 300mA$_{dc}$ based on its coil resistance of 33Ω. Both devices show reasonable constancy of inductance with DC current, which is exactly what we would hope for from a proper, gapped core.

4.3.4: Power Transformers as Chokes

A question most valve enthusiasts ask at some point is: can a power transformer be used as a cheap and dirty smoothing choke? The answer is: sort of. Since they have no air gap, even a tiny DC current is enough to push the core into saturation, causing inductance to drop rapidly. To see this effect in action, fig. 4.25 shows the inductance of a junk-box 6VA transformer when used as a makeshift choke in the test circuit of fig. 4.23 (using its 120V primary coil as the choke winding). The device is clearly not designed for the task. Nevertheless, for DC current less than about 10mA it still has sufficient inductance to be useful, and many

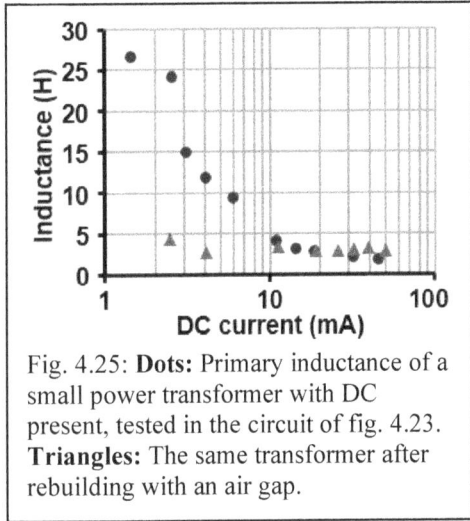

Fig. 4.25: **Dots:** Primary inductance of a small power transformer with DC present, tested in the circuit of fig. 4.23. **Triangles:** The same transformer after rebuilding with an air gap.

preamp applications fall within this range. Bear in mind that power transformers have no stated DC voltage rating, so it is best to assume it is less than 350V$_{dc}$, i.e. less than peak wall voltage. All secondaries should be left open circuit.

A further option is to dismantle the core and reassemble it with an air gap. This was done with the previously tested transformer. The U-clamp was pulled off and a thin blade was slid between the laminations to crack the varnish. The first lamination was pulled out with needle-nose pliers, destroying it in the process (see fig. 4.26), after which the rest were more easily removed. With the E and I pieces collected together, the core was reconstructed with a piece of paper between the two stacks to create a gap. Without the original interleaving the laminations will not naturally hold

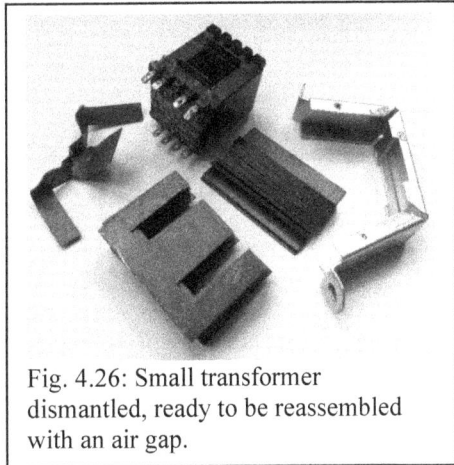

Fig. 4.26: Small transformer dismantled, ready to be reassembled with an air gap.

together, so a generous coat of varnish was used to glue the lot together before the U-clamp was refitted, squeezing it in a vice. Fig. 4.25 shows the measured result was a pleasingly-constant 3 henries up to 50mA and probably beyond; a useful smoothing choke had for practically no cost.

Another possible way to use a transformer, if it has identical dual primary windings, is as a high-inductance common-mode choke. With the correct phasing as illustrated in fig. 4.27, the same DC current flows in opposite directions through each winding, cancelling out core flux and avoiding saturation. Note that this also means it offers no

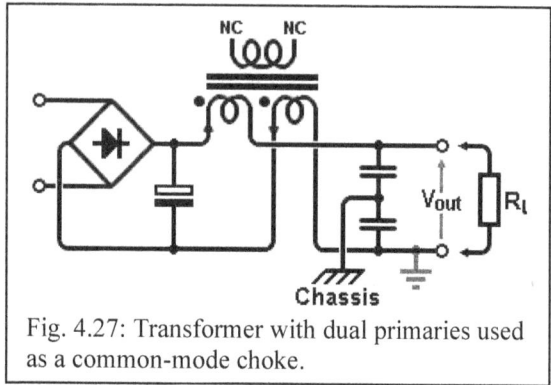

Fig. 4.27: Transformer with dual primaries used as a common-mode choke.

inductance to ordinary power supply ripple, although the coil resistance may still provide some useful smoothing. *Common-mode* signals on the other hand, such as those which leak from the wall supply to the HT through stray capacitance around a power transformer, will try to flow to ground in the *same* direction through each winding of the common-mode choke. Since these signals do not cancel, they encounter the full inductance of both windings, and may be considerably attenuated. The author has never used this setup so it is left as an experiment for interested readers.

4.4: Active Smoothing

Passive filter circuits can be augmented with active devices to enhance behaviour in various ways. Active smoothing circuits sit at the nexus between pure active filter design, and voltage regulator design (see for example the various ripple-suppressor circuits in chapter 6).

4.4.1: The Capacitor Multiplier

Among the simplest of active smoothing circuits is the capacitor multiplier. Its simplicity is deceptive, however, as it remains one of the most useful tools for managing power supply noise even in modern IC circuits, and there are some important design caveats that become important at high voltages. Fig. 4.28 shows the basic concept. It is really just a passive RC filter followed by a unity-gain buffer. The buffer or follower does its best to reproduce the smoothed voltage presented to the gate/base/grid, minus any bias voltage. In doing so, the follower uncouples the passive filter from any difficult loading, allowing it to achieve excellent attenuation and (except for the triode) minimal DC voltage drop too.

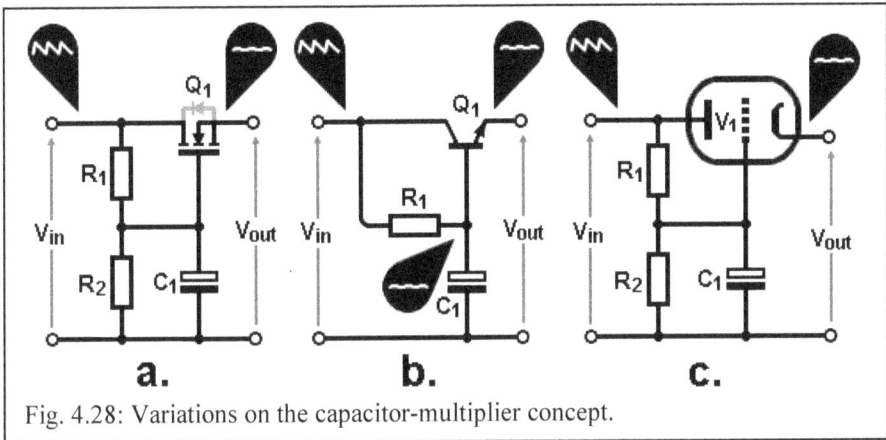

Fig. 4.28: Variations on the capacitor-multiplier concept.

This circuit is not a voltage regulator or stabiliser, it is simply an enhanced RC smoothing filter. The smoothed gate voltage is still free to rise and fall with slow changes in input voltage, e.g. wall fluctuations, and the output voltage must follow likewise. This 'rising tide that raises all boats' can be an advantage as it means dissipation in the pass device does not greatly increase if the wall voltage happens to increase. Unlike with a voltage regulator, this seriously reduces design headaches when it comes to heatsinking and makes the circuit an economical solution if we don't need a critical DC output voltage. It can also be used as a sort of pre-regulator to unburden a proper voltage regulator further downstream, or even be used *after* a voltage regulator to reduce broadband noise.[3]

Fig. 4.29 shows a practical circuit using a MOSFET. The purpose of R_2 is to form a potential divider with R_1, so the gate voltage is pulled below the input voltage, otherwise the transistor would switch off during the valleys of the incoming ripple

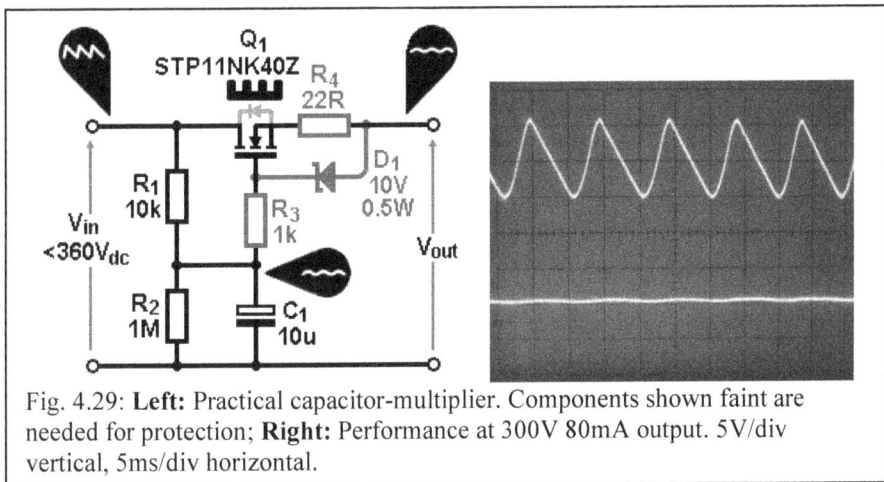

Fig. 4.29: **Left:** Practical capacitor-multiplier. Components shown faint are needed for protection; **Right:** Performance at 300V 80mA output. 5V/div vertical, 5ms/div horizontal.

---

[3] www.edn.com/simple-circuits-reduce-regulator-noise-floor

voltage since it would drop below the smoothed voltage stored on C_1. As shown, the circuit will tolerate about 3% input ripple. During normal use the voltage across the transistor is equal to the voltage across R_1 plus V_{GS}. For load currents greater than about 80mA a TO-220 package will need heatsinking. Practically any MOSFET with sufficient V_{DS} rating will work in this application.

D_1 protects the gate-source junction from overvoltage and reverse voltage and, in combination with R_4, provides an essential soft current limit. Since the voltage across R_4 cannot exceed $V_Z - V_{GS}$, or about $10 - 4 = 6V$ in this case, if the load current tries to exceed $6V / 22\Omega = 0.27A$ the Zener will turn on and pull the MOSFET gate voltage down, preventing any further increase in current. This is necessary to protect the MOSFET from inrush currents at start-up, and from careless oscilloscope probing! The drain-source junction is inherently protected from reverse voltage by the MOSFET's internal body diode, shown faint. The gate stopper R_3 additionally protects D_1 from transient current being dumped out of C_1 if the output is accidentally shorted.

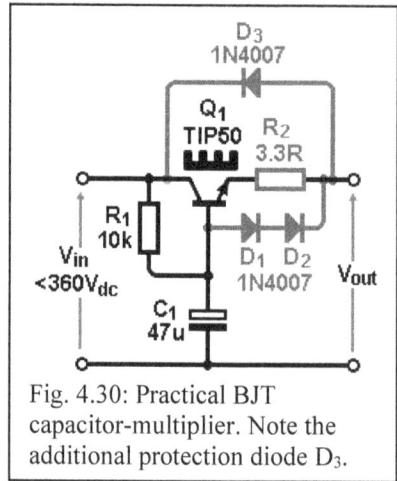

Fig. 4.30: Practical BJT capacitor-multiplier. Note additional protection diode D_3.

Fig. 4.29 also demonstrates the performance of the circuit when supplying $300V_{dc}$ at 80mA from a $310V_{dc}$ raw supply. The input ripple was $\sim 11V_{pp}$ while the output ripple is only $\sim 180mV_{pp}$ and has much less harmonic content. This is exactly the amount of ripple rejection we would expect from a $10k\Omega/10\mu F$ smoothing filter. However, to achieve the same DC drop at 80mA from an RC filter *alone* would have required 113Ω and $880\mu F$ (i.e. the same time constant as R_1C_1), hence the name 'capacitor multiplier'.[*]

A BJT can also be used as capacitor multiplier, as in fig. 4.30, but BJTs have lower gain and voltage ratings than MOSFETs. Also, they cannot simultaneously withstand as much collector current and collector-emitter voltage as similar MOSFETs, owing to their 'secondary

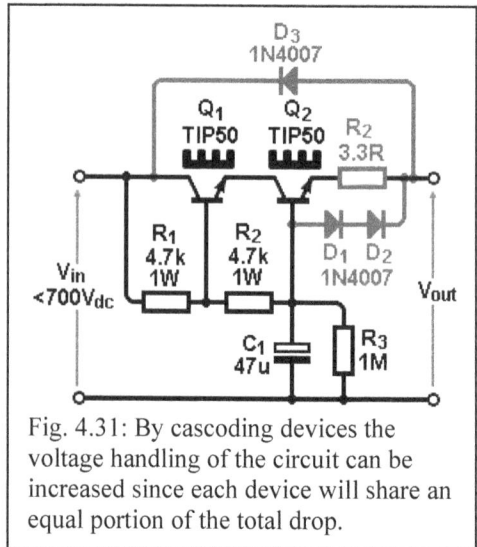

Fig. 4.31: By cascoding devices the voltage handling of the circuit can be increased since each device will share an equal portion of the total drop.

---

[*] Or capacit*ance* multiplier if you prefer.

breakdown' phenomenon, making them more likely to fail at start up. Also, a trap awaits if the power supply is switched off and the collector voltage falls faster than the voltage stored across C_1, leading to a large reverse collector-base voltage which may destroy the BJT. This is not a problem for the MOSFET circuit since the body diode provides a discharge path in combination with the gate-source protection diode. It is essential to include a similar diode in the BJT circuit as shown here, or alternatively a diode pointing from base to collector. However, a resistor from base to ground is not essential here since the inherent base current in the (fairly low h_{FE}) BJT will naturally pull the base voltage down below the collector.

Overall voltage and power handling can be increased by cascading or cascoding devices, as in fig. 4.31. The base-supply resistor (R_1 in previous circuit) has now been split into two equal parts, R_1-R_2, which ensures the collector-emitter voltage of both transistors is equally shared, most importantly at switch-on. Total dissipation is therefore also shared between the two transistors. The same principle can be applied to MOSFETs too, of course.

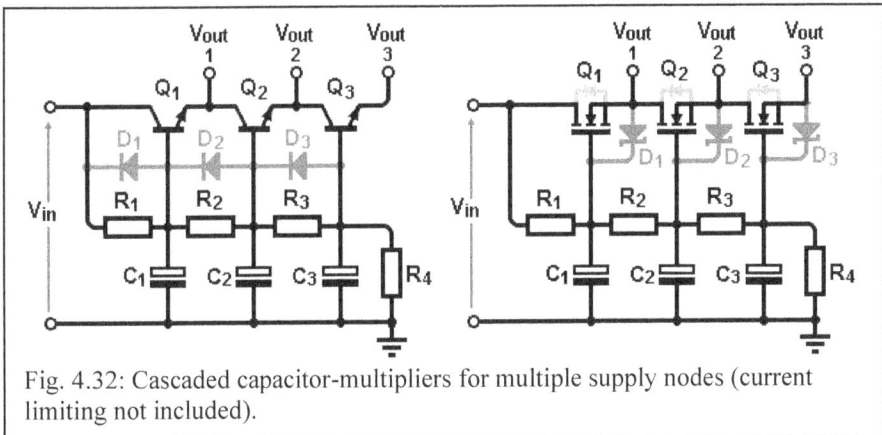

Fig. 4.32: Cascaded capacitor-multipliers for multiple supply nodes (current limiting not included).

Similarly, a cascade of capacitor multipliers could be used to supply every stage of an amplifier, as illustrated in fig. 4.32. Three stages are shown, but this could be extended at will. If each transistor directly supplies a typical valve gain stage (not a capacitive load) then over-current protection is not essential; even the humble little MPSA44 would be sufficient for designs consuming less than about 10mA. Note that BJTs must still have base-collector protection diodes added, and MOSFETs should have gate-source protection Zeners included, as indicated in the figure. Also remember that each transistor

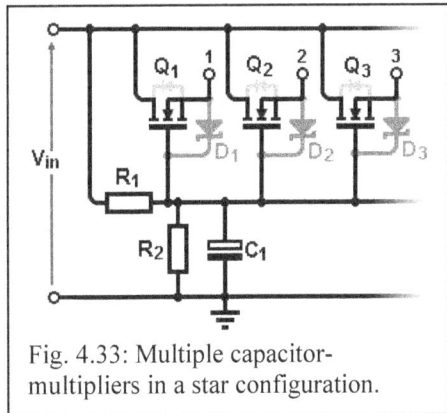

Fig. 4.33: Multiple capacitor-multipliers in a star configuration.

must pass the load current for its own node *plus* all those further downstream. Alternatively, we could use a star arrangement with the transistors supplying individual valve stages but sharing the same drain supply and smoothing filter, as illustrated in fig. 4.33.

It is also possible to use an ordinary opamp buffer in a capacitor multiplier, even in a high-voltage environment, as illustrated in fig. 4.34. The supply voltage for the opamp is clamped to a safe level by Zener diode D_1 –which might typically be a 25V device– and the Zener current and opamp quiescent current flow down through R_3 to ground. The opamp therefore floats within a safe 'window', suspended below the HT rail. R_1 and R_2 set a suitable reference voltage which must be about mid-way within the 'window', and this

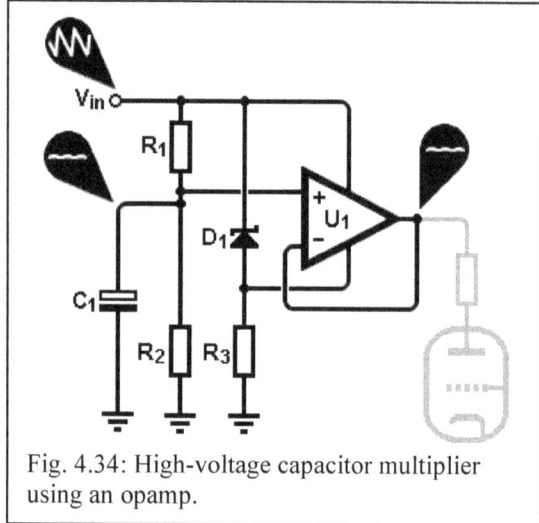

Fig. 4.34: High-voltage capacitor multiplier using an opamp.

is smoothed by C_1. The opamp dutifully follows this smoothed voltage and supplies the valve stage. A typical opamp like the TL07x might only be able to supply about 5mA, but since they come in dual or quad packages, several valve stages could be supplied from one IC.

There was once a price advantage to this topology,[*] compared with using discrete, high-voltage MOSFET source followers, but the cost of MOSFETs has fallen sufficiently that the opamp approach has lost most of its utility.

4.4.2: The Gyrator or Simulated Inductor

Another sort of active filter is the gyrator or 'simulated inductor', which simulates the reactance of an inductor by 'inverting' the reactance of a capacitor. It deliberately exploits the fact that the output impedance of a cathode follower becomes inductive owing to grid-cathode capacitance, although these days a transistor does the job more efficiently. A practical circuit is shown in fig. 4.35, and the result is equivalent to an inductance:

$$L_{eq} = C_1 \times R_1 \times R_3 \qquad (4.8)$$

---

[*]Overgrown versions of this circuit can be found in the Audio Research SP10 preamplifier, for example.

In series with a damping resistance:

$$R_o = R_3 \left(\frac{R_1 + R_2}{R_2} \right) \tag{4.9}$$

In this case amounting to 155H in series with 66Ω –much more impressive than the average smoothing choke.

The voltage across R_2 is equal to the gate-source voltage (typically around 4V) plus the drop across R_3. In this case $R_1 = R_2$, so the total DC drop across the gyrator is simply $2(V_{gs} + iR_3)$ and the total dissipation in Q_1 is $i(2V_{gs} + iR_3)$. If the load current i is 100mA, say, then this would result in about 17.4V total drop and 1.27W dissipation in Q_1, the remaining 0.47W being dissipated by R_3. A TO-220 transistor package can dissipate little more than 1W by itself, so a clip-on heatsink would be needed.

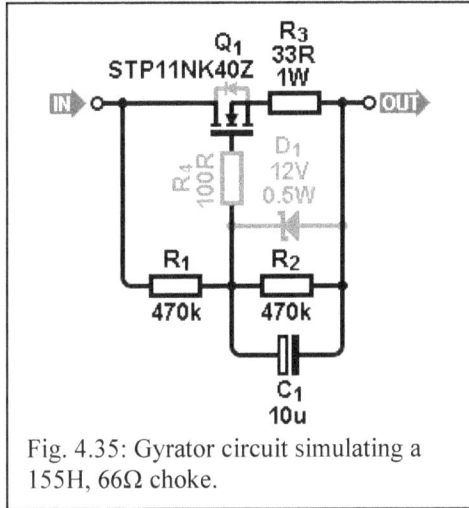

Fig. 4.35: Gyrator circuit simulating a 155H, 66Ω choke.

There must be enough drain-gate voltage –i.e. the voltage across R_1– to accommodate the peak incoming ripple voltage, otherwise the MOSFET will switch off during the valleys of the ripple waveform. In other words, the peak-to-peak ripple voltage applied to this circuit must be less than $2(V_{gs} + iR_3)$. R_4 is the usual gate stopper and must be mounted very close to the MOSFET to discourage oscillation. D_1 protects the gate-source junction and C_1 from overvoltage, which also allows C_1 to be a small, low-voltage device. Most high-voltage MOSFETs will work in this application; the IRF820 and STP4NK50Z are 500V-rated options. The STP- types are especially handy as they have built-in gate-source protection and are available in the all-plastic TO-220FP package, so can be bolted to the chassis without the need for insulation hardware.

The gyrator can be tuned or damped just like a real choke (section 4.3.1). However, unlike a real choke the gyrator cannot store energy. For smoothing this is normally an advantage as it means the gyrator cannot produce unwelcome flyback voltage, or magnify the output voltage above the input voltage. But, for the same reason, a gyrator cannot replace a real choke in a choke-coupled or parafeed amplifier.

Chapter 5: Fusing, Switching, and Other Refinements

It is essential that a power supply be reasonably protected against short circuits and other serious failures. This is for reasons of general reliability and ease of repair, and more importantly to protect the user against electric shock and fire. Switching and fusing might seem like overtly simple subjects, but like anything in analog electronics there are subtleties that emerge if you dive deep enough into the subject.

5.1: Fuses

When current flows in a fuse, the wire heats up, and with enough current it will melt. Once the wire breaks, an arc will be sustained across the gap between the two broken ends. This will continue until the gap becomes too large or the voltage drops too low to sustain the arc, and the current finally stops. The **melting time** plus the **arcing time** together make the total **clearing time** of the fuse, as illustrated in fig. 5.1. The clearing time depends on many factors, the most significant being by how much the instantaneous current exceeds the rated fuse current.

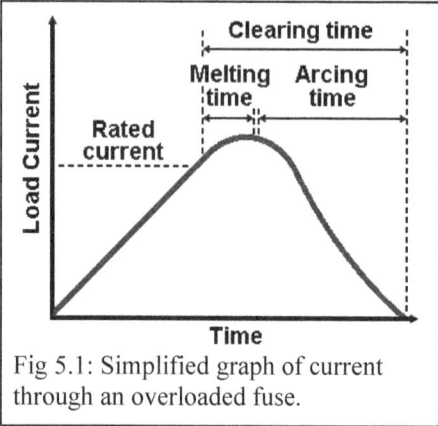

Fig 5.1: Simplified graph of current through an overloaded fuse.

Since fuses work by heating, they will blow faster in elevated ambient temperatures. A wire fuse is typically derated to about 80% of its normal current rating when the ambient temperature is 50°C, which can occur inside a hot chassis. However, fuse selection is not an exact science. In most applications the fuse rating will need to be considerably larger than the normal operating current to prevent nuisance blowing, so temperature derating is not usually something we consider.

The most common cartridge-fuse sizes are 20×5mm and 32×6.3mm (1¼×¼ inch). The 20×5mm variety is now the most common choice for current ratings up to a couple of amps.[*] Wire-ended fuses are available for direct soldering to a printed circuit board, but this of course makes replacement more difficult. Larger fuse sizes are available for high-voltage and high-energy applications.

---

[*] Ratings up to 10A are available, but with several amps flowing, the small end caps suffer more from heating than 1¼×¼ inch fuses, leading to more nuisance blowing.

5.1.1: Current Rating

In Europe there are two principal sorts of fuse to consider: **time delay** (T) and **fast acting** (F). Near equivalents in America are **slow-blow**[†] and **quick-blow**. European fuses fall under IEC rules and are specified by the maximum current they can handle *without* blowing. American fuses fall under UL/CSA rules and are specified by the minimum current which *will* cause them to blow. In other words, a 1A IEC fuse will not blow when the current reaches one amp (at 25°), but a UL 1A fuse *will*, though it may take a few hours to do so. This can be deduced from fig. 5.2 which shows the typical clearing time against overload current for some typical IEC and UL fuses. However, in normal consumer applications this difference between ratings is of little practical consequence because, as noted earlier, we do not select fuses to operate right on the verge of failure under normal running conditions.

Most power supplies suffer a brief period of inrush current when first switched on, before the transformer is fully magnetised and before the capacitors have charged, etc. Therefore, fuses used anywhere in a power supply will nearly always be of the time-delay/slow-blow type, since these can withstand the inrush. Nevertheless, every time the fuse is subjected to inrush current it will flex due to the powerful magnetic field generated by the current surge –in a glass fuse you can often see the wire 'jump' when power is applied. During operation the fuse will also warm up and expand, causing further

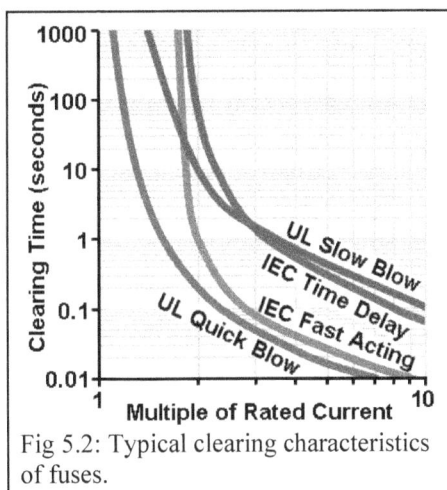

Fig 5.2: Typical clearing characteristics of fuses.

mechanical stress. Over many cycles of operation, the fuse wire will eventually fatigue and break; it is quite normal for an old fuse to fail even though there is nothing wrong with the circuit. This sort of nuisance blowing is just something we must live with. Of course, if the fuse is replaced and immediately fails again, then further investigation is called for.

Fuses for use on the mains supply or in other potentially high-energy environments are designated **high breaking capacity**, or HBC. The breaking capacity is the maximum current overload the fuse can safely interrupt without exploding or rupturing. The maximum fault current that might flow in a $230V_{ac}$ environment is much larger than in a $120V_{ac}$ system, so IEC (European) HBC fuses are made from ceramic and filled with sand to dissipate heat and discourage vapourised metal from leaving a conductive path inside the cartridge. Sand also helps to quench arcing more

---

[†] The alternative spelling 'slo-blo®' is not a technical term but a registered trademark of Littlefuse.

109

rapidly. By contrast, UL/CSA fuses can be made from glass while still achieving an impressive 10kA breaking capacity up to $125V_{ac}$.

5.1.2: Voltage Rating

A fuse also has a maximum voltage rating. Since it is just a piece of wire there is negligible voltage across it during normal running, so the rating refers to the maximum voltage allowed across the fuse after it has blown, that is, the open-circuit voltage. Note that this may be higher than we expect during normal running, due to the power supply 'un-sagging' when the load current is interrupted.

As explained earlier, when the fuse melts, an arc will be sustained until the gap sufficiently widens, or the voltage drops enough to quench it. With alternating current this happens very quickly since the voltage is continually passing through zero; it is self-extinguishing. But with DC the arc may be sustained for much longer. In other words, it is much easier to interrupt AC than DC. Commonly, a fuse rated for $250V_{ac}$ might only be rated for $30V_{dc}$, if it carries a DC rating at all. If the fuse is used beyond its voltage rating then its clearing time may exceed the manufacturers guarantee, and at excessive voltages it could arc over or allow creepage current between the terminals after blowing.

5.1.3: Primary Fusing

A high breaking capacity fuse in series with the live feed, where it enters the appliance, is mandatory. This protects against live-to-neutral faults which might cause a fire, and against live-to-earth/chassis faults which would also be a hazard to the user. In Britain, mains wall plugs have a built-in fuse, but they are large 3A to 13A types which principally protect against a faulty power cable rather than a faulty appliance. Since most appliances now use detachable IEC cables ('kettle leads') there is no telling what cable might be plugged in, and what plug fuse might be installed in it, so a primary fuse inside the appliance is still mandatory.

On account of nuisance blowing, the primary fuse is usually made user accessible via an integrated IEC-inlet assembly or a separate panel-mounted fuse holder. When using the latter it is good practice to connect the incoming live wire to the end terminal of the fuse holder, so the fuse breaks contact with the live connection as soon as it is withdrawn from the holder, reducing the possibility of an accidental shock when handling.[*]

It should be obvious that we get maximum circuit protection when the fuse is only *just* large enough to allow normal operation. However, because of the somewhat unpredictable inrush current, choosing the right fuse rating often requires some experimentation. We have little choice but to estimate a suitable value, and if the inrush causes it to blow then a larger rating is required, within reason. Obviously, the fuse must not be so grossly over-rated that it no longer offers much protection against genuine faults.

---

[*] This is not a requirement under UL 60065

In a typical transformer-based power supply the primary fuse will usually be rated for about 1.5 to 2 times the normal mains operating current. It must never be larger than the mains voltage divided by the transformer primary winding resistance, otherwise it cannot blow at all. If a fuse three times greater than the normal operating current is still too small to prevent nuisance blowing, then some form of inrush limiting should be considered (e.g. section 5.4.1), rather than a bigger fuse. This is often the case for large toroidal transformers which can have brutal inrush current. Off-line switch-mode power supplies also have high inrush current but usually have appropriate fusing built in as standard.

Sometimes fuses are placed in both the live *and* neutral feeds. This is only necessary in countries which still use reversible wall plugs, since either wire could end up being the live one. However, a valid argument against this practice is that if the neutral fuse blows first then it will leave the circuit still physically connected to the live wire, which is less safe for the user. A compromise

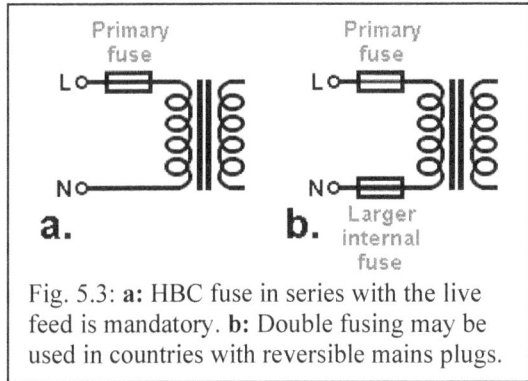

Fig. 5.3: **a**: HBC fuse in series with the live feed is mandatory. **b**: Double fusing may be used in countries with reversible mains plugs.

solution is to fuse the live wire as normal –which may be user accessible– but to use a fuse with a larger current rating in the neutral wire, which is not user accessible, as in fig. 5.3b. The smaller fuse protects against all line-to-neutral faults, and against live-to-ground faults when the plug is correctly polarised. Only if the wall plug is incorrectly polarised and there is a neutral (now live)-to-ground fault does the larger fuse blow, forcing a proper investigation. To emphasise, this should only be considered in countries with reversible wall plugs.

In countries with non-reversible plugs, a similar principle might still be useful for guitar amps, but with both fuses in series with the live feed only. If a careless or desperate musician is tempted to replace the accessible fuse with one of the wrong value (or worse, with an old nail or whatever can be jammed in to make it through the performance) the larger fuse hidden inside the chassis still protects against gross genuine faults, again forcing a proper investigation. This approach was used in some Garnet amplifiers.

5.1.4: Secondary Fusing

A fuse on the primary side of the mains transformer is a necessity as a precaution against fire, but it might not provide much protection for the transformer or amplifier circuit itself, since heavy fault currents on the secondary side may translate to quite small currents in the primary. Most builders therefore add fuses on the secondary side too, and some regulatory agencies may actually require this. Secondary fuses can be non-HBC glass fuses since secondary fault currents are greatly limited by the transformer itself. Glass fuses also make it easier to see which one has failed. They should be rated for around 1.5 to 3 times the rated transformer current, unless the transformer is capable of a lot more current than you're using, in which case you may want to use a smaller fuse more appropriate to the load current.

Most cartridge fuses are rated for $250V_{ac}$ since this accommodates all the world's wall voltages. Unfortunately, the HT voltage in a valve amp is often much higher than this. High-voltage fuses are available (such as the Littlefuse 477 and Bussmann S505H series rated for $500V_{ac}/400V_{dc}$) but they are less widely stocked, so it is tempting to use $250V_{ac}$ fuses everywhere, even in the HT supply. Fortunately, the safety implications for secondary fusing are much more benign compared with primary fusing, so although such use would not pass 'official' certification, it is not a great crime for a homemade, non-commercial amp.

It is harder to interrupt direct current than alternating current, so it is preferable to place secondary fuses in the AC (or pulsating DC) part of the circuit, i.e. before the reservoir capacitor. The best option is directly in series with the (AC) transformer winding, as shown in fig. 5.4c, so that it protects the transformer against a shorted rectifier.

Vintage amplifiers using two-phase rectifiers often used a single fuse in series with the transformer centre tap, as in fig. 5.5a.[1] However, if both rectifier diodes were to fail short then the whole transformer winding will be shorted –the fuse in the centre tap cannot protect against this. This sort of fault is unlikely in a valve

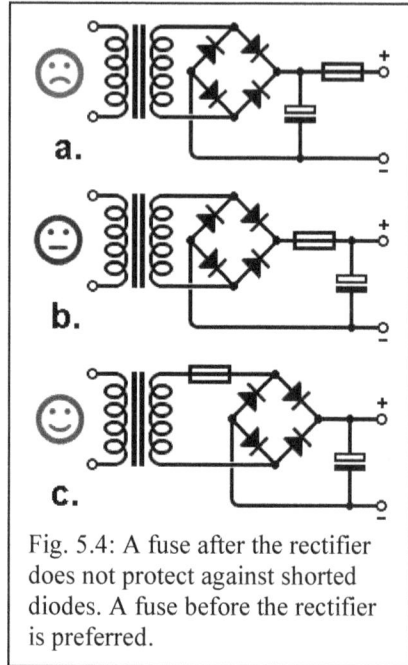

a.

b.

c.

Fig. 5.4: A fuse after the rectifier does not protect against shorted diodes. A fuse before the rectifier is preferred.

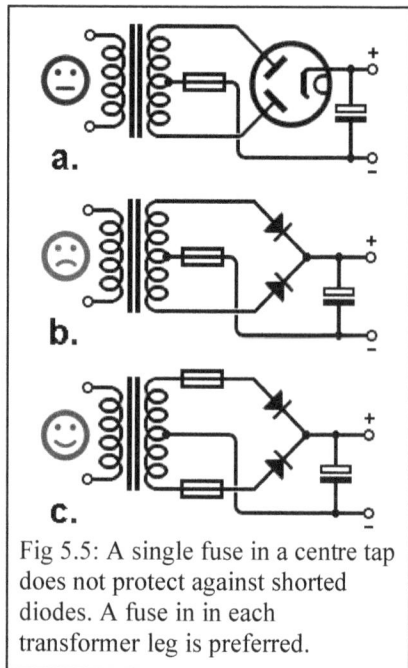

a.

b.

c.

Fig 5.5: A single fuse in a centre tap does not protect against shorted diodes. A fuse in in each transformer leg is preferred.

---

[1] Delaney, W. J. (1951). More Power Supply Problems, *Practical Wireless*, January, pp32-3.

rectifier but ageing silicon diodes have a nasty habit of doing exactly that. Therefore, when using silicon diodes, modern practice demands that two fuses be used, one in each leg of the winding as in fig. 5.5c.

One place where such fuses might *not* be used is on a purely AC heater supply, since there is very little to go wrong and the likelihood of a fault creating a permanent short is slim. Also, cold valve heaters have very low resistance, so they cause severe inrush current at switch on which can last several seconds (much longer than it usually takes to magnetise a transformer or charge a reservoir capacitor). As a result, any fuse used on the heater supply will need to be generously overrated. Moreover, valves from different manufacturers can vary wildly in their heater cold resistance, which can lead to unexpected fuse failure. For example, the 1993 issue Vox *AC30* uses a pair of T6.3A fuses in the heater supply, which are happy with the four Sovtek EL84s that are supplied as standard, but most other brands of EL84s have smaller cold-resistance than the Sovteks, so replacement valves can cause the fuses to blow unexpectedly unless they are replaced with 10A types, even though the normal working current is the same either way.

It is also important to consider what might happen to the rest of the circuit when a fuse blows. For example, a blown fuse in the bias supply of a power output stage could lead to red-plating and the destruction of the output valves. Similar problems could occur in a DC-coupled amplifier if a heater fuse blows. This may require the addition of fuses or other protective measures to other parts of the circuit, too.[*] Resettable 'polyfuses' fuses can be useful for this.

5.1.5: Resettable Fuses

A resettable fuse or polyfuse[†] is a type of thermistor with an unusual positive-temperature coefficient (PTC). It is designed to have very low resistance for currents up to its **hold current** rating, but will snap into a high-resistance state when its **trip current** rating is exceeded, with a similar response time to a slow-blow fuse. Typically, the trip current rating is about twice as large as the hold current rating. In between these two figures is a grey area where the device may or may not trip, depending on thermal details beyond our control. When the overload is removed and the device cools down, it returns to its low-resistance state, although this figure can degrade over time if the fuse is tripped many times.

When the device trips and goes high resistance, it will maintain *constant power dissipation* within itself, as specified on the datasheet. This means some leakage current will continue to flow through it during the tripped state; exactly how much depends on the open-circuit voltage. A resettable fuse is therefore not like a traditional fuse which cuts the current completely, and not quite like a foldback circuit either (section 7.2), but something in between.

---

[*] *Quis custodiet ipsos custodes* –who watches the watchers?
[†] A trademark of Littlefuse

Resettable fuses are available with trip ratings from tens of milliamps to several amps. However, the maximum open-circuit voltage rating is usually 60V at most, so they cannot be used directly in the HT supply.

5.1.6: Fuse Annunciation

Occasionally it is useful to have a lamp or LED to indicate when and which fuse has blown. Fig. 5.6 shows an old trick for doing this. Normally the LED is shorted by the fuse, so it does not light. If the fuse blows, fault current is diverted through the LED, limited to a few milliamps with a suitable resistor. The same principle can be applied to a resettable fuse too, of course. For an AC circuit an

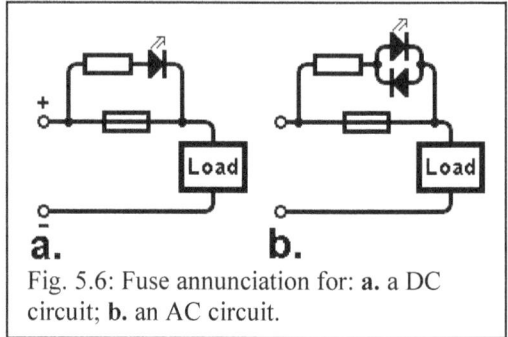

Fig. 5.6: Fuse annunciation for: **a.** a DC circuit; **b.** an AC circuit.

extra diode may be needed in anti-parallel with the LED to protect it from excessive reverse voltage, as in b. For safety reasons these circuits must *not* be used on the main primary fuse; the primary fuse must always result in total isolation after blowing.

5.2: Switches and Switching

A mechanical switch endures similar conditions to a fuse. If the switch is opened while it is carrying current, an arc will be drawn across the gap. The switch contacts may also bounce a few times before finally coming to rest, which may induce further spurts of arcing. This corrodes the switch contacts and may eventually lead to intermittency. In extreme cases (usually in high-current DC circuits) the switch can become welded shut. Switches therefore have maximum current and voltage ratings much like fuses. Again, the DC voltage ratings are typically much smaller than AC voltage ratings, so it is desirable to use a switch in an AC circuit rather than a DC one, if possible.

5.2.1: Primary Switching

A requirement of UL 60065[*] is that any appliance powered from the wall supply must include some form of mains 'disconnect device'. In practice this means either a switch, or a detachable cable such as an IEC cable or 'kettle lead'. If the mains cable is 'captive', i.e. permanently wired to the amplifier, then both the live *and* the neutral *must* be switched simultaneously using a double-pole switch as in fig. 5.7b, and the 'on' position must be indicated with a label or by illumination. However, if a detachable mains cable is used then a switch is not essential, though most users will want one anyway (it can be single- or double-pole). Usually the switch will be placed after the fuse rather than before, although the author is not aware of any legal

---

[*] UL 60065 section 8.19.1

rules either way. Note that the mains safety earth conductor must *never* be switched under any circumstances.

The switch must be rated for use on mains voltage, and its current rating should be generously larger than the operating current. Most readily-available mains switches are rated for several amps –far more than most valve amplifiers consume– so this should not be a problem. As a practical matter, an unofficial convention is to mount the switch so that, if something were to fall on it from above, it would tend to knock it into the 'off' or 'safe' position.

5.2.2: Secondary Switching

Switching of low-voltage circuits is simple and requires hardly any thought beyond the appropriate voltage and current ratings. Arcing can be suppressed to some extent by placing a small (100nF say) capacitor in parallel with the switch, as in fig. 5.8. When the switch opens, the capacitor presents a brief low-impedance path until it is fully charged, by which time the switch has (hopefully) already opened.

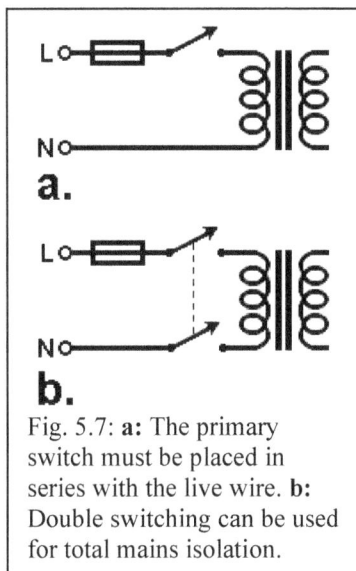

a.

b.

Fig. 5.7: **a:** The primary switch must be placed in series with the live wire. **b:** Double switching can be used for total mains isolation.

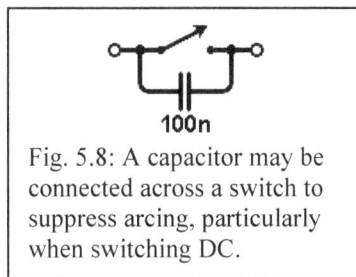

100n

Fig. 5.8: A capacitor may be connected across a switch to suppress arcing, particularly when switching DC.

5.3: Standby Switching

Occasionally we may want to provide a separate switch for the HT supply. This commonly takes the form of a so-called standby switch, which allows the heaters to warm up before the HT is applied. Really it would be better to call it a 'pre-heating switch', as 'standby' has since become an accepted term for the power-saving feature found on many modern appliances. Nevertheless, the term 'standby switch' is firmly entrenched in guitar amp lore, somewhat unfortunately, as explained in the next section.

There are some valves that genuinely *do* need a warm-up period before the HT is applied, but this will be explicitly stated on their data sheets. This includes mercury-arc rectifiers (the mercury must be vaporised before use), valves used in DC-coupled applications such as voltage regulators, and some special purpose valves with unusual getters that must be pre-heated. Such valves are not normally used in audio amplifiers, much less guitar amplifiers.

The trouble with implementing a standby (pre-heating) switch is that the HT voltage will often exceed the voltage ratings of readily available switches by a fair margin, so the hobbyist is coerced into using an ordinary mains-rated switch and hoping for the best. Admittedly, if the switch is suitably heavy-duty and over-rated for current then this should not be a problem. A heavy-duty switch will have large dimensions to support the heavy metal parts, implying large creepage distances and thick insulation. But in a commercial product this would not pass official certification, so an alternative is to let a high-voltage optocoupler or solid-state relay (SSR) do the job, which could also perform automatic start-up, shut-down, and fail-safe procedures; whatever we care to dream up.

In theory, when a two-phase rectifier is used, a switch can be placed in the centre tap of the transformer and will always have zero volts across it, so it might be tempting to use a cheaper, low-voltage switch. In practice, however, depending on when and exactly how the switch is opened, a hair-raising flyback voltage can be generated across the transformer winding. This could lead to arcing across the switch, so a high voltage rating is needed after all. Moreover, there are other potential problems that standby switches create, so we should think very hard about whether to include one at all. It may turn out to be more trouble that it is worth.

5.3.1: Standby-Switch Origins

Standby switches are now almost ubiquitous in valve guitar amplifiers. They are occasionally found in hi-fi amplifiers too, but this seems to be cross contamination from the much larger market for guitar amplifiers, where they have attained a kind of mythical value among guitarists, some of whom become infatuated with the idea of adding standby switches to amplifiers that neither have them nor need them. Even Gibson and Vox, who ought to know better, have added standby switches to some reissue products which never had them originally, presumably in response to customer expectation rather than engineering need. A notable howler was the addition of hot switching to the Vox *AC30CC* which immediately led to instances of valve rectifier failure.

The reasons why standby switches have become such a common feature of guitar amps are obscure; such switches were never found on similar appliances from the valve era. The trend seems to have originated in 1954 with the Fender *5D6* Bassman, which introduced a new hot-switching standby. Although hot switching is bad practice it apparently 'got away with it' by using a pair of 5U4G rectifiers in parallel, with a modest 16µF reservoir capacitance. However, in 1957 the *5F6* Bassman, from which so much guitar amplifier lore is descended, took this to the extreme by using a single GZ34 with 40µF capacitance. Rectifier failure became a known problem with this model and its many imitators[2] such as the Marshall JTM45, and this design flaw has become even more noticeable with modern-production rectifiers.

But why was a standby switch added to the *5D6* Bassman in the first place? One theory is that it was intended as a kind of mute switch, allowing the amp to be

---

[2] Doyle, M. (1993). *The History of Marshall: The Illustrated Story of "The Sound of Rock"*. Hal Leonard Corporation, p17.

silenced between songs without letting the heaters go cold. The amp could then be switched on with no waiting time when the next performance demanded it. This is not a compelling argument since many similar amps in the Fender catalogue had no such switch, and neither did those of competing manufacturers Gibson and Vox at the time. Moreover, stage amplifiers were not mic'd or wired to house PA during this period, so it is difficult to imagine a little hum being noticed by audiences between songs. There are also much simpler ways to mute the audio.

Another theory suggests the switch was added when silicon rectifiers started to be used, because they did not provide the natural soft start of the original valve rectified designs. This cannot be true, however, since the first amps fitted with standby switches still had valve rectifiers.

Probably the most popular explanation is that the standby switch was included to increase the useful lifespan of the valves. This is partially true, as we shall see, but not in the way most commentators suppose. The principal cause of valve decay is evaporation of the cathode material, which is entirely determined by cathode temperature. Since the traditional standby switch leaves the heaters running at full temperature it does nothing to slow down this process. Other forms of cathode deterioration are possible, as explained in the following sections, but they are either not improved by the use of a standby switch, or may actually be made worse.

There are three more plausible reasons why Fender added the standby switch:
- To allow time for the bias voltage to reach its normal value and so prevent a current surge in the power valves when the HT is applied. This is at best only a partial explanation since many fixed-biased Fender amps continued to be produced with no standby switch.
- To protect the HT reservoir/smoothing capacitors from overvoltage. Large, high-voltage capacitors were both expensive and of poor reliability at the time. Costs could have been kept to a minimum by using capacitors which were only just rated for the normal working voltages in the amp. But at start-up, before the valves are hot and conducting, the HT may rise above these ratings.[*] Some original schematics are clearly labelled with no-load voltages that exceed the smoothing capacitors' ratings.[†]
- To protect the DC-coupled cathode follower from arcing. If this valve is not warmed up before the HT is applied, the full voltage will appear between grid and cathode, often resulting in grid-cathode arcing. This is a persuasive explanation because this valve stage was newly implemented in the *5D6 Bassman*. This rare amplifier was produced very briefly, in small numbers, and can be regarded as a developmental prototype for the later, more established models.

---

[*] Vintage electrolytic capacitors tend to become accustomed to operating at a particular voltage. Suddenly subjecting them to a higher voltage can cause them to fail even when it is within the supposed rating.

[†] e.g. *Bassman 135*.

The standby switch would alleviate these stresses by preheating the valves. Nowadays, of course, we have ready access to cheaper, better-quality capacitors which can be properly rated for the off-load voltages, and DC-coupled valves are better protected by placing a diode or neon lamp between grid and cathode, so a preheating switch can be entirely dispensed with.

5.3.2: Cathode Stripping

It must be impressed again that the life of a receiving valve is not measurably improved by pre-heating the cathode. If it were, we would expect to see it mentioned in classic textbooks and valve manuals, yet such remarks are conspicuous by their absence. Indeed, some texts explicitly exclude receiving valves from discussion of similar pre-heating procedures.[3] Despite this, promotional materials glorifying the supposed life-prolonging effects of standby switches are often published –on the internet and in non-technical magazines, but never in academic work– even by well-known guitar amp manufacturers, presumably with good-but-misguided intentions. Such myths become self-propagating. The claims invariably include specious references to supposed cathode stripping or poisoning effects, described below.

Cathode stripping, in particular, seems to have been a subject of great confusion, conjecture, and misinformation among the audio community in recent years. There really is no excuse for this as plenty of material was published on it during the 20th century, and it presented no special concerns to circuit designers of the day. In its purest form, cathode stripping occurs when particles of the oxide coating are physically torn from the surface of the cathode when it is exposed to a powerful electrostatic field from the anode. This would happen if the valve were operated at saturation, without a usual space-charge of electrons to protect it. Now, it is important to note that this effect *does not exist in receiving valves*, even when operated at saturation, because it requires an electric field strength of at least 4MV/m (4 million volts per metre!).[4] Taking an extreme example of a 5V4GA rectifier which has an unusually close anode-cathode spacing of 0.5mm, operating at its maximum peak voltage of 1400V, the electric field strength would reach a maximum of only 2.8MV/m. In an amplifying valve the distances tend to be greater, the voltages lower, and the cathode shielded from the electric field of the anode by the low-voltage grid. Pure cathode stripping is therefore a non-issue.

What most people mean by cathode stripping is more properly known as **cathode sputtering**. This occurs when stray gas molecules in the valve become ionised by the electron stream, and the positive ions accelerate towards the more negative grid and cathode. If the cathode is working at saturation and not protected by the space charge, and provided the ions are not re-neutralised along the way, they may crash

---

[3] Harrison, N. J. (1956), *RCA Transmitting Tubes: Technical Manual No. 4.* RCA, Tube Division. p65.
[4] Herrmann, G. & Wagener, S. (1951). *The Oxide-Coated Cathode.* Chapman & Hall, Ltd. London. p111.

into the cathode and physically damage its surface. This does not require unusually high voltages but it is nonetheless rare in receiving valves.

Good receiving valves have excellent vacuums, of the order of 10^{-9}mmHg / 130nPa, so gas content is extraordinarily low. Traces of gas may evolve from the various parts of the valve over its lifetime, but these should be effectively absorbed by the getter. Even when a valve is started from cold so that for the first few seconds the cathode is saturated, the emission current will be very low and consequently the rate of ionisation will be very small too. Furthermore, the cathode normally possesses far more reserves of activation than are required for use in ordinary (space-charge limited) audio circuits, so minute loss of cathode material due to sputtering is of little consequence. The valve we would most expect to suffer from cathode sputtering is the rectifier, since it operates with the heaviest currents and closest to saturation, yet hot switching of rectifier valves is expressly *not* recommended by manufacturers. Sputtering might be more troublesome in a valve with an unusually poor vacuum, but then, it has bigger problems –a gassy valve will most likely be suffering from heavy reverse grid current, cathode poisoning, or even arcing. In short, cathode stripping/sputtering effects are a concern only in hot-cathode gas tubes, and in general only manifest under very high voltages or high-current saturation conditions such as might be found in cathode ray tubes, high-power rectifiers, and transmitting tubes.

5.3.3: Cathode Poisoning

Cathode poisoning refers to chemical –rather than mechanical– processes occurring at the cathode. There are several forms of cathode poisoning, including absorption of gas into the oxide coating, but the most pernicious type is the growth of **interface resistance**. When a valve cathode is fully heated but no anode current is allowed to flow for long periods of time (hours), an interface resistance may grow between the cathode tube and the oxide coating.[5] This resistance acts like an unbypassed cathode resistor, reducing the g_m of the valve even though it may test normal for emission. It also greatly increases flicker noise.[6] Moreover, once the interface resistance has formed it cannot be removed, and it was once said that the useful life of a valve could be almost completely defined by its rate of growth.[7] Late-generation computer valves which *had* to spend much of their time in cut-off were built with high-purity cathodes which virtually eliminated the problem, but whether the same material quality was applied to the popular receiving valves –let alone modern production valves– is unclear. For this reason, a traditional standby switch may actually shorten the life of valves, contrary to popular myth.

---

[5] Waymouth, J. F. (1950). Deterioration of Oxide-Coated Cathodes Under Low Duty-Factor Operation. *Massachusetts Research Laboratory of Electronics, Technical Report No.159.* pp1-15.
[6] Lindemann, W. W. & Van der Ziel, A. (1952). On the Flicker Noise Caused by an Interface Layer, *Journal of Applied Physics,* 23, pp1410-1411.
[7] Eaglesfield, C. C. (1951). Life of Valves with Oxide-Coated Cathodes, *Electrical Communication,* June, pp95-102.

The safest way to avoid this alternative form of cathode poisoning is to ensure that at least *some* anode current flows whenever the cathode is heated. A reduction of heater power also provides a considerable improvement.[8] If you have no choice but to leave the valves in standby, reducing the heater voltage by up to a third of its normal value, while allowing a trickle of anode current to flow, is a good compromise.

5.3.4: Least-Worst Standby Switches

Despite all the discouragement of the previous section, if a standby switch is still wanted then we might at least make a better job of it than was attempted in the Vox *AC30CC*. Most importantly, if a valve rectifier is used then we must avoid hot switching, that is, the rectifier must be allowed to charge the reservoir naturally from cold, as the heater warms up. This means a plain standby switch must not be placed between the valve rectifier and reservoir, or in the centre tap of the transformer,* as illustrated in fig. 5.9a and b. This is not a concern with silicon rectifiers, of course. A better way to standby switch a valve rectifier is by switching off its filament supply, as shown in fig. 5.9c. Be sure to use a heavy-duty switch since it has to carry the filament current, and there will be high DC voltage between the switch and its casing.

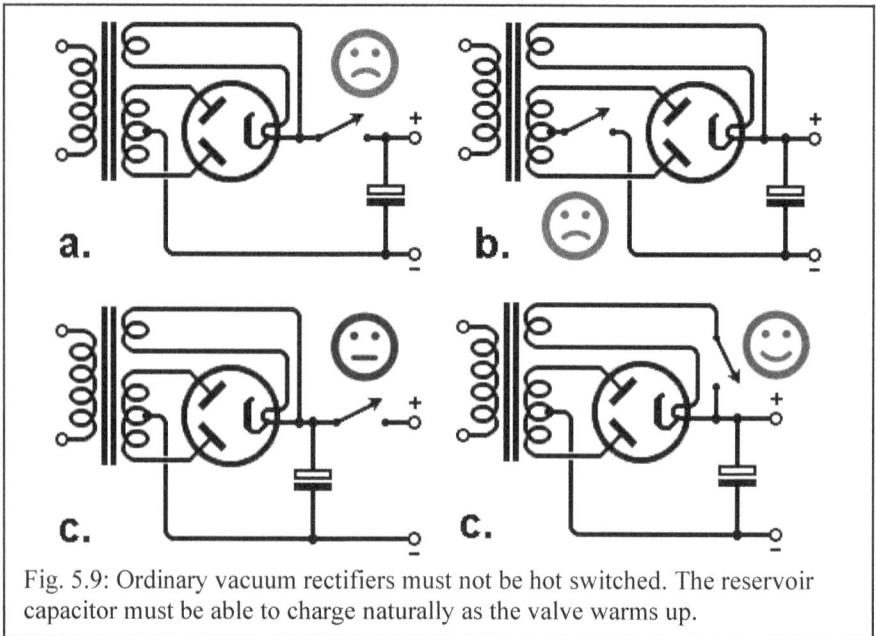

Fig. 5.9: Ordinary vacuum rectifiers must not be hot switched. The reservoir capacitor must be able to charge naturally as the valve warms up.

---

[8] Metson, G. H. (1956). On the Electrical Life of an Oxide-Cathode Receiving Tube, *Advances in Electronics and Electron Physics*, 8, pp403-46.
* It is telling that Fender did this in the *5D5* and *5E5 Pro-Amp* models in 1954/55, but then removed the standby switch from the 1956 *5E5-A* model.
120

It is also worth noting that a centre-tap standby switch should not be used if a bias supply is derived from the same transformer winding, as in fig. 5.10. As indicated by the arrow, when the switch is open, a current path still exists through the whole of the transformer winding. Depending on the component values this can mean up to twice the normal HT voltage is produced during standby! Another position not to put a standby switch is directly in series with a smoothing choke, as shown in fig. 5.11. This will result in an almighty flyback voltage across the choke when the switch is opened, which may damage the switch or the choke insulation.

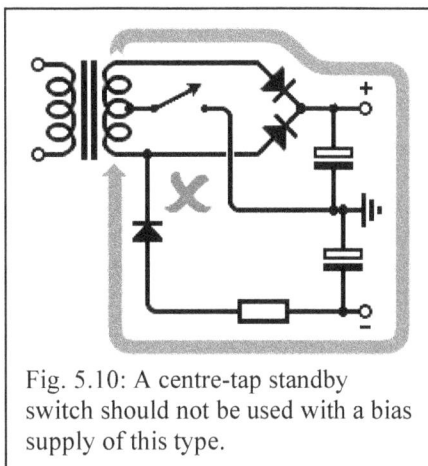

Fig. 5.10: A centre-tap standby switch should not be used with a bias supply of this type.

A traditional standby switch leaves the valves heated but without anode current, which encourages the growth of interface resistance. A compromise is to put a resistor in parallel with the switch to allow a trickle current to flow during standby.[†] This has the added advantage that it provides a soft start for the whole amplifier by allowing the power supply

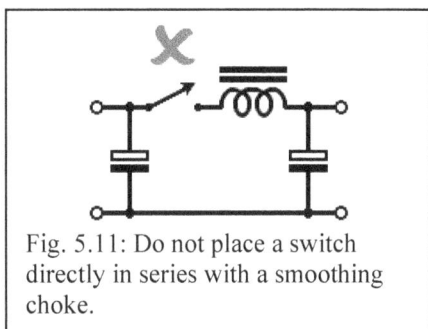

Fig. 5.11: Do not place a switch directly in series with a smoothing choke.

capacitors to partially charge, which should also reduce speaker thump when the switch is finally closed. A disadvantage is that the amplifier might not be completely silenced during standby but may produce a thin, strangled sound when attempting to play music. Whether this is a legitimate criticism is left to the reader to decide. The value of resistor is not critical; a 10kΩ 5W to 47kΩ 2W device will suit most circumstances.

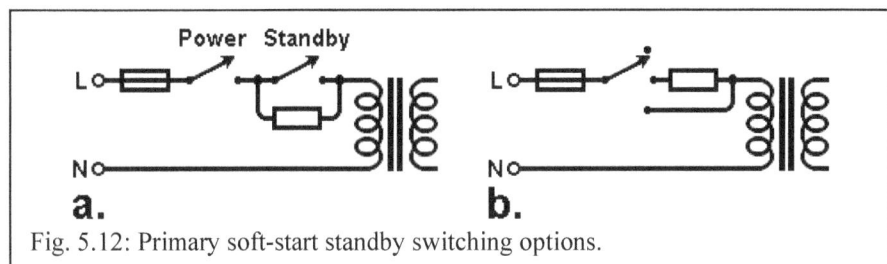

Fig. 5.12: Primary soft-start standby switching options.

[†] Vox added such a 39kΩ 5W resistor to later runs of the *AC30CC* to alleviate the problem with hot-switching.

121

A further standby option is to move the switch to the transformer primary, as in fig. 5.12. It will then provide a soft start for everything, including magnetising inrush current. With this arrangement a mains-rated switch can be used with confidence, and a 4.7kΩ 5W resistor is a good place to start. Fig. 5.12b shows how a three-position switch offers a more ergonomic transition from off, to standby, to on, without the user having to worry about the correct procedure. During standby, all secondary voltages will be reduced; this is potentially advantageous since reducing the heater voltage by about a third during standby will slow down cathode evaporation and discourage the growth of interface resistance, while still retaining a relatively quick warm-up when the amp is eventually switched from standby to on. On the other hand, this approach might not be suitable if it causes some essential secondary circuit –such as a microprocessor– to operate marginally or drop out.

5.3.5: Automatic Delay

Going further, soft start can made fully automatic. It is possible to buy self-delaying relays (they are very expensive), but fig. 5.13 shows an old and delightfully simple approach.[9] At start-up the coil is not energised and the relay is open, so the reservoir capacitor C_1 charges slowly through R_1. Meanwhile the valve heaters are warming up, and once the amplifier draws enough current through the coil the relay will close, shorting the resistor and

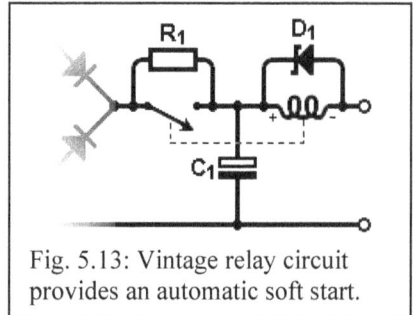

Fig. 5.13: Vintage relay circuit provides an automatic soft start.

bringing the power supply into normal operation. An obvious requirement is that the load current must be large enough to energise the relay. Most 12V power relays need about 30mA for the coil, which is well within the range of valve power amps, but some experimentation may be needed to find a value for R_1 that ensures reliable operation every time. Zener diode D_1 passes any excess load current which would otherwise overload the coil (it also quashes flyback voltage), so it should be appropriate to the voltage rating of the coil, e.g. 12V 5W. Notice that the relay armature is in the path of the pulsating ripple current rather than the smooth DC, allowing a conventional $250V_{ac}$ relay to be used in a power supply up to $350V_{dc}$.

A more modern option is to use a MOSFET to switch the HT. Indeed, an optically controlled MOSFET is effectively a solid-state relay, which makes switching high-voltages trivial (see section 7.6.3). Alternatively, a self-delaying MOSFET circuit is shown in fig. 5.14. The MOSFET is inserted into the negative side of the rectifier as this allows the gate to be pulled up above the drain, so there is no loss due to V_{GS} (this should not be used with a conventional bias supply, however, remember fig. 5.10). At switch-on the MOSFET is off and the capacitor C_2 will begin to charge

---

[9] Clements, D. (1952). High-Tension Delay Circuit. *Wireless World*, (April), pp163-4. A similar principle was used for overload protection in the HP 711A valve-regulated bench power supply, for example.

through R_1. Once
the voltage across it
reaches about 4V
the MOSFET will
switch on and the
HT will ramp up to
its full value. This
is the same
principle as the
conduction-angle
regulator in section
7.6.2. Using the
component values
shown, the delay
before the
MOSFET switches
on will be about 10
seconds for a
typical HT voltage.
The ramping itself
can take as much as
one second to
complete, so it also
qualifies as a soft
start.

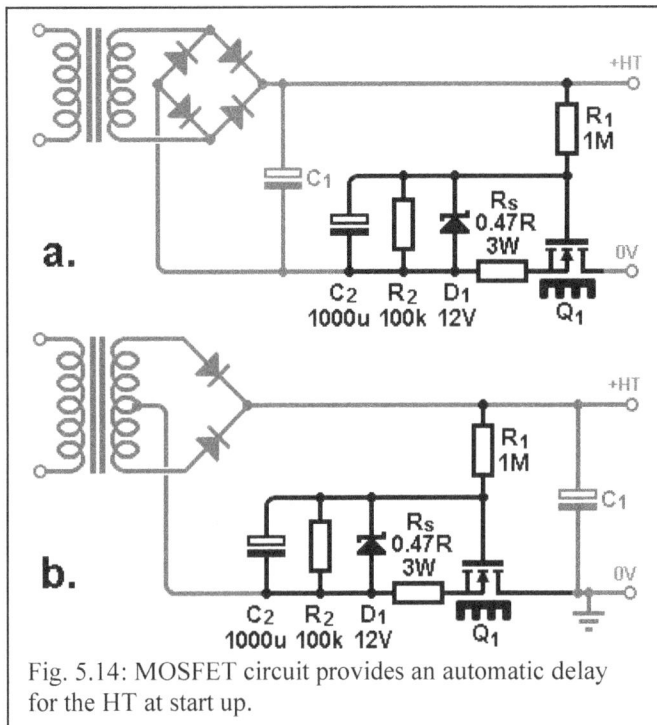

Fig. 5.14: MOSFET circuit provides an automatic delay for the HT at start up.

Zener diode D_1 protects the gate-source junction and C_2 from overvoltage (so C_2 can be a low-voltage part), and in combination with R_s also protects the MOSFET against inrush transients if the amplifier is switched off and on again rapidly. R_2 allows C_2 to discharge after switch-off. R_1 could be replaced with several resistors in series to share the voltage burden, allowing ¼W devices to be used. If the delay circuit is placed after the reservoir capacitor as in fig. 5.14a, then a heatsink is not usually necessary, and almost any N-channel MOSFET rated to withstand the full HT voltage should work. But if the MOSFET must handle ripple current as in b. then it must have sufficient drain current rating to handle the peaks, and a clip-on heatsink (or bolting to the chassis) may be necessary. For even longer delay times it is probably better to incorporate some type of timer IC into this circuit,[10] rather than use more R or C which might introduce unreliability due to capacitor leakage.

Turning our attention to the primary side of the power supply, fig. 5.15 shows how a low-voltage DC supply can be obtained directly from the mains, which can then be used for any number of automatic features (this is permissible only when the circuit to be powered is isolated from the user). The live feed passes through a 1µF capacitor which drops voltage without dissipating any heat, so it is sometimes called

---

[10] Didden, J. (2019). A High-Voltage Delay for Tube Amplifiers: the Sequel. *Audio Xpress*, (November), pp 50-3.

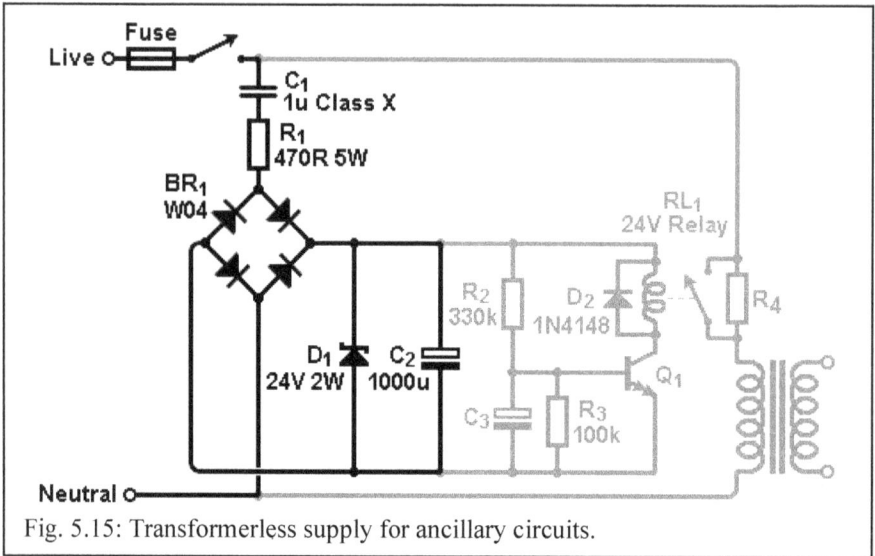

Fig. 5.15: Transformerless supply for ancillary circuits.

a **wattless dropper**. This must be a class-X safety capacitor as only these may legally be connected across the mains.[*] Since the capacitive reactance is much larger than the effective load resistance, the capacitor behaves like a constant-current source, always limiting the RMS current to less than:

$$I = \frac{V_{rms}}{2\pi fC} \tag{5.1}$$

Where V_{rms} is the wall voltage and f is the mains frequency.

R_1 is essential to limit the peak charging current at switch on. The bridge rectifier directs the current into a Zener diode to provide a stabilised DC voltage which can be anything up to about 30V. The Zener diode should be capable of handling the same amount of current as calculated above. The maximum DC current available for the load will be about twenty percent less; trying to draw more will steal too much from the Zener, causing the DC voltage to droop. With a 1μF wattless dropper the circuit can provide up to about 55mA$_{dc}$ on 230V mains, or 30mA$_{dc}$ on 120V mains.

The components shown faint represent a typical circuit to be powered –in this case an inrush-limiting relay. At switch on, C_2 will charge up to the Zener voltage. C_3 then charges much more slowly through R_2 until the voltage across it is enough to turn on the Darlington transistor Q_1, energising the 24V relay and shorting out the inrush-limiting resistor R_4. As drawn, the time delay is approximately one second per hundred microfarads used for C_3, and a second or two should be more than enough to suppress inrush. C_1 discharges again through R_3 at switch-off. Q_1 can be any general-purpose Darlington such as 2N6427, MPSA13, etc. Remember, this circuit is allowable only because there is no way for the user to come into contact

---

[*] Technically, class-Y is also allowed, as they are even more robust than class-X.

with it. It should be considered part of the mains supply and must have no galvanic connection to the audio circuit.

5.4: More Power Supply Refinements

The following is a collection of further refinements that are typically found in modern, professional power supplies. For hobby purposes these are optional, but they are certainly deserving of adoption into the old-fashioned world of valve amplifiers. Although they may be omitted from circuit diagrams in this book to save space and keep things clear, this does not mean they cannot be added as the reader's discretion.

5.4.1: Inrush Current Limiting

Inrush current limiting has already been touched upon in the previous section. When power is first applied to a transformer there will be a brief but possibly enormous current surge that lasts for a few mains cycles. There are two reasons for this. Before power is applied, the transformer is not magnetised, i.e. it has zero core flux. If we are unlucky enough to switch on when the mains voltage is at the zero crossing where the voltage is changing fastest, the core flux will increase to the point of saturation and beyond. Momentarily, the only thing limiting the primary current is the winding resistance, so the primary current will be abnormally large for the first few cycles until the core flux settles to its steady state. This effect will be worse for high-power transformers since they have little primary resistance. Conversely, if the circuit is switched at the peak of the mains cycle, then there will be less magnetising inrush, but the near-instantaneous change in voltage across the reservoir capacitor will cause it to take several large gulps of charging current instead. Switched-mode power supplies often have fierce inrush current for the same reason. How ever you slice it, inrush current is inevitable when switching on a power supply.

Transformers smaller than about 10VA use such fine primary wire that it will flex under the magnetic force field of the inrush current. After a few hundred switching cycles this will lead to metal fatigue and eventually a break, so small transformers benefit enormously from some form of inrush limiting. Larger transformer can easily withstand the current, but we may need inrush limiting measures for a different reason –to prevent otherwise reasonably-sized fuses and circuit breakers from blowing and tripping every time we switch on.

Big power amplifiers commonly use a resistor to limit the current, which is automatically shorted out by a hefty relay or triac after a second or two (e.g. fig. 5.15). A less elaborate option is to use a negative temperature coefficient (NTC) thermistor in series with the mains input, as shown in fig. 5.16d. NTC thermistors have relatively high resistance when cold, but quickly fall to a fraction of an ohm as the device heats up with normal current flow. The thermistor should be chosen so its cold resistance is a significant fraction of the transformer primary resistance, e.g. one third to one half. It will get hot during normal running so be careful with its physical placement. It is also worth noting that it takes time for the device to cool down after

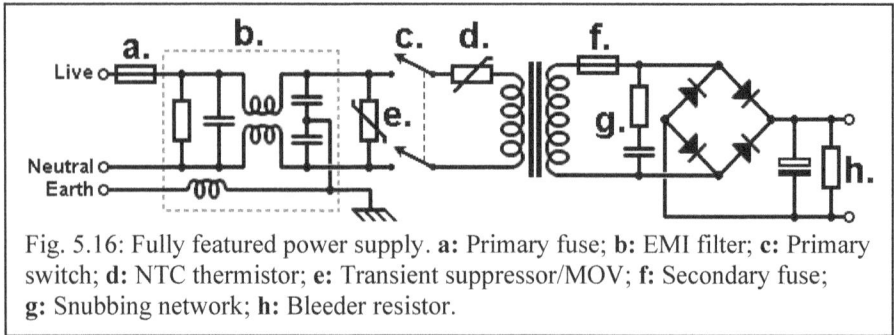

Fig. 5.16: Fully featured power supply. **a:** Primary fuse; **b:** EMI filter; **c:** Primary switch; **d:** NTC thermistor; **e:** Transient suppressor/MOV; **f:** Secondary fuse; **g:** Snubbing network; **h:** Bleeder resistor.

use, so a thermistor cannot protect against rapidly turning the supply off and on again.

5.4.2: Transient Suppression

The mains supply is occasionally polluted with transient voltage spikes caused by industrial motors, lightning strikes and so on. A protective measure against these spikes is a transient suppressor or metal oxide varistor (MOV), connected between live and neutral after the primary fuse, as in fig. 5.16e. 'Varistor' is short for variable resistor, although the device behaves more like a pair of back-to-back Zener diodes. Under normal circumstances the MOV has a very high resistance (several megohms), but if the voltage across it reaches a certain threshold –called the breakdown or clamping voltage– its resistance suddenly drops, quashing the voltage spike and possibly blowing the fuse. Unlike a Zener diode, however, a MOV can withstand extremely large (but brief) current pulses.

Selecting a MOV is a simple matter of choosing one with a continuous RMS working voltage that is 1.2 to 1.5 times the nominal mains wall voltage. However, sustained or repeated overloads can –and frequently do– cause MOVs to burn out, so they must be suitably positioned so as not to ignite anything else and create a fire hazard. For UL-certified equipment a thermal fuse must be connected in series with the MOV and mounted side-by-side with it, to detect overheating and interrupt the supply to the MOV. Thermally protected MOVs are also available, such as the Littlefuse TMOV® series.

5.4.3: EMI Filtering

A common addition to any modern appliance is an electromagnetic interference (EMI) filter, where the mains enters the unit. Such filters are often built into IEC inlets and have become so ubiquitous that the author has never had to buy one, only scrounge them from discarded equipment. They usually consist of a common-mode choke with capacitors on each side to create a balanced CLC filter, as shown in fig. 5.16b. An inductor is sometimes included in the earth feed which helps to suppress earth loop induction at high frequencies. A bleeder resistor may also be included to ensure the capacitors discharge after the power cable is removed.[*]

Beginners sometimes assume that an EMI filter is supposed to turn an ugly mains supply into a sparkling clean sine wave, but we should be so lucky. They only do anything useful in the range of hundreds of kilohertz to megahertz and, even then, they are largely intended to attenuate noise generated *inside* the appliance (particularly from an SMPS) from leaking *out* onto the mains (the mains cable can act as quite a good transmitting antenna, jamming local radio reception). Valve circuits generally use linear power supplies and seldom contain any heavy switching currents, so they have minimal problems with EMI, either reception or transmission, but it can do no harm to include a filter. Be sure to use one that can handle the maximum expected current. This should be no problem in a valve amp as even the most modest filters are rated for a couple of amps.

5.4.4: Bleeder Resistor

When a high-voltage power supply (>50V) is switched off it is desirable that the reservoir/smoothing capacitors have some means of discharging, otherwise they pose a shock hazard during repair or when replacing valves. Sometimes a discharge path is already provided by a heater-elevation divider or by equalising resistors across

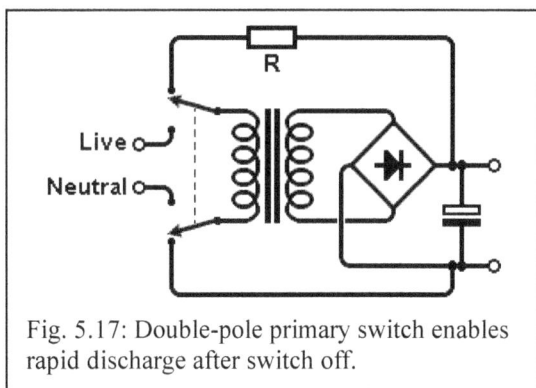

Fig. 5.17: Double-pole primary switch enables rapid discharge after switch off.

series-connected smoothing capacitors etc., in which case nothing more is required. But if not, it is a simple matter to add a bleeder resistor in parallel with one of the capacitors, as in fig. 5.16b. A value of 330kΩ to 1MΩ is typical, determined mainly by how much current we are willing to waste during normal running. Adding an LED in series with the resistor, visible from inside the unit to give some indication of discharge, is a nice touch for the repairman.

If the discharge time needs to be especially rapid, as in a particularly dangerous high-energy power supply, then the bleed resistance will need to be so small that it wastes excessive current during normal running. In such cases the spare pole of a double-pole main primary switch can be used to switch in the discharge resistance only when needed. If there is no spare pole because both live and neutral are switched, the variation in fig. 5.17 could be used.[11]

---

[*] According to UL 60065 the voltage must discharge to <60V within two seconds after switch off.
[11] Woodward, S. (2001). Quickly Discharge Power-Supply Capacitors, *EDN*, July, p132.

Chapter 6: Shunt Stabilisers and Regulators

Any circuit designed to hold some parameter –voltage, current, power etc.– within certain predetermined limits can be described as a **stabiliser**. Traditionally, circuits which achieve stabilisation by using negative feedback get promoted to the title of **regulators**. Over time, however, the name 'stabiliser' seems to have drifted out of use; everything is called a regulator these days, whether or not it uses feedback. You can't fight progress.

The raw DC voltage provided by the conventional transformer and rectifier circuit is unstabilised. It includes rectifier ripple, noise, and is free to fluctuate and drift in response to changes in wall voltage and load current. Often this does not matter a great deal for valve amps. Wall voltage variations are usually modest; power valves are often pentodes or tetrodes which are fairly insensitive to changes in anode voltage; cathode-biased stages naturally adjust to changes in HT.

But there are also situations where stabilisation is wanted. Sometimes we really do want true voltage regulation, i.e. a DC voltage that stays within strictly controlled limits to ensure consistent performance at all times. Sometimes we want the hum and noise reduction that a regulator can offer, whether or not we care much about the DC voltage. And sometimes we need a way to adjust a voltage over a wide range for the purposes of experimenting or bench testing, and this often turns into a regulator circuit.

6.1: Shunt Voltage Regulation

This book is mostly concerned with linear regulators. These are the kind that achieve their regulation by continuously burning away excess voltage as heat. In essence, the regulator forms one arm of a potential divider, as illustrated in fig. 6.1. The shunt regulator places the power-controlling device in parallel with the load, that is, in the 'shunt' arm of the divider; it

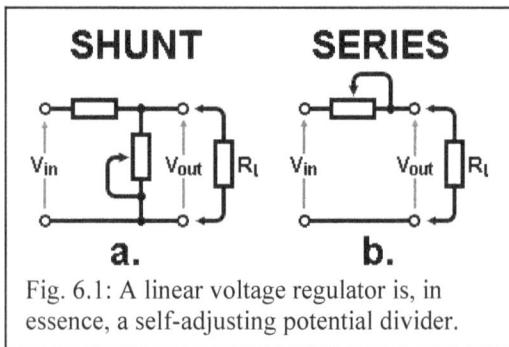

Fig. 6.1: A linear voltage regulator is, in essence, a self-adjusting potential divider.

attempts to draw current from the power supply in such a way that it counterbalances unwanted perturbations in the power supply voltage. It therefore has the full load voltage imposed across it, and it continually wastes some current in its operation. This wastefulness is the main shortcoming of any shunt regulator and is the reason they play second fiddle to series regulators, covered in the next chapter.

The advantage of shunt regulation is that it can be as simple and cheap as a Zener diode, and the voltage lost across the series arm can, if necessary, be made very small, i.e. a near-zero drop-out voltage. Shunt regulators can also be more tolerant of

128

output short circuits since the only part that might burn out is the series arm, which could be a nice, robust resistor. Shunt regulation is also prized by audiophiles because it localises signal currents in the same manner as a local bypass capacitor (section 4.1.2), but can achieve it without frequency-dependence, all the way down to DC, unlike a mere capacitor.

When it comes to high-voltage shunt regulation, valves are a natural fit. They are at home with high-voltage across themselves, and need no heat sinking, whereas transistors need more care and attention in this regard. Also, in a shunt regulator the cathode will normally be fairly close to ground potential, so there isn't the problem of heater-to-cathode leakage that series valve regulators present. Of course, valves cannot handle nearly as much current as solid-state devices, but shunt stabilisers are normally reserved for low-current applications anyway, owing to their wastefulness. Shunt regulators therefore present an interesting way to use those unpopular television valves you can't bear to throw away, but which aren't well suited for audio amplification. Almost any valve with enough electrodes can be pressed into service as a shunt regulator for a small project, if you can afford to burn the extra watts that is.

6.1.1: Principles of Shunt Regulation

Fig. 6.2 shows the functional parts of a feedback shunt regulator. A fraction of the output voltage –the feedback voltage– is compared with a fixed reference voltage. Any difference between the two is called the error voltage, or simply error, and indicates that the output voltage of the regulator is not what it should be. This error is amplified and sent to the shunt control device which does the heavy lifting; conducting more current to try to pull the output voltage down, or conducting less to allow the output voltage to un-sag back up (we really need a word for this). These functional blocks are not always obvious in a practical regulator circuit, but they can be found if you stare at it long enough. For example, the error amplifier and shunt control device might be one-and-the-same component, or the series dropping resistor might be formed by the implicit source resistance of the transformer/rectifier.[1] In a Zener diode it all takes place within the invisible world of atoms and electric fields.

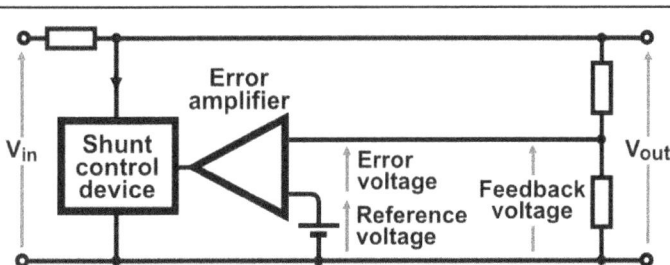

Fig. 6.2: Functional block diagram of a feedback shunt regulator.

---

[1] Guy, H. (1956). A Compact Stabilised High Voltage Supply, *Practical Wireless*, June, pp242-4.

A feedback regulator is just like any feedback amplifier, except that its 'input signal' is a DC reference voltage rather than an audio signal. Note, therefore, that the cleanliness of the output can be no better than that of the DC reference itself. The output is an amplified version of the reference, plus any line fluctuations that still squeeze through. In general, the more signal we feed back, and/or the more gain the error amplifier has, the better the regulation. This is called the **loop gain**; it is the gain we would measure if we broke the feedback path and measured the gain between the two broken ends. However, the higher we make the loop gain the harder it is to make a feedback circuit stable, owing to unwanted phase shifts in the feedback signal. If such phase shifts become too great then the feedback signal sent to the control device will be of the wrong polarity, and the control device will reinforce the error rather than cancel it, which means ringing or oscillation. As a rule of thumb, the smaller the number of active amplifying stages inside the loop, the easier it is to make stable. Simple circuits have simple shortcomings, and the circuits explored in this book try to adhere to this maxim.

6.1.2: Line and Load Regulation

The degree to which a change in the input (line) voltage causes a change in the output (load) voltage is called **line regulation**. A perfect voltage stabiliser would hold the output voltage constant no matter what the input voltage happens to be. It would eliminate any trace of rectifier ripple and noise, as well as reject longer term DC drifts caused by changes in wall voltage. As we shall see later, feedforward nulling is the most potent way to achieve high line rejection, but it requires careful tuning which makes it a frustrating and capricious technique to rely on.

The degree to which a change in load current causes a change in load voltage is called **load regulation**. A perfect stabiliser with infinite load regulation would not care how much load current is demanded, it would maintain a constant output voltage regardless. This is another way of saying it would have zero output impedance. The output impedance of the stabiliser forms a potential divider with the load impedance. The more current the load demands, the greater the internal voltage drop across the output impedance, meaning the voltage remaining across the load will sag. Negative feedback is the most potent way to achieve good load rejection.

6.2: Passive Shunt Regulators

As mentioned earlier, a shunt regulator can be as simple as a diode. A diode has a steep I/V characteristic, meaning a large change in current through it will cause only small change in voltage across it. For example, the forward voltage across an ordinary silicon diode is about 600 to 700mV whatever the current through it, within practical limits of course. Since no amplifying devices are involved, this is a purely passive form of voltage regulation (strictly *stabilisation*).

Fig. 6.3 shows an archetypal passive shunt regulator using a series resistor R_s to limit the current through a diode. If the supply voltage rises, more current will flow into

D$_1$ but, being a diode, the voltage across it will hardly change. The increase in supply voltage is therefore taken up entirely across the series resistor. If the supply voltage rises by 1V, say, then the drop across R$_s$ is now 1V greater, since the other side is clamped at about 700mV by D$_1$. The total current therefore increases by 1V/R$_s$.

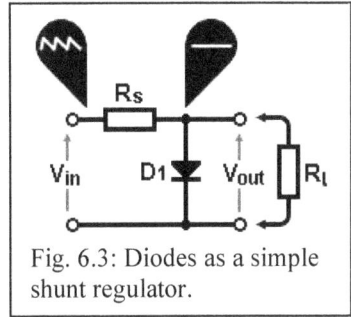

Fig. 6.3: Diodes as a simple shunt regulator.

If we now attach a load in parallel with D$_1$, some current will be diverted into it, away from the

diode. If the load current increases further then more current will be 'stolen' from the diode, and if the load current relaxes, the diode must take up the deficit. Fig. 6.4 illustrates how the total current flowing through R$_s$ is shared, see-saw fashion, between the diode and load. The diode dissipates maximum power when there is no load since it has to sink the full quota of current. Conversely, if the load draws all the current then there will be none left for the diode and the circuit 'drops out' of regulation; any further increase in load current will drag the output voltage down. Therefore, when we design this kind of shunt regulator, we must ensure the quiescent current in the shunt is large enough to pay for any expected increase in load current, and

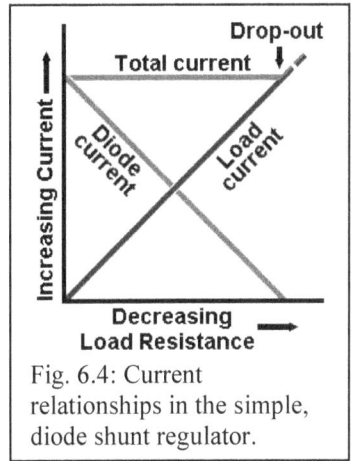

Fig. 6.4: Current relationships in the simple, diode shunt regulator.

conversely that the shunt can safely swallow the extra current if the load relaxes. As a rule of thumb, be prepared to make the idle current in the shunt device at least equal to the nominal load current. This of course means the power consumption of the entire appliance is effectively doubled! Such is a price of shunt regulation.

6.2.1: Zener Diodes

We rarely need stabilised voltages as low as 700mV, especially when building valve amplifiers. However, we can stack any number of diodes in series so their forward voltages sum. We can also use other kinds of diode such as LEDs, whose forward voltage covers the range between about 1.2V (infrared) to 5V (hyper white/blue). But the most common sort of stabiliser diode is the **Zener diode**.

Whereas ordinary diodes and LEDs can be used to provide constant voltages equal to their forward drop, Zener diodes are connected 'the other way round'; they are used in reverse-bias mode, as shown in fig. 6.5 (when forward biased they behave like ordinary diodes with about 700mV drop). The voltage drop across the diode in this orientation is the **breakdown voltage**, which is specially engineered to be some particular value. Zener diodes are available with nominal breakdown voltages from 2.7V to about 280V, with power ratings from milliwatts for tiny surface-mount devices to mighty 75W metal-clad devices that must be bolted to a heatsink.

131

There are actually two kinds of reverse breakdown effect. We need not trouble ourselves with the underlying physics but simply note the following. In devices below about 6V the **Zener effect** dominates, which has a negative temperature coefficient. This means the voltage across a low-voltage Zener diode will tend to decrease with increasing temperature. Above about 6V the

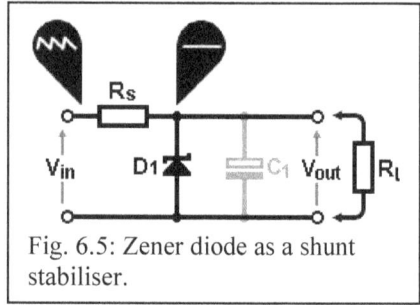

Fig. 6.5: Zener diode as a shunt stabiliser.

avalanche effect takes over, which has a positive temperature coefficient, so the voltage will tend to increase with increasing temperature. Diodes in this range are therefore sometimes called **avalanche diodes**, but generally the name Zener is used for both types. Between about 6V and 7V both mechanisms operate simultaneously and their characteristics tend to cancel. Diodes in this range therefore tend to give the best temperature stability, lowest noise, and smallest internal resistance. Sometimes they are sold under the name of 'precision Zener' or something similar.

6.2.2: Internal Dynamic Resistance

In simple terms, the output impedance of a voltage regulator is a measure of how well it can hold the load voltage constant despite changes in load current. A perfect voltage regulator would have zero output impedance at all frequencies from DC to light; it would hold the voltage across itself utterly constant no matter how much load current was demanded. Practical circuits, like people, fall short of perfection.

In the case of a diode shunt regulators the output resistance is determined mainly by the internal **dynamic resistance** or **slope resistance**, r_d, of the diode itself. This can be represented by a resistance in series with an idealised diode, as shown in fig. 6.6. The output resistance of the whole stabiliser is then the Thévenin (parallel) combination of R_s and r_d, but usually R_s is large enough by comparison that it can be ignored. The overall output impedance for *AC* signals

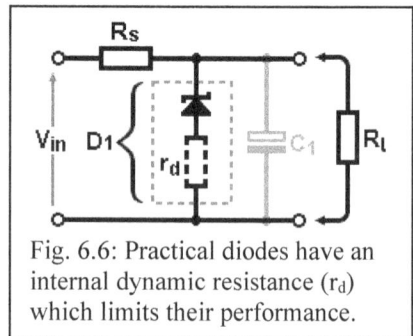

Fig. 6.6: Practical diodes have an internal dynamic resistance (r_d) which limits their performance.

can always be made smaller by adding a capacitor in parallel with the diode (shown faint), but the DC output resistance we are stuck with.

Fig. 6.7 shows the conductance characteristics of some Zener diodes measured by the author, plus a couple of cheap LEDs for comparison. The internal dynamic resistance is equal to the inverse of the gradient of the slope. The steeper the slope, the lower the internal resistance. A general trend is evident: Low- and high-voltage Zeners have the most curved characteristics and need at least 5mA anode current for best regulation, avoiding the 'bottom bend' as it used to be known. Devices between

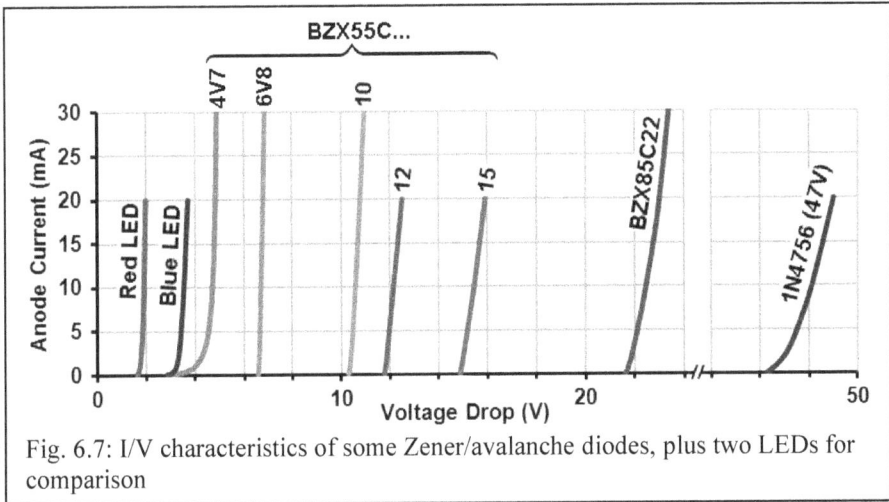

Fig. 6.7: I/V characteristics of some Zener/avalanche diodes, plus two LEDs for comparison

about 6V and 10V (where Zener and avalanche effects are in balance) have smaller and more linear internal resistance and can be operated with less current if necessary.

6.2.3: Noise in Zener Diodes

Whenever current has to cross a potential boundary –as it must do inside a diode– noise is generated. This can be modelled as a noise current generator in parallel with the diode, as shown in fig. 6.8. For an ordinary forward-biased diode such as a rectifier or LED, the noise current riding on the average anode current is quite predictable. It is given by Schottky's theorem:

$$i_{shot} = \sqrt{2qI_{DC}B} \quad \text{amps RMS} \qquad (6.1)$$

Where:
q = charge on an electron, 1.6×10^{-19} coulombs
I_{DC} = mean anode current
B = noise measurement bandwidth in hertz

This is a type of white noise. The same applies to true Zener diodes even though they are operated in reverse bias, remembering that true Zener diodes are those with breakdown voltages below about 6V. Higher voltage (avalanche) diodes generate much higher levels of white noise, plus another sort called multistate noise which is mainly pink, neither of which is so easy to calculate. Diodes also exhibit excess noise, though it is often negligible in the audio band.

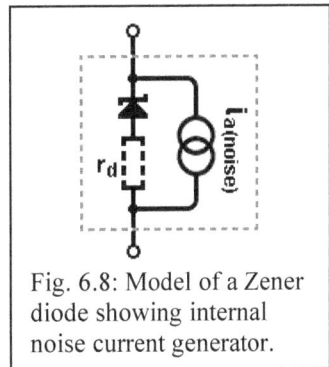

Fig. 6.8: Model of a Zener diode showing internal noise current generator.

However, we are more interested in how much noise *voltage* appears across the diode, that is, on our stabilised supply. From fig. 6.8 it is clear the noise current flows in a loop through the diode's internal resistance r_d, as well as through any

133

external impedance we choose to put in parallel. From Ohm's law, the noise voltage across the diode will be equal to the noise current multiplied by the total parallel impedance.

Now, the noise current increases with DC bias current, but the internal resistance of the diode falls, and usually at a faster rate. In other words, the more bias current you shove through the diode, the quieter it gets, at least when you get above 1mA. A capacitor added in parallel with the diode will shunt the noise even further; with most diodes it will need to be at least 100nF or larger to have any effect in the audio band. In most cases a smoothing/decoupling capacitor will be added anyway, which will fulfil the same job.

Gas reference diodes / glow discharge tubes are the valve version of modern avalanche diodes. They are usually regarded as being noisy, but data presented by Jones[2] appear to show that they generate mainly shot noise. If true, this makes them just as quiet as LEDs and Zeners, technically speaking. It is instead their much higher internal resistance that gives rise to their greater output noise *voltage*. They also have severe limitations on how much capacitance can be added in parallel to suppress the noise, which no doubt contributes further to their noisy reputation.

6.2.4: Amplified Zener

For high-power applications, high-power Zeners exist, but being a specialist item they are expensive and not widely sold. It is often cheaper and more convenient to use an ordinary, low-power Zener, and augment it with a readily-available power transistor. This arrangement is called an amplified Zener, shown in fig. 6.9.

An NPN or PNP transistor can be used, according to convenience, as shown in a. and b. respectively. Either way the operation is the same: Zener diode D_1 stabilises the base-collector voltage of the transistor, while the base-emitter is essentially a diode so its voltage is quite constant by default (typically 0.7V for a power transistor). The total output voltage is therefore $V_Z + V_{be}$. If the base current alone is not enough to bias the Zener then more can always be furnished by adding R_1. Since Q_1 does the hard work of dissipating most of the heat it will normally require a heat sink.

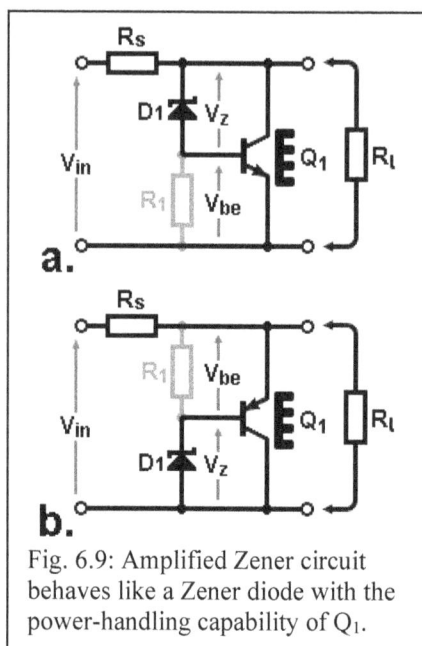

Fig. 6.9: Amplified Zener circuit behaves like a Zener diode with the power-handling capability of Q_1.

---

[2] Jones, M. (2012). *Valve Amplifiers*, 4th ed., Elsevier, Oxford.

A MOSFET could alternatively be used in place of the BJT, but the gate-source voltage of a MOSFET is far less predictable than the V_{be} of a BJT, so the stabilisation voltage will be less precise. This matters less at high voltages since the gate-source voltage is then only a tiny fraction of the total, and high voltage is probably the reason for using a MOSFET in the first place. But we're getting ahead of ourselves. Although the amplified Zener appears to behave like a big Zener diode, it is not actually a passive circuit; it is a feedback regulator. We will work our way back to this by the end of the chapter.

6.3: Glow-Discharge Devices

Glow-discharge devices are the ancestors of modern Zener diodes, and they are used in the same general way. A glow-discharge device is in essence a fancy spark gap, consisting of a bulb containing one or more noble gases* and a pair of electrodes. If a high voltage is applied between the electrodes then some electrons will be emitted by the negative cathode and will accelerate towards the anode, colliding with gas molecules on the way. If they gather sufficient velocity they will succeed in knocking electrons out of orbit from the gas molecules, ionizing the gas. These new electrons will also accelerate towards the anode and in turn collide with more gas molecules, potentially ionizing them too. Under the right conditions this process reaches an avalanche point and becomes self-sustaining; all the gas becomes ionized and a continuous current will flow between the electrodes. Some gas-discharge devices are bi-directional, e.g. neon lamps and surge arresters, while proper gas stabiliser tubes have a specific cathode which is chemically treated to encourage emission, in which case polarity must be observed.

During operation, some of the electrons moving inside the tube will collide with previously-ionized molecules and will be re-captured, neutralizing the ions again. Each time this happens, a photon is emitted, and this is what causes the gas to glow. The colour of the glow depends on the gas used: red/orange for neon, pink/purple for argon, blue for krypton, and blue/purple for xenon. Since it is mainly the low-velocity electrons which are re-captured, the glow will be concentrated around the cathode (electrons near the anode are travelling too fast to be re-captured). The intensity of the glow varies in sympathy with the current, which in some circumstances creates a pleasing light show.

There will be a particular voltage at which the initial electrons reach sufficient average velocity to trigger the avalanche effect and cause the gas to ionize fully. This is called the **striking voltage, breakdown voltage** or **ignition voltage**. The voltage across the device must reach this level before it will switch 'on' and conduct. However, once the tube strikes, the current will generate some heat, and this helps to keep the gas ionized. The voltage required to *keep* the current flowing therefore drops to a lower level which is fairly constant over a wide current range. It is this region that behaves like a Zener diode, and is called the **maintaining voltage** or

---

* The dot inside the circuit symbol indicates the presence of gas rather than vacuum.

135

stabilisation voltage. It may also be called the **extinguishing voltage** since the tube will switch 'off' if the supply voltage drops below this value.

6.3.1: Gas Stabiliser Tubes

Gas stabiliser tubes were widely used as voltage references and stabilisers before Zener diodes and other solid-state solutions became available. Many types were produced over the years, with different stabilisation voltages. Although they look rather like valves, no heater is required, so they are sometimes called **cold-cathode** valves. Many American gas tubes have designators beginning with zero, such as the 0A2, indicated zero heater voltage is required (frustratingly, the rest of the code gives no clue to the stabilising voltage).

Tube type	Stabilisation voltage	Striking voltage	$I_{a\ (min.)}$	$I_{a\ (max.)}$	Dynamic resistance	Base
0A2	150V ±8%	200V	5mA	30mA	240Ω	B7G
0B2	108V ±3%	150V	5mA	30mA	140Ω	B7G
0C2	75V ±6%	130V	5mA	30mA	500Ω	B7G
0A3	75V ±5%	105V	5mA	40mA	200Ω	IO
0B3	90V	110V	5mA	30mA	200Ω	IO
0C3	108V	115V	5mA	40mA	100Ω	IO
0D3	150V ±10%	160V	5mA	40mA	160Ω	IO
5651	87V ±6%	110V	1mA	3.5mA	1500Ω	B7G
83A1	83V ±2%	125V	3.5mA	6mA	300Ω	B7G
85A2	85V ±4%	125V	1mA	10mA	300Ω	B7G

Table 6.1: Some voltage-stabiliser tubes and their nominal operating conditions. Tubes in bold type are voltage reference tubes.

Table 6.1 lists some of the more common gas tubes and their operating conditions. Note that the values for striking voltage and dynamic resistance may vary considerably between samples. Voltage *reference* tubes are listed in bold; these are not rugged power supply stabilisers but precision voltage references designed to operate at small, relatively constant current, with only light loading. Then *can* be used as low-current stabilisers but the performance may not be as impressive as anticipated. Furthermore, the maintaining voltage of gas tubes may wander over time, and they also tend to become accustomed to running under fixed conditions. If the operating current changes, or the tube is plugged into some other circuit, there may be a lot of drift and fluctuation until it eventually settles down again.

Whole books have been written on gas tubes and their uses,[3] but we really only need the essential operating information. Gas tubes are used in more-or-less the same way as Zener diodes. The datasheet should state the range of current over which it can safely operate; exceeding this rating will cause overheating and eventual damage (they are normally rated for a maximum running temperature of only 90°C). At currents below the recommended minimum, the voltage across the tube becomes unreliable and noise will become excessive, resulting in poor stabilisation. Gas tubes should never be connected directly in parallel with one another without limiting resistances, otherwise one will inevitably strike first and take all the current, possibly burning out in the process. So far, so much like Zener diodes. But there are further limitations for gas tubes.

With a Zener diode we usually add an arbitrarily large capacitor in parallel to shunt noise and improve transient response. However, this is not possible with a gas tube, because it can induce oscillation, particularly at small operating currents. This is caused by the tube exhibiting a negative resistance region and a hysteresis effect.[4] There is no need to delve deeply into the exact mechanism, but simply imagine that when the tube strikes, it immediately drains so much current out of the capacitor that it puts itself out again. The capacitor charges up again through R_s and the process repeats endlessly, forming a relaxation oscillator rather than a stabiliser. The datasheet will therefore quote a maximum safe value for parallel capacitance, which is usually 100nF. If more is needed then it should be isolated from the tube by some resistance (try 100Ω).

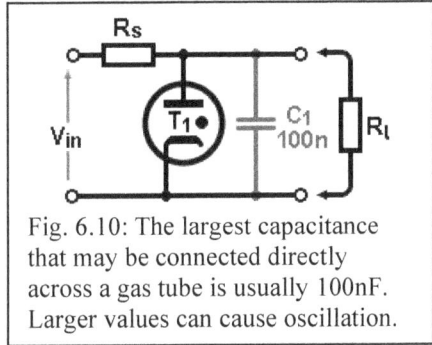

Fig. 6.10: The largest capacitance that may be connected directly across a gas tube is usually 100nF. Larger values can cause oscillation.

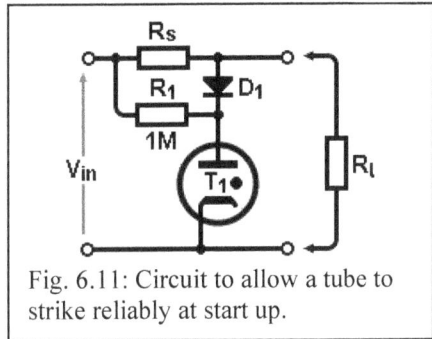

Fig. 6.11: Circuit to allow a tube to strike reliably at start up.

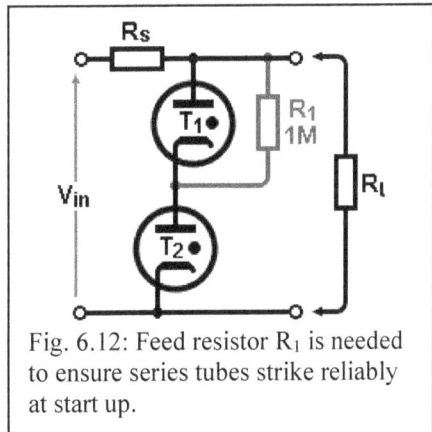

Fig. 6.12: Feed resistor R_1 is needed to ensure series tubes strike reliably at start up.

---

[3] Benson, F. A. (1965). *Voltage Stabilization*. Macdonald and Co, London.
[4] Edwards, P. L. (1952). Relaxation Oscillations in Voltage-Regulator Tubes. *US Naval Ordnance Report No. 2698.*

The voltage across the tube must reach the striking voltage before it will start regulating. It is therefore no use trying to stabilise a raw 180V supply down to 150V with an OA2, since that tube

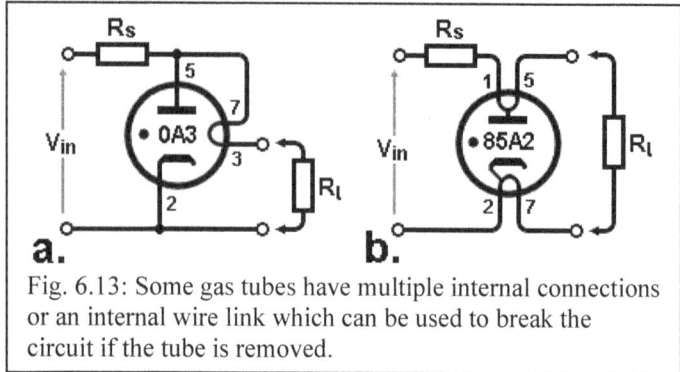

a. **b.**

Fig. 6.13: Some gas tubes have multiple internal connections or an internal wire link which can be used to break the circuit if the tube is removed.

needs 200V to strike in the first place. A related conundrum can arise at switch-on if the load immediately pulls the voltage below the striking voltage. A classic solution is to add a feed resistor R_1 and diode, as shown in fig. 6.11, so the tube receives the full input voltage at start up, allowing it to strike. The feed resistance can be very large $-1M\Omega$ say– so it does not interfere much with normal operation. A similar problem arises when two or more tubes are stacked in series to sum the voltages as in fig. 6.12. At switch-on T_1 cannot conduct until T_2 does, but if T_1 is not conducting then there can be no voltage across T_2 to make it strike in the first place –a chicken and egg scenario. Again, the solution is a feed-resistor R_1, added so the supply voltage initially appears across T_2, allowing it to strike. Alternatively the resistor could be connected in parallel with T_2, so T_1 strikes first.

Some gas tubes have special pin connections which can be used to disable the circuit if the tube is removed from its socket. American tubes such as the 0A3 in fig. 6.13a provide a wire link between two pins that can be used like a switch to break the circuit. With many European tubes there are multiple connections to the anode and/or cathode which can be used similarly, as illustrated in b.

6.3.2: Neon Lamps

A neon lamp is a cheap and simple example of gas-discharge device which can be used as a voltage stabiliser, reference, or voltage clamp. Note, however, that neon lamp assemblies advertised as indicators for mains equipment usually have a current-limiting resistor built in, so these are not suitable as stabilisers unless you can get access to the naked bulb.

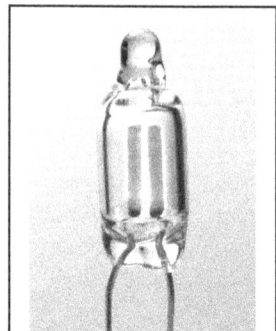

The striking voltage of a neon lamp is normally about 90V, and the maintaining voltage will be somewhere between 60V and 70V, varying between samples. The normal operating current for most neon lamps when used as indicators is 1–2mA, and the maximum for use a stabiliser is probably 10mA. For lamps measured by the

Fig. 6.14: A neon lamp can be used as a cheap 60V stabiliser.

author, the dynamic resistance was around 400Ω. One property which is sometimes useful is that the characteristics are symmetrical, unlike a Zener diode or gas regulator tube. In other words, it doesn't matter which way round you connect a neon lamp, it always stabilises at the same voltage. Like gas tubes they may oscillate at low currents with parallel capacitance, and there is usually a burn-in period of perhaps 100 hours over which the striking and maintaining voltages will drift before settling down.

6.4: Shunt Stabilisers and Load Lines

When designing simple shunt stabilisers using diodes or gas tubes etc., it is necessary to allow for variations in line and load, that is, variations in input voltage and output current demand. Input-voltage variations are normally due to variations in wall voltage. In most parts of the word this can be expected to be within ±10% of the nominal value, although it is usually safe to assume ±5% for developed, populous areas. If you live on a remote hillside farm, however, then your experience may be worse. Changes in load current are of course caused by changes in the effective load resistance. The effects of both can be explored in a visual way using load lines –a technique which will already be familiar to anyone designing valve amplifiers.

If the load current is expected to be substantially constant then the shunt stabiliser only needs to fight variations in line (wall) voltage. Fig. 6.15 shows an example using a 47V 1W Zener (avalanche) diode, together with its I/V characteristic and load lines. Note that there is no load current in this scenario, or it is assumed to be negligibly small compared with the diode current. The nominal input voltage happens to be 80V, and the solid load line shows the situation when V_{in} is right on the money; the intersection with the I/V characteristic is the operating point and indicates about 7mA through the diode. The dashed lines show what happens if V_{in} rises or falls by ten per cent, causing the operating point to move also, covering a range of about 5mA to 9mA. The smaller the series resistor, the steeper the load line would be and therefore the larger the current variation would be –the harder the

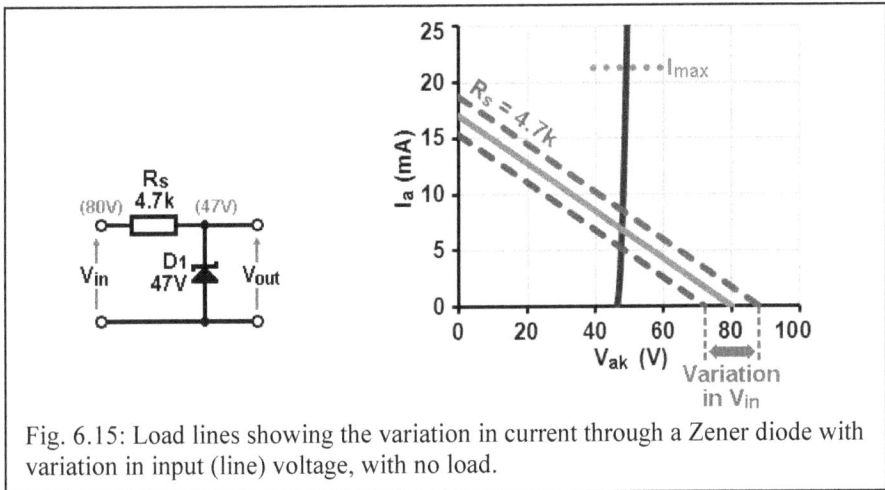

Fig. 6.15: Load lines showing the variation in current through a Zener diode with variation in input (line) voltage, with no load.

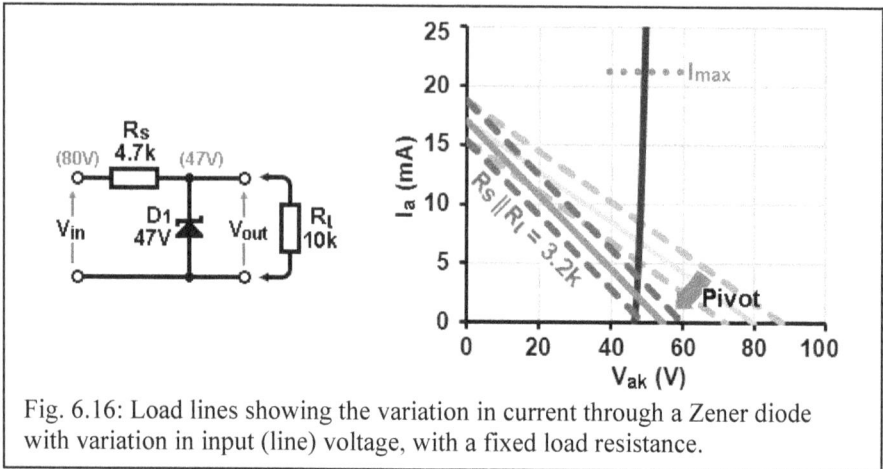

Fig. 6.16: Load lines showing the variation in current through a Zener diode with variation in input (line) voltage, with a fixed load resistance.

diode would have to work to fight the change, if you like. The dotted line I_{max} shows the maximum permissible current beyond which the diode would dissipate more than its rated 1W. Of course, for long term reliability we would probably want to keep dissipation below half this figure.

With no load current this circuit is really only acting as a voltage reference. The situation gets a little more involved when a load is attached, as shown in fig. 6.16. Adding the 10kΩ load resistance causes the load line to pivot about its apex from its original slope of $-1/R_s$ to a new slope equal to $-1/(R_s||R_l)$. The diode current has reduced since the load is stealing a portion of the total. We can also see that when the input voltage falls to its lowest level the operating point gets dangerously close to bottom of the characteristic, i.e. close to dropping out of regulation. This suggests the series dropping resistance needs to be smaller, to increase the quiescent current in the diode and so create more margin for variation.

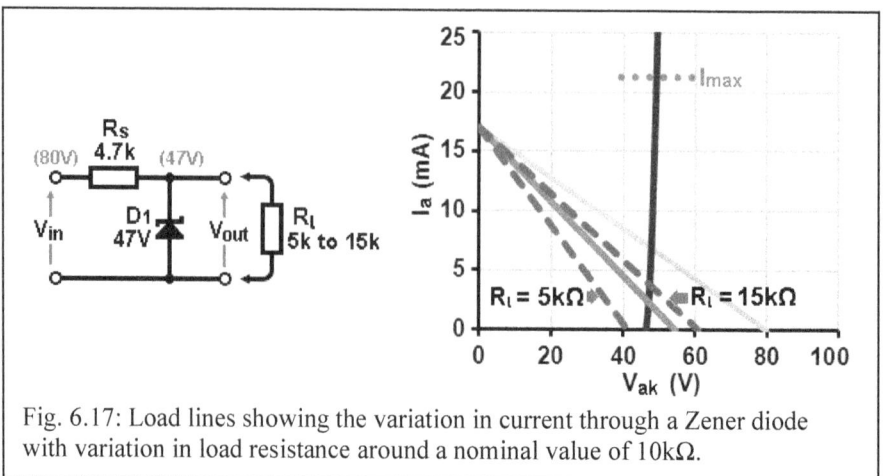

Fig. 6.17: Load lines showing the variation in current through a Zener diode with variation in load resistance around a nominal value of 10kΩ.

When the load current is not constant then the picture is complicated still further. In fig. 6.17 the solid load line is reproduced from the previous figure, showing the situation when the input voltage is 80V and the load resistance is 10kΩ. The dashed lines show the effect of varying the load resistance by fifty percent, i.e. from 5kΩ to 15kΩ. The larger value means less current is stolen from the diode, so the operating point climbs up the I/V characteristic, but when the load resistance drops to 5kΩ it steals over 8mA –more than original 7mA no-load diode current– causing the diode to drop-out completely. In other words, we need to go back to the drawing board and increase the no-load current enough to accommodate the maximum expected load current to more than 8mA. But remember, this figure only shows the case when the input voltage is at its nominal value of 80V! If the input voltage falls then all the load lines shift to the left, pushing them even closer to –or beyond– drop out. Hence when designing a stabiliser we must assume the worst-case (low) input voltage simultaneously with the worst-case (maximum) load current.

6.5: Feedforward Shunt Valve Regulators

Having covered passive shunt regulators we can move on to some more interesting, active circuits. Feedforward shunt stabilisers work by feeding unwanted input (line) voltage variations to the shunt control device. The control device then adjusts its conductance in order to produce a counterbalancing voltage across a series resistor. With careful user adjustment it is possible –in theory– to achieve a perfect null; total cancellation of unwanted line variations.

A feedforward system is anticipatory rather than reactionary; the control device is told what the error is going to be before it arrives. But the control device has no knowledge of what is actually happening 'downstream' at the output, and therefore cannot self-correct, unlike a feedback system. In other words, it cannot provide load regulation. Indeed, feedforward stabilisers usually *require* the load current to be constant or the nulling will be impaired. This reliance on user trimming, and on the load current being constant, are big shortcomings of the feedforward approach, so it is rarely used, at least on its own. We will hastily explore only two feedforward ideas here, partly to learn why to avoid them!

6.5.1: Feedforward Ripple Suppressor

Fig. 6.18 shows a very simple feedforward shunt regulator. The purpose of this circuit is not to produce a fixed DC output voltage but only to reduce power supply ripple, giving a cleaner supply to the load. It is fairer to call it an active smoothing circuit or ripple suppressor, i.e. it is the shunt version of the capacitor multiplier from section 4.4.1.

The method of operation is simple. R_s and the triode form a potential divider, so any ripple voltage at the input will appear at the valve's anode (slightly attenuated). Meanwhile, the input ripple is also 'fed forward' to the grid through C_1, and the triode will amplify and invert this signal. If the inverted signal can be made exactly equal to the ripple already at the anode, the two will cancel out, producing a clean, ripple-free output. In other words, if the input voltage rises, the valve simultaneously conducts more current, increasing the voltage drop across R_s in perfect opposition to the rise, and vice versa.

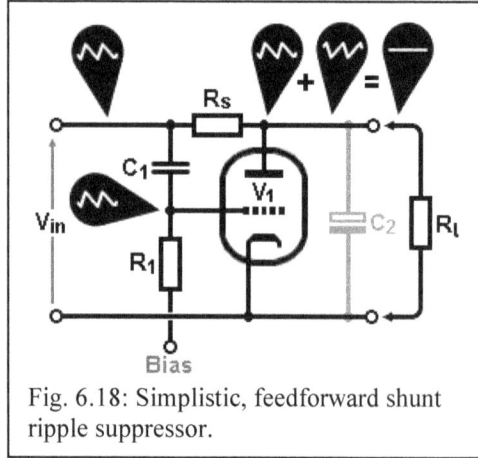

The circuit only rejects *input* ripple; it provides no load regulation, so signal-induced fluctuations are unaffected. The output impedance of the regulator is equal to R_s in parallel with the valve's internal resistance, but the latter is usually much larger so can be ignored. In

Fig. 6.18: Simplistic, feedforward shunt ripple suppressor.

other words, as far as the load is concerned this circuit looks like a simple dropping resistor R_s, so an ordinary decoupling capacitor C_2 is normally needed too, but this does not influence the operation of the regulator.[5]

The valve will produce a change in current equal to the grid voltage multiplied by the effective transconductance. The optimum value for R_s is:

$$R_s = \frac{1}{g_m} \tag{6.2}$$

In theory, design is straightforward. R_1 is the usual grid leak which might be 1MΩ, and C_1 must be large enough to pass the lowest ripple frequency with negligible phase shift, e.g. the cut-off frequency should be at least ten times lower than the ripple frequency. The valve must have enough grid base (input headroom) to accept the peak-to-peak ripple voltage, which implies a low-μ device. The negative supply biases the valve to the desired quiescent current. And remember, to avoid clipping, the bias voltage must be greater than the peak ripple voltage.

This would be all well and good if we had perfectly linear triodes, but reality bites. Performance is at the mercy of g_m which varies between samples and decays over the lifetime of the valve, so the circuit will need occasional trimming. Furthermore, g_m is non-linear and therefore not exactly complemented by the linear resistor R_s. In other words, the amplified waveform is slightly distorted, so it will not be a perfect mirror-image of the original ripple waveform, so we cannot get truly perfect cancellation no matter how carefully we trim. Providing a negative bias supply may also be a minor inconvenience.

These problems can be alleviated to some extent by using cathode bias. This eliminates the need for negative supply and, if it is left unbypassed, will degenerate g_m. Degeneration linearises g_m, stabilises it against ageing, and increases the

---

[5] Langford-Smith, F. (1957) *Radio Designer's Handbook* (4th ed.), p1200, Iliffe and Sons ltd., London.

effective headroom. The original g_m will be degenerated down to a new effective value of:

$$g_m' = \frac{g_m}{1+g_m Rk} \tag{6.3}$$

This is the figure that should be entered into formula (6.2). Nevertheless, any changes in load current will cause the voltage across the valve to change, which will vary its transconductance. This circuit therefore works best when the average DC load current is constant. This is normally the case for preamps, and this stabiliser is not much use for anything else.

Even with (or without) cathode degeneration we are sure to need some way to trim the circuit for optimum ripple rejection. There are many ways this could be arranged, but perhaps the simplest is to make R_s a little larger than necessary, and then to feed a smaller fraction of the ripple to the grid to compensate. For example, if R_s is made twice as large as $1/g_m'$ then we would need only half the ripple to be fed to the grid. In practice we replace R_1 with a trimpot.

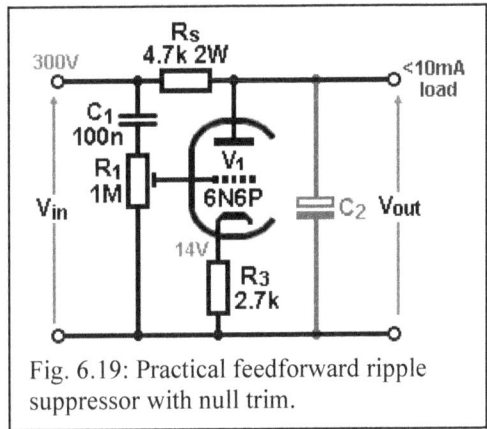

Fig. 6.19: Practical feedforward ripple suppressor with null trim.

Fig. 6.19 shows a practical example using one half of a Russian 6H6П/6N6P, which is a beefy little dual triode with fairly high g_m (it can be considered a mini power valve). The JJ ECC99 is essentially the same but with a different pinout. This valve is overkill for this particular application, but when it comes to DC regulation later, dissipation in the valve can double between quiescent and maximum workload. This valve's 4.8W rating will come in useful then.

In this case R_3 biases the stage to about 5mA and degenerates the g_m from 2.6mA/V to about 2.6/(1+2.6×2.7) = 0.3mA/V. This means R_s should theoretically be 1/0.3 = 3.3kΩ, but a larger value of 4.7kΩ has been used to give some margin for trimming. The input voltage was set to 300V nominal, but the circuit should still work with voltages from 250V to at least 400V. The actual load current should be limited to perhaps 10mA at most, otherwise the drop across R_s becomes excessive.

The circuit was supplied by a conventional full-wave rectifier and trimmed while monitoring the output ripple on an oscilloscope. Trimming must be carried out with the intended load already attached, after which C_2 can finally be added to taste. With the cathode resistor bypassed by a capacitor (not shown), the peak-to-peak ripple rejection was about −34dB, but a figure of −40dB was possible when unbypassed, owing to the improved linearity. This is roughly what we would get from a plain RC filter using the same 4.7kΩ dropping resistor and a 30μF smoothing capacitor. Such

143

performance would have been impressive one hundred years ago when circuits like this were invented and thirty microfarads was expensive, but times have changed. Six watts of combined heater and anode power is rather a lot just to avoid thirty microfarads. This can be improved with minor modifications, as we shall see later.

6.5.2: King Regulator

Taking the previous design a little further, what if we supply both AC *and* DC to the grid? Such an arrangement was described by King in 1923[6], making it one of the oldest valve stabiliser circuits. Fig. 6.20 shows the basic principle; the feedforward capacitor in the previous circuit is now a resistor, R_1. Since the grid voltage is now elevated by DC, the cathode voltage must be raised too to achieve a suitable bias. A Zener diode is shown here, but any other voltage reference might serve. This has the advantage of reducing the voltage across the valve, meaning less anode dissipation for a given current. Since both AC and DC are fed forward to the grid, the circuit provides a measure of AC *and* DC line regulation, which is something a simple smoothing capacitor cannot do. In practice, however, it is difficult to make this topology work usefully.

The voltage fed to the grid by the potential divider is attenuated by a factor of: $B = R_2/(R_1+R_2)$, so the ideal value for R_s is:

$$R_s = \frac{1}{Bg_m'} = \frac{R_1+R_2}{g_m'R_2} \qquad (6.4)$$

The quiescent current in the triode, multiplied by R_s, is the largest DC voltage change the circuit can stabilise against. In other words, if we expect the line voltage to vary by ±25V, say, then the idle drop across R_s must be *at least* 25V even before we account for load current. Load current will increase the drop still further, so this is going to be a rather lossy circuit. Moreover, degeneration is not really an option

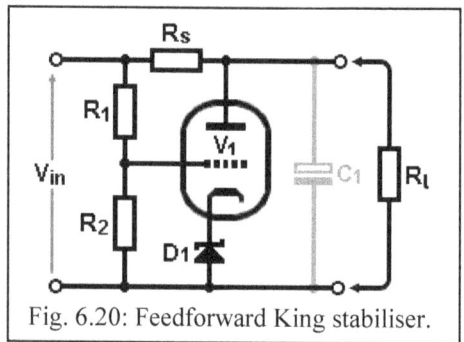

Fig. 6.20: Feedforward King stabiliser.

here since it is likely to lead to an excessively large value for R_s, yet without degeneration the circuit is extremely sensitive to bias trimming! All this makes the circuit difficult to design and annoyingly sensitive to component values. The potential divider determines the bias, which determines g_m, which determines the optimum value for R_s, which might produce too much DC drop to be a useful circuit, or not enough to cope with the expected input variations. You can end up chasing your tail! Nevertheless, fig. 6.21 shows a practical example for any sadist who wishes to experiment.

---

[6] King, (1923). *Bell Systems Technical Journal*. **2**, p98.

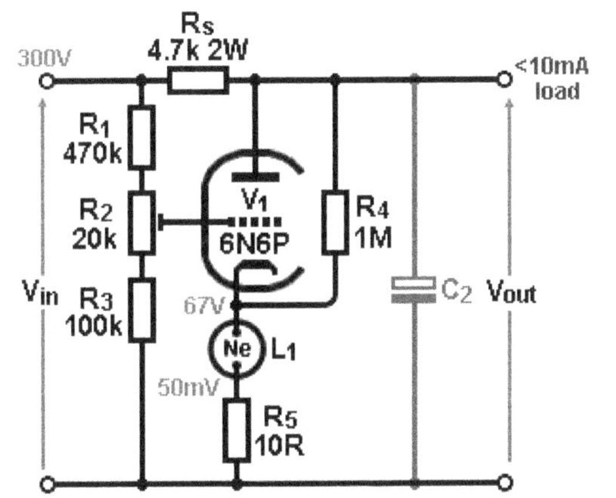

The neon lamp provides a relatively fixed cathode voltage which, after everything else, conveniently allowed a 4.7kΩ dropping resistor to be used with 5mA shunt current as before. This is not the true optimum value (which will also depend on load current since that also affects the anode voltage) but it is close enough to prove the point. R_4 ensures the neon always strikes, as explained in 6.3.1 earlier. The purpose of R_5 is to allow the quiescent current to be monitored and

Fig. 6.21: Practical King stabiliser. Not recommended for general use.

set to 5mA, i.e., adjust the trimmer R_2 until the voltage across R_5 is 50mV. This can be a frustrating exercise; the circuit is *very* sensitive and will drift for at least five minutes as everything warms up.

Since R_s is not optimised, ripple reduction is not worth mentioning –output capacitor C_1 can do that job. The main feature of this circuit is its ability to stabilise DC. This turned out to be better than the author expected, considering the compromises made. Once fully warmed up, it can cope with an input voltage variation from 280V (drop out) to 320V (10mA maximum neon current) while maintaining the DC output voltage within ±1.5%. Note that the heater voltage was supplied from the same transformer and so followed the same line variation. Not bad for one modest triode. Really though, the circuit is interesting as an academic exercise. It is too sensitive to be useful outside the laboratory, so we shall leave it in 1923 and move on to another 1920s development: negative feedback.

6.6: Feedback Shunt Valve Regulators

Feedforward shunt stabilisers provide line regulation but not load regulation. With feed*back* we get load regulation *and* some line regulation. Negative feedback brings with it all the well-known advantages of lower output impedance and greater independence from component variation and ageing. This means it can eliminate the need for trimming. Feed*forward* circuits can deliver better line regulation with the same parts, but the attendant difficulties in making this happen are usually enough to make anyone choose the feedback route instead, trading some line regulation in exchange for load regulation and less swearing. It is also possible to use a little of both, as we shall see later with mixed-mode regulators.

145

6.6.1: Feedback Ripple Suppressor

The simplest, triode feedback regulator is shown in fig. 6.22. It is the twin-brother of the stabiliser from section 6.5.1, with the feed capacitor connected to the output (load) instead of the input (line). Again, it does not regulate DC, it is only a ripple suppressor. In case this circuit does not immediately look like the block diagram in fig. 6.2, the voltage reference is the cathode voltage. The valve amplifies the difference between the cathode and feedback (i.e. grid) voltage. Thus the triode does double-duty as error amplifier

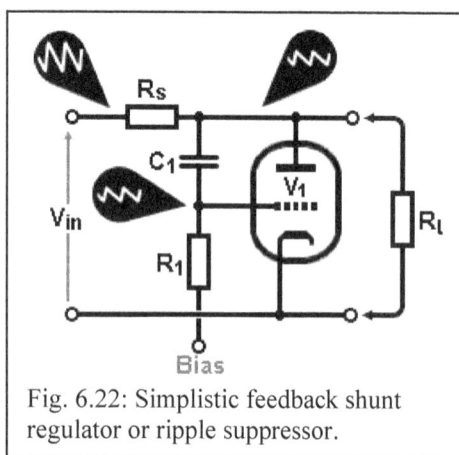

Fig. 6.22: Simplistic feedback shunt regulator or ripple suppressor.

and control device. In short, it is an ordinary gain stage with a 100% local feedback factor.

Unlike the feedforward version, we can now make R_s any value we like, depending mainly on the DC drop we can tolerate. A larger resistance increases the gain of the stage, i.e. more loop gain. With the feedforward version of this circuit, cathode degeneration was an advantage because the regulator needed linearity more than it needed gain. Here it is the reverse: more loop gain means more feedback, and therefore better regulation and lower output impedance. Since the valve has 100% negative feedback, the impedance when 'looking into' the anode will be approximately equal to $1/g_m$. This impedance forms a potential divider with R_s, so by applying the formula for a potential divider, the ripple reduction factor will be roughly $1/(1+g_mR_s)$. The denominator is the loop gain –the bigger the better. The output impedance of the whole circuit will be $1/g_m$ in parallel with R_s.

Fig. 6.23 shows a practical circuit using the same component values as the feedforward version, for easy comparison. The cathode is now bypassed to maximise gain, and note that this regulator requires no trimming –we simply feed back all of the output noise. With $10V_{pp}$ of input ripple the output ripple was found to be $0.7V_{pp}$. This is a reduction factor of 0.07 or –23dB. From this the output

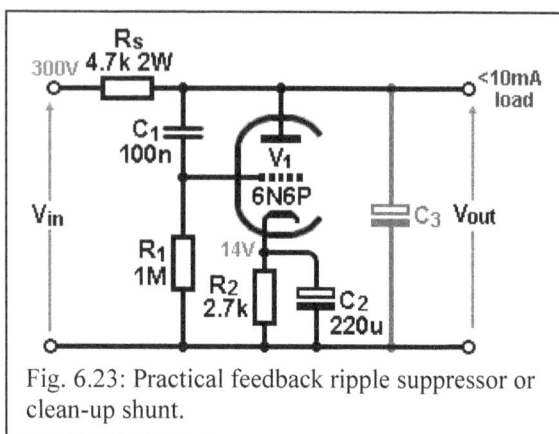

Fig. 6.23: Practical feedback ripple suppressor or clean-up shunt.

146

impedance of the regulator can also be calculated:
$Z_o = R_s \times 0.07 = 4700\Omega \times 0.07 = 329\Omega$.

The ripple reduction is much worse than we achieved with the feedforward version earlier. In peak-to-peak terms we would get the same ripple reduction from the same 4.7kΩ dropping resistor and a 2μF smoothing capacitor. But the reactance of a smoothing capacitor continues to fall with frequency, so would provide more filtering of ripple harmonics and lower output impedance above 240Hz. Since we could just as easily use, say, a 22μF capacitor, which would deliver even better performance than this regulator across the whole audio spectrum, does this circuit have any use at all?

Perhaps you spotted the clause in the last sentence. The one useful thing this circuit can do is suppress *sub*-audible fluctuations without needing an enormous, high-voltage smoothing capacitance. If you've ever looked at power supply ripple on an oscilloscope with AC-coupling, and watched the trace meander up and down the screen, then you have been watching those murky rumblings that can yet intermodulate with hi-fi audio signals. In other words, this regulator can be regarded as a 'clean-up shunt' or 'helper' for a conventional decoupling capacitor, C_3. Of course, C_1 and C_2 can be increased to push the rejection range down even further. And don't forgot to consider using an unloved TV valve for the job; whatever it is, it can be run in triode-mode by connecting any screen grids to the anode. There is usually no point using pentode-mode for a shunt regulator since the voltage across the valve is already quite constant (almost by definition), and triode-mode maximises g_m.

6.6.2: Back-Biased Ripple Suppressor

A variation on the previous circuit allows the cathode bypass capacitor to be eliminated without causing degeneration. The change is to use *back biasing*, that is, to connect the load directly across the valve as in fig. 6.24. Both the regulator current and load current now flow in the bias resistor R_2, which makes setup more critical, but a nice bonus is that R_2 now effectively adds to R_s which increases the loop gain a little. Back biasing can only be used when the 'upstream' part of the power supply does not require a direct ground connection; a typical rectifier arrangement is shown to make this clear. This proviso, and the

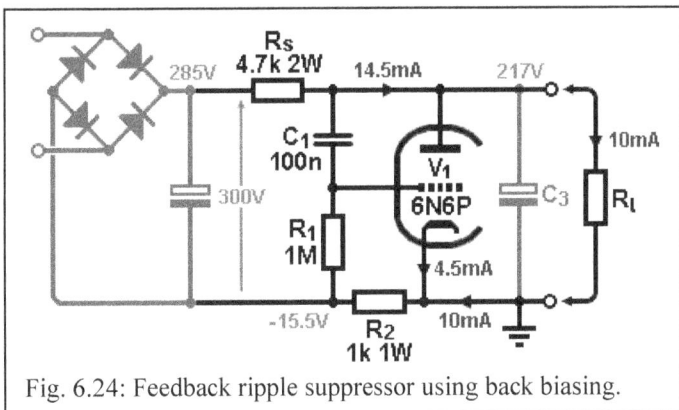

Fig. 6.24: Feedback ripple suppressor using back biasing.

dependence on load current, make back-biasing less universally adaptable than previous circuits, which is why it gets only an honourable mention here. Back biasing can be used with feedforward regulators too of course, if you have the patience.[7]

6.6.3: Mixed-Mode Ripple Suppressor

Feedback is largely self-correcting, but cannot completely eliminate ripple. Feedforward can completely eliminate ripple (in theory), but the outcome rests on a user-trimmed knife-edge. Can we take a little from both columns? Fig. 6.25 shows one possible approach. Here a pot allows the feed signal to be swept continuously from input to output, that is, from purely feedforward to purely feedback. The optimum mixed-mode condition occurs when R_2 becomes equal to $1/g_m$. The version in b. avoids the need for a high-power pot, in which case the optimum condition occurs when R_2 is equal to $(R_1+R_2)/g_mR_s$, e.g. if R_s is three times larger than $1/g_m$ then the pot must be set one third from the 'feedback end'. Of course, the pot can be replaced with fixed resistors if you're not worried about exact trimming.

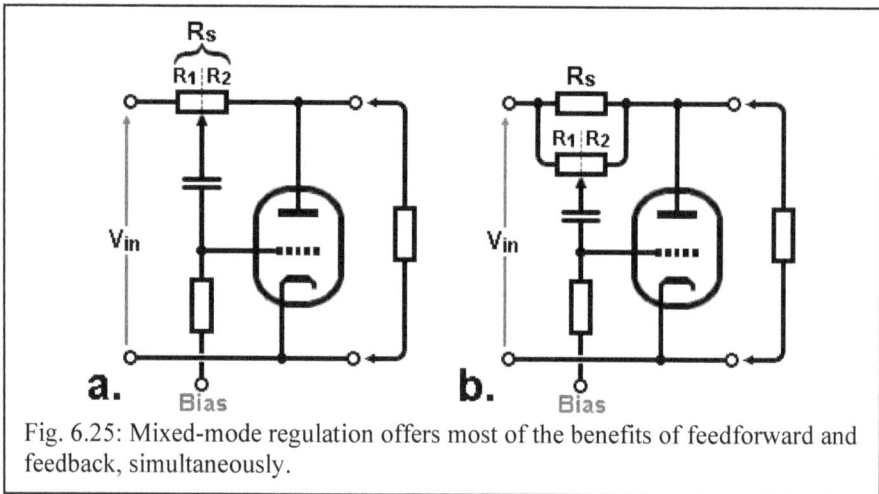

Fig. 6.25: Mixed-mode regulation offers most of the benefits of feedforward and feedback, simultaneously.

When optimised, from the point of view incoming ripple we have a feedforward stabiliser, but from the point of view of the load we have a feedback regulator, albeit one with less than 100% feedback. With this combination we get excellent line regulation plus at least *some* load regulation, and an output impedance of $1/g_m$ either way. What's more, as the valves ages and g_m declines, the circuit will drift more towards feedback-mode, rather than falling off the knife-edge as a purely feedforward circuit would. It grows old gracefully.

Fig. 6.26 shows a practical example, modified from fig. 6.23 earlier. When trimmed, the ripple reduction reached −44dB, which is roughly what we would get with R_s and

---

[7] Anon. (1937) An Amplifier Without Phase Distortion. *Electronics*, June, pp26-7.

a 47µF smoothing capacitor. This is an excellent result; the main downside is the need for a trimpot to find the true optimum. Some further examples of mixed-mode operation are shown later.

6.6.4: King-Type Feedback Shunt Regulator

Having seen the simple feedforward shunt ripple suppressor converted into a feed*back* shunt ripple suppressor, it should come as no surprise to learn that we can do the same with the King regulator. Fig. 6.27 shows the new circuit (compare with fig. 6.21). As previously, this is no longer a nulling circuit, so there is no particular optimum value for R_s; we can use whatever value satisfies our requirements for DC voltage drop or our lust for loop gain.

Fig. 6.26: Practical mixed-mode ripple suppressor

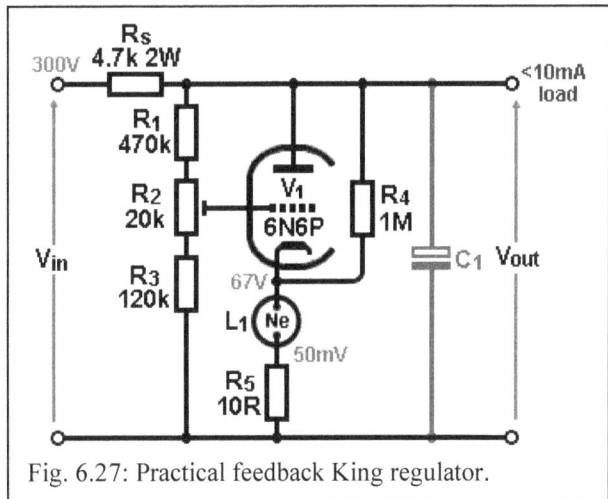

Since this circuit is DC-coupled and has no (intentional) degeneration at DC, a bias trimmer is useful, but thanks to negative feedback the adjustment is *much* less tiresome than with feedforward. If the valve tries to draw more current, the anode voltage is pulled down, which is fed back, which counteracts the increase in current. This makes the bias adjustment much easier and more stable over time. Note that R_2 has been increased slightly to maintain a suitable range of grid voltage adjustment, since the feedback divider is now being supplied from the (lower voltage) output.

This regulator does not feed back the whole of the output fluctuations, only a fraction of them, so regulation will not be spectacular. Testing produced a figure of −8.4dB ripple suppression, which is an output impedance of 1.8kΩ. One the other hand, this is maintained all the way down to DC. Indeed, with 5mA quiescent current at a nominal input of 300V, the regulator can cope

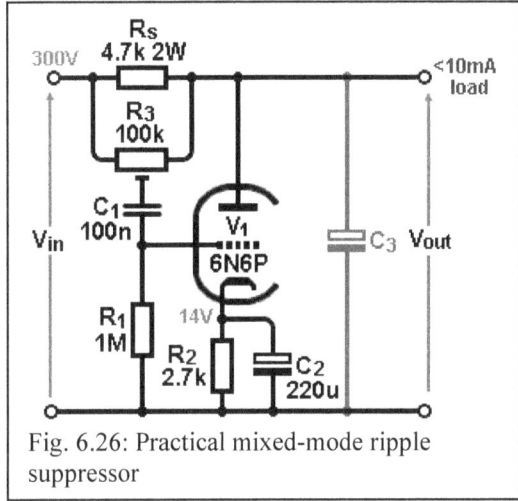

Fig. 6.27: Practical feedback King regulator.

149

with a practical input range from 275V to 330V, which more-or-less covers all likely wall variation. The output voltage stayed within ±3% of nominal. The brightness of the neon lamp even gives a visual indication of how hard the regulator is working.

A nice little circuit. Can we improve it? Since there is no longer a special relationship between the divider and R_s, an easy change is to put a capacitor in parallel with the upper part of the feedback divider. This allows AC ripple to bypass the divider and queue-jump directly to the grid, meaning we get 100% feedback at

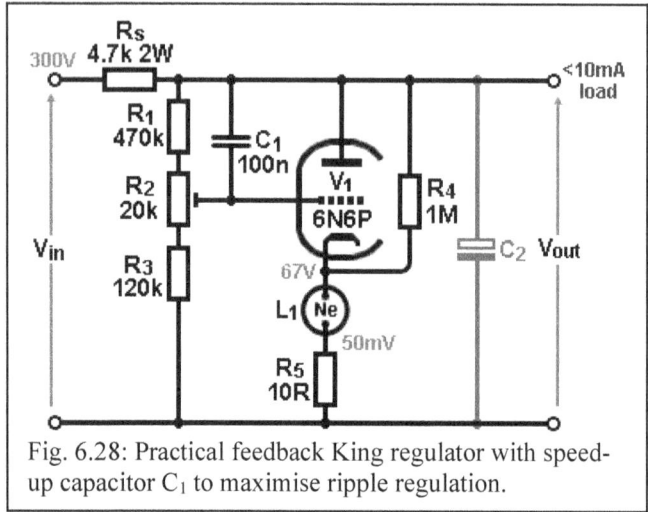

Fig. 6.28: Practical feedback King regulator with speed-up capacitor C_1 to maximise ripple regulation.

those frequencies where the capacitive reactance is small. This is sometimes called a speed-up capacitor. Most discrete voltage regulator circuits adopt the same approach, because not only does it maximise regulation for frequencies above DC, it also (usually) improves the stability of the feedback circuit.

Fig. 6.28 shows the revised circuit. C_1 is the speed-up capacitor and could be made arbitrarily large, but 100nF is enough to cover mains frequency and higher, since the cut-off frequency of a bypassed potential divider like this is:

$$f = \frac{1}{2\pi C_1 (R_{upper} || R_{lower})}$$

The ripple reduction of this circuit improved to −16.5dB. The reason this is not as good as fig. 6.23 earlier, despite apparently similar AC operating conditions, is due to the internal resistance of the neon lamp causing some degeneration. It could be replaced with a Zener diode of course, but if we use one Zener then we might as well replace the whole regulator with Zeners and we are on the slippery slope to siliconville. Besides, a Zener does not glow so nicely. Another option would be to supply C_1 from a pot in parallel with R_s to create a mixed-mode regulator (section 6.6.3), but a circuit that needs *two* timpots is getting out of hand.

6.6.5: Enhanced King-Type Regulators

The previous circuit employed a speed-up capacitor to bypass the upper arm of the feedback divider to achieve 100% feedback at ripple frequencies. But at very low frequencies the reactance of the capacitor must rise, and the feedback falls back to

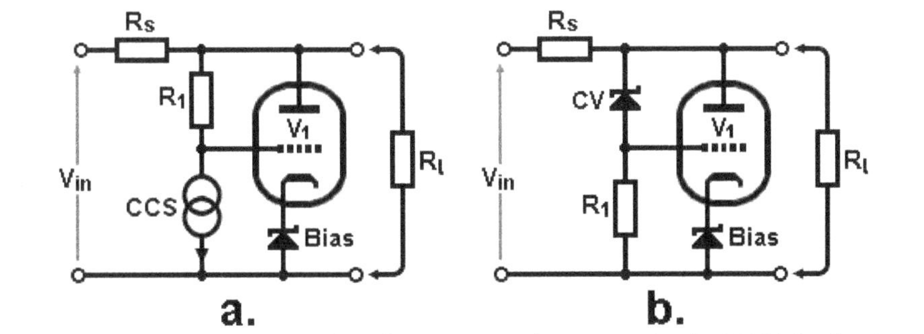

Fig. 6.29: Enhancements on the King-type regulator to maintain 100% feedback down to DC **a.** Hunt-and-Hickman arrangement; **b.** Amplified Zener arrangement.

the DC level determined by the resistor divider. But to those skilled in the art it is not hard to see how the idea can be extended down to DC, as illustrated by fig. 6.29.

The topology in a. replaces the lower arm of the divider with a constant-current sink (CCS) which has near infinite impedance down to DC. With a constant current flowing in R_1, the voltage across it must also be constant; if the top of the resistor rises, the bottom must rise by the same amount, relaying the full voltage change to the valve grid. A (feedforward) version of this topology was described by Hunt and Hickman in 1939.[8]

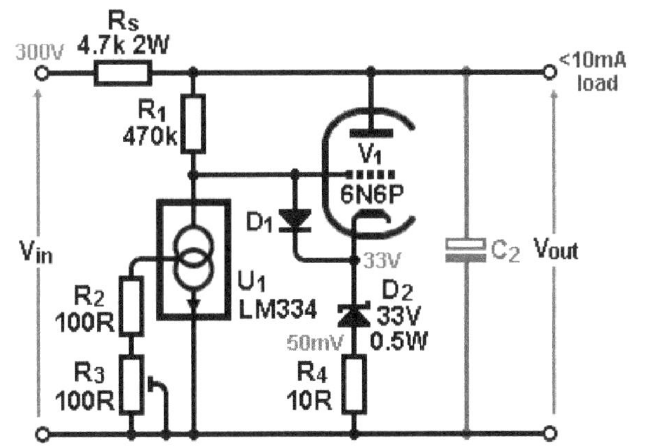

The topology in b. is the 'dual' of this idea; it replaces the upper arm with a constant-voltage (CV) device with near zero impedance down to DC. A Zener diode is shown for familiarity, but a gas tube or other cunning artifice would do as well. This can also be thought of as the valve version of the amplified Zener from section 6.2.4, an idea which goes back at least

Fig. 6.30: Practical Hunt-and-Hickman regulator.

to the 1940s.[9] Both approaches accomplish the same thing, the choice is really between what technology you want to throw at a practical design. Indeed, there is

---

[8] Hunt, F. V. & Hickman, R. W. (1939) On Electronic Voltage Stabilizers. *Review of Scientific Instruments*, 10, pp. 6-21

[9] Ledward, T. A. (1942) Constant Voltage Supply. *Wireless World*, April, pp33-5.

nothing to stop you using both the CCS *and* the CV together, if you are exploring that field of precision. But here we will keep things simple.

Fig. 6.30 shows an example of the Hunt-and-Hickman arrangement, using an LM334 adjustable current sink. The current sink is adjusted with R_3 which allows the voltage across R_1 –and therefore the bias voltage at the grid– to be set. The IC can only handle 40V across itself, so the neon lamp in previous circuits has been replaced with a 33V Zener, plus a protection diode between grid and cathode. At startup, when the valve is still cold, D_1 will conduct and safely clamp the voltage across U_1 to just over 33V. Once adjusted to a quiescent current of 5mA, the circuit delivered about −24dB of ripple reduction. The input voltage could be varied from 280V (dropout) to 330V while the output varied by less than ±2%. This circuit extracts all the juice the valve itself has to give, but with a Zener diode and a modern current-regulator IC doing all the squeezing, it is starting to look like a solid-state circuit being photobombed by a triode.

6.6.6: Cascode Ripple Suppressor

Since triodes frequently come two in a bottle, there is a rational urge to use both of them in the same circuit block. Various configurations can be dreamt up, but perhaps the most obvious is to connect them as a cascode. This can be highly advantageous in a shunt regulator since it means less voltage appears across each device, so dissipation is shared. This could allow them to be operated at higher current, which leads to greater g_m. On the other hand, more current also means more voltage drop across R_s, and if we make R_s smaller then we undo the advantage of having more g_m. Everything is a compromise.

Fig. 6.31 shows a practical mixed-mode example, adapted from earlier circuits. The upper grid is fed from the input-side of the circuit to reduce unnecessary further drop across R_s. Technically, this also means the upper triode provides a small measure of feedforward regulation, but the contribution is miniscule since it is so heavily degenerated by the lower triode. The heater should be elevated to about one quarter of the HT to avoid exceeding $V_{hk(max)}$ of the upper triode.

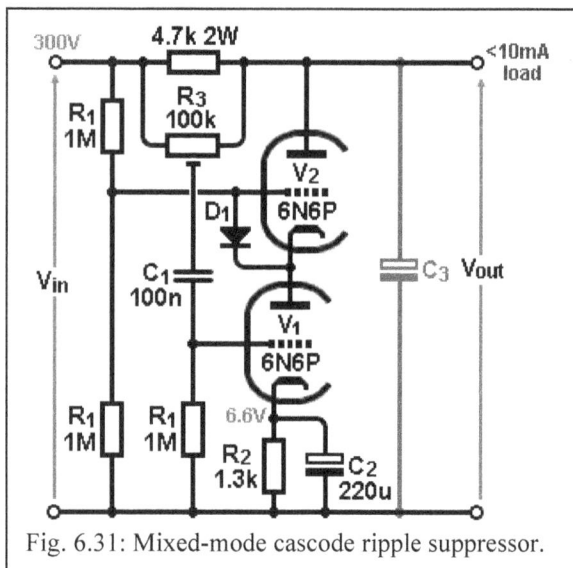

Fig. 6.31: Mixed-mode cascode ripple suppressor.

With the pot set to pure feedback mode, the circuit produced -28dB of ripple reduction, with an output impedance of 180Ω. But when optimally tuned, this increased to -46dB with an output impedance of 190Ω. Even though the circuit is running at the same quiescent current as the single-triode shunt stabilisers shown earlier, cascoding has nonetheless provided us with a little more g_m. A $\pm10\%$ change in input voltage resulted in practically the same percentage change in output, so the upper triode is delivering negligible feedforward line rejection; this is only a ripple suppressor, but still an effective one.

6.6.7: Janus Ripple Suppressor

The Janus regulator is an interesting example of feedforward-feedback design by Broskie.[10] The circuit is shown in fig. 6.32 and is aptly named after the Roman god with two faces, who looks to the past and the future simultaneously. For incoming ripple, R_1 and V_1 form a

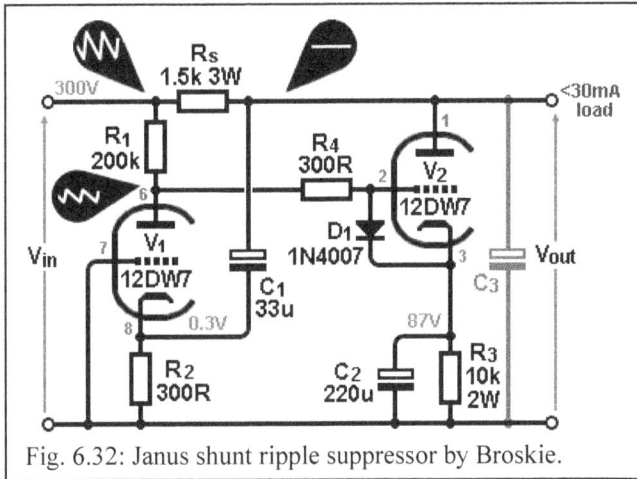

Fig. 6.32: Janus shunt ripple suppressor by Broskie.

potential divider that feeds a portion of the ripple forwards to the shunt regulator valve V_2. Meanwhile, C_1 feeds output noise back to the cathode of V_1 which (now acting as a common-grid amplifier) amplifies it and feeds this signal to V_2. Thus V_1 does double-duty as a potential divider for feedforward signals, and error amplifier for feedback signals.

Note that the circuit does not regulate DC; it is another ripple suppressor. This at least means relatively little idle current is needed, so a small valve can be used, in this case an ECC832 / 12DW7 / 7247. This valve contains one half of an ECC83/12AX7 (which provides high gain for the error amplifier) and one half of an ECC82/12AU7 (which has more g_m, making it best suited to the control device). The two triodes are directly coupled together, so V_2 must have a fairly large cathode resistor to raise its cathode voltage up high enough to bias it to a suitable idle current. R_4 is simply a grid stopper, and D_1 protects the triode against grid-cathode arcing at startup. Input voltages in the range of 280V to 350V can be handled, and heater elevation is not essential.

---

[10] http://www.tubecad.com/2007/06/blog0112.htm
Compare with fig. 6 in: Hunt, F. V. & Hickman, R. W. (1939). On Electronic Voltage Stabilizers. *Review of Scientific Instruments*, 10, p11.

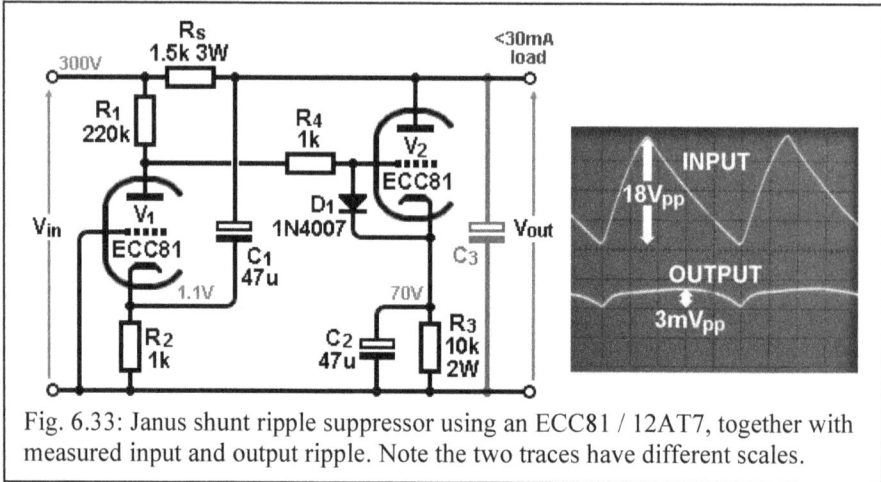

Fig. 6.33: Janus shunt ripple suppressor using an ECC81 / 12AT7, together with measured input and output ripple. Note the two traces have different scales.

The high-gain error amplification greatly increases the feedback loop gain. When combined with the feedforward contribution, the ripple reduction is excellent; testing produced a figure of about −75dB at 100Hz and above. If any criticism can be levelled at this circuit at all, it is that it produces a small amount of *boost* around 0.4Hz, i.e. it slightly magnifies wall fluctuations. The use of an unusual valve, E24 resistors, and two high-voltage capacitors with different values are also minor irritants.

Fig. 6.33 shows a modified version of the circuit using a more common ECC81/12AT7. Small-signal analysis suggests this valve ought to produce a slightly poorer result than the 12DW7, but actual testing produced practically the same figure of −75dB. This is likely due to large-signal nonlinearity being the limiting factor in either case. The combination of C_1 and C_2 here also eliminates the unwanted low-frequency boost. Also shown are the ripple waveforms produced on test. Notice that the output ripple is measured in *milli*volts; if both traces were shown with the same scale, the lower trace would appear be a flat line! In short, the Janus regulator is one of the most effective ways to use two triodes as a shunt ripple suppressor.

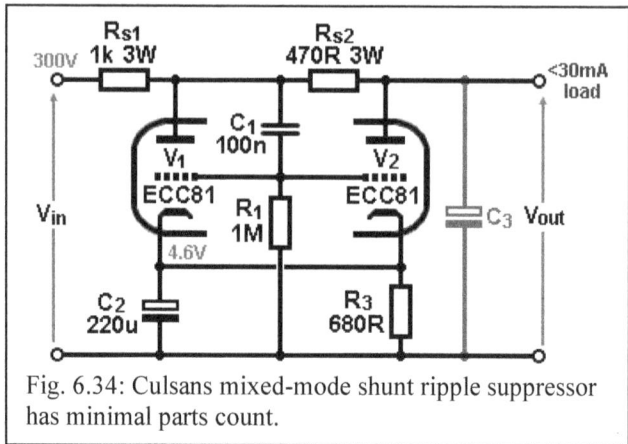

Fig. 6.34: Culsans mixed-mode shunt ripple suppressor has minimal parts count.

6.6.8: Culsans Ripple Suppressor

Fig. 6.34 shows another two-valve ripple suppressor. Here, V_1 is a feedback ripple suppressor while V_2 is effectively working in mixed mode, but sharing the same divider and biasing components, which neatly economises on parts. You can think of the first stage as a pre-regulator, reducing ripple initially so that it does not overwhelm the headroom of the second stage. The value of R_{s1} is not critical since it is part of the feedback stage, but R_{s2} should be equal to $1/g_{m2}$ for maximum nulling in the second stage. There is nothing to stop both valves being given individual biasing or back biasing, or adding a means for critically tuning V_2, but the point of this circuit is minimalism. Using an ECC81/12AT7 with the component values shown, ripple reduction was found to be -14dB after the first stage and -29dB after both stages.

6.7: Transistor Shunt Regulators

The foregoing valve regulators can of course be reconfigured using transistors, which offer far more g_m and current handling ability. MOSFETs are the obvious choice as substitutes for valves. A depletion MOSFET like the DN2540 can be substituted for a triode almost directly, with only a small change in bias needed, but depletion MOSFETs are rare whereas enhancement devices are available in great abundance. BJTs offer more g_m than MOSFETs, but BJTs are less readily available with high voltage ratings and can raise problems with thermal runaway. In either case, we can expect better regulation performance from a transistor compared with a valve under similar conditions. Moreover, transistors make it much easier to build a proper DC regulator rather than a mere ripple suppressor. The main design constraint when using transistors is heat, and getting rid of it.

6.7.1: King-Type Regulator

The King regulator is a common choice for a MOSFET shunt regulator. As usual, the quiescent current, multiplied by R_s, is the largest input change that can be suppressed. If the input voltage is 300V, and we expect as much as a 10% (30V) increase due to wall variation, when we must drop at least 30V across R_s at idle. We could use a small R_s and a large shunt current, which will mean large dissipation in the shunt device and also in the voltage reference device. Alternatively, we could use a large R_s and a small shunt current, which means the load itself will cause significant additional drop across R_s, which may leave us with too little output voltage for our needs.

Fig. 6.35 shows a practical example. A 12V, 0.5W Zener has been chosen as the voltage reference (because that's what the author had in stock). This can handle at most $500mW/12V = 42mA$. The quiescent shunt current must therefore be no more than half this value, since we can expect it to double when the input rises by the expected 30V. Having settled on 20mA of quiescent shunt current, R_s must be at least $30V/20mA = 1.5k\Omega$. We will also apply the rule of thumb that the output load current should be no greater than the quiescent shunt current itself. Any power MOSFET with sufficient voltage rating should work here; the author used an

STP11NK40Z. D_2 protects the gate-source junction from overvoltage at start up. R_3 is a standard gate stopper which suppresses any possibility of oscillation. R_5 allows the shunt current to be monitored and set to 20mA by adjusting R_2. The worst-case dissipation will happen if there is no load current and the

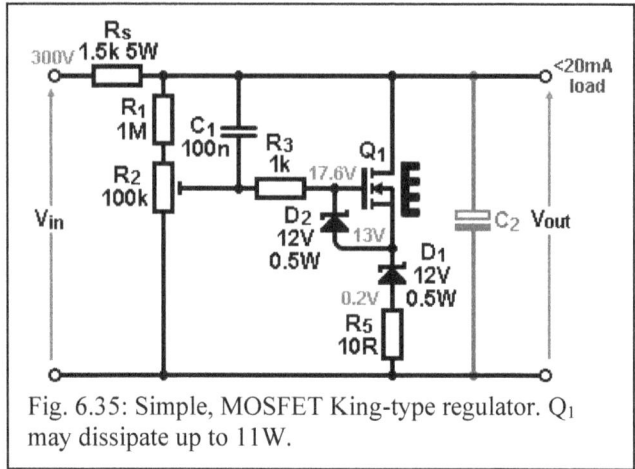

Fig. 6.35: Simple, MOSFET King-type regulator. Q_1 may dissipate up to 11W.

input voltage rises to its highest value of 330V. Estimating the MOSFET current to double under these conditions, it will dissipate almost 11W, so a heatsink is certainly needed.

The ripple reduction was found to be ×0.016 or −36dB, which corresponds to an output impedance of only 24Ω. The DC regulation is useful but not impressive, since the small reference voltage imposes a small feedback factor at DC. A ±10% change in input voltage caused the output voltage to vary by about ±5%. With no load and the maximum 330V input, the shunt current increased to 43mA –a little *more* than double the quiescent, owing to MOSFET nonlinearity. This actually exceeds the limit of D_1, so for a long-term solution we ought to upgrade to a 1W Zener or better. Incidentally, there is little point in bypassing it with a capacitor since any Zener noise is insignificant compared with the power supply ripple that will always be present to some degree, and a capacitor invites instability at very low frequencies.

6.7.2: Cascode Regulator

In the previous circuit, all the power dissipation is handled by one MOSFET, and with a suitable heatsink it would certainly manage this. But it is often convenient to use the chassis itself as a heatsink, and the non-ideal geometry of a sheet-metal chassis tend to result in a practical dissipation limit of about 10W per device. If we expect to burn off more than 10W then we ought to spread the dissipation over a larger area by using more than one transistor. In any case, sharing dissipation between more than one device is conservative design practice. As in section 6.6.6, one way to do this is to use a cascode, with two devices (or more) each sharing a roughly equal fraction of the overall voltage drop.

Fig. 6.36 shows a practical example in which R_1 from the previous circuit has been split into two equal parts, with the junction feeding the gate of a cascoded MOSFET Q_2. No other change is needed, but by way of example, a second dropping resistor R_{s2} has been added. It is roughly equal to the output impedance measured for the

156

previous circuit; in other words, it is roughly equal to $1/g_m$, and therefore converts the circuit to mixed-mode. It is such a small value compared with R_{s2} that it has negligible effect on DC conditions –it can be added without a second thought– yet it bestows a considerable performance upgrade. Even if you don't know the g_m of the transistor in use,

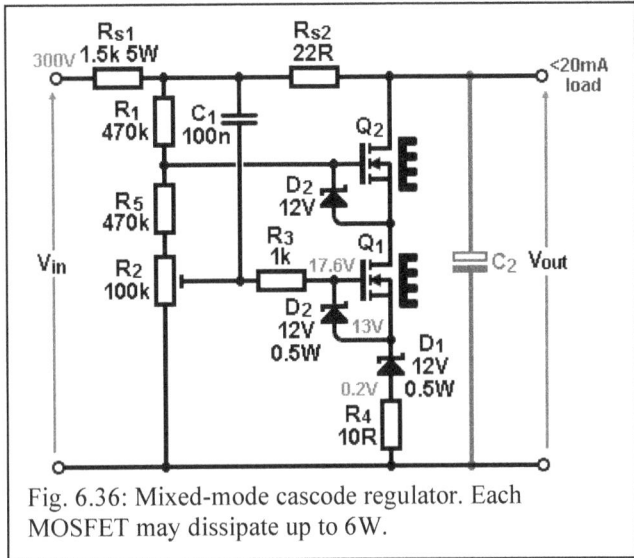

Fig. 6.36: Mixed-mode cascode regulator. Each MOSFET may dissipate up to 6W.

simply adding a 10Ω resistor in this position will normally buy at least *some* extra ripple reduction at practically no cost, by introducing a little feedforward action. In this case the peak-to-peak ripple reduction was found to be almost −48dB.

6.7.3: King-Type Regulator with Amplified Zener

With the cascode regulator in the previous section, the upper device acts mainly as a power sharing device; it doesn't improve the actual regulation. It was added simply to offload some of the power dissipation that would otherwise be concentrated entirely in the main amplifying MOSFET. What if we reverse our approach and build the totem pole from the top down, rather than the bottom up? Remember, the simple King-type regulator from section 6.7.1 put nearly all the dissipation burden on the MOSFET because the voltage reference (Zener diode) could not take it. We could replace that little Zener with an amplified Zener. This gives us a freer choice over the

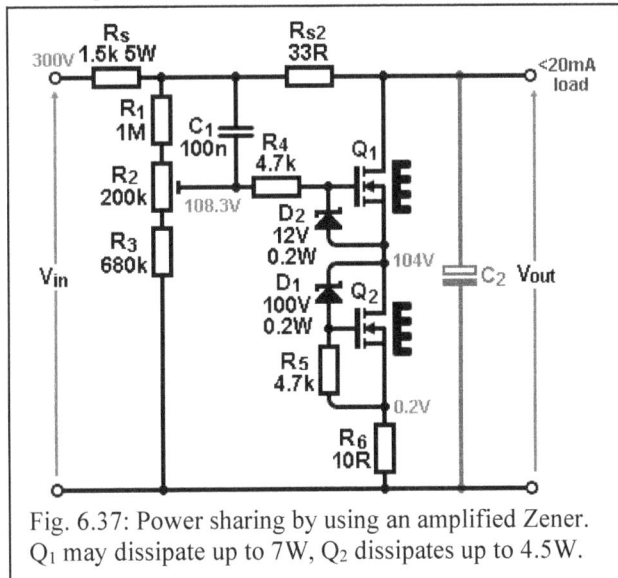

Fig. 6.37: Power sharing by using an amplified Zener. Q_1 may dissipate up to 7W, Q_2 dissipates up to 4.5W.

reference voltage, so there is little to stop us splitting the total dissipation roughly equally between the two transistors just as with the cascode regulator. Moreover, a higher reference voltage also means a larger feedback voltage, which leads to more loop gain and therefore *better regulation too*.

Fig. 6.37 shows a practical example. The reference is D_1 which is now a convenient 100V device –not exactly half the output voltage (which could be anywhere between about 240V and 270V depending on loading) but enough to offset a large fraction of the total circuit dissipation. R_5 determines the current through D_1, which is equal to $V_{GS}/4.7k\Omega$ or about 1mA in this case. Q_2 completes the amplified Zener circuit and passes the bulk of the shunt current. The optimum value for R_{s2} was found on test to be 33Ω, a little more than previously because Q_1 is degenerated a little more by the internal resistance of the amplified Zener sitting below it.

Since the reference voltage is nearly half the output voltage, the feedback fraction must follow suit. After adjusting R_2 for 20mA quiescent shunt current (once again by measuring the voltage across the current-sense resistor R_6) the ripple reduction was found to be $-54dB$; twice as good as the previous cascode regulator. What's more, the DC regulation is better too. A $\pm5\%$ variation in input voltage caused less than 0.5% variation in output voltage. With a larger $\pm10\%$ variation in input, the DC regulation degraded to $\pm3\%$ because the regulator comes close to dropout, but still a good result. The advantage of this topology over the previous cascode is clear.

6.7.4: Amplified Zener Cascode

Given the success of the previous topology, we might well wonder how much benefit the upper device is really delivering. What if we reconfigured the circuit into one, big amplified Zener? Fig. 6.38 shows exactly that. A pair of identical 120V Zener diodes, plus V_{GS}, make up the total regulated voltage. R_1 sets the idle current in the Zener string to V_{gs}/R_1. Since this current is small, low-power Zeners can be used. Q_1 and Q_2 do the hard work, dissipating as much power as we demand of

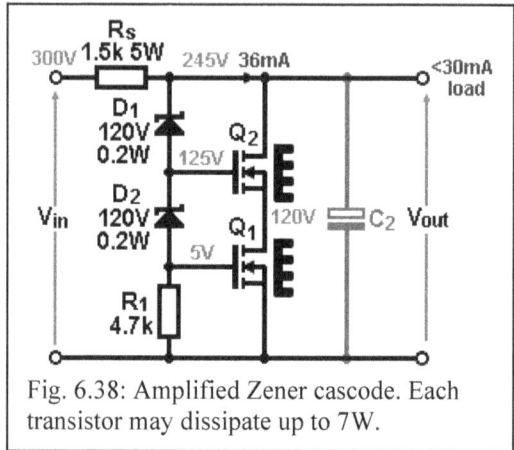

Fig. 6.38: Amplified Zener cascode. Each transistor may dissipate up to 7W.

them, effectively creating a 245V power Zener diode. Again, STP11NK40Zs were used but any other suitably rated MOSFETs would do. Because the output voltage is now reliably set by our choice of Zeners, it is easy to anticipate the amount of shunt current that will flow, given our choice of R_s, so no current-sampling resistor is needed.

The peak-to-peak ripple reduction of this circuit was just over -43dB, indicating an output impedance of about 10Ω, which is effectively the internal resistance of the amplified Zener. This ripple reduction is not quite as good as a mixed-mode regulator using similar parts, but the output impedance is lower and the DC regulation is better –holding the output within 1V for a $\pm5\%$ variation in input voltage. The simplicity and lack of any need for trimming are also significant advantages over previous circuits. Remember, although we can treat this circuit like a big Zener diode, it is not a passive regulator but an enhanced King regulator (remember section 6.6.5). On balance, this topology is the default, high-voltage solid-state shunt regulator of choice, unless you are chasing a very specific performance metric.

6.8: Constant-Current-Enhanced Shunt Regulation

Ignoring feedforward regulation which is a type of nulling or bridge circuit, the line regulation (ripple rejection) of a typical shunt regulator is a consequence of the potential divider formed by the series dropping component and the internal resistance of the shunt. It is therefore maximised by making the former as large as possible, and the latter as small as possible. Most of this chapter has dealt

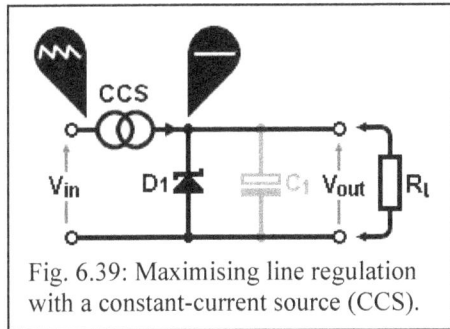

Fig. 6.39: Maximising line regulation with a constant-current source (CCS).

only with the shunt part of the regulator, but what about maximising the series part? The traditional option would have been a choke, but the modern approach is to use a constant-current source, as illustrated in fig. 6.39. The shunt arm is shown as a Zener diode, but any other sort of shunt regulator could in theory take its place. That leaves the details of the series CCS to fill in.

Fig. 6.40 shows some current sources suitable for HT applications. The first is a low-cost option using an MJE573x PNP transistor. Bipolar junction transistors are cheap but seldom rated for more than 350V, hence the input voltage limit. The voltage across R_e is equal to the LED drop minus the V_{be} of the transistor. When using a red LED this conveniently amounts to about 1V, so the set current will be roughly equal to: $i \approx 1/R_e$.

The second circuit uses a P-channel enhancement MOSFET. Such devices are readily available with much higher voltage ratings than BJTs. The set current is equal to the V_{be} of Q_2 divided by R_s, or roughly: $i \approx 0.6/R_s$. A $1k\Omega$ base resistor protects Q_2 against transient currents at start up, and the 12V Zener (faint) protects both the gate-source junction and Q_2 from over voltage, meaning any general purpose PNP transistor can be used for Q_2.

The third circuit is a Maida-type circuit. An LM317 voltage regulator IC is configured as a constant current source; it will maintain 1.25V across R_1, so the set current is: $i = 1.25/R_1$. The job of the MOSFET Q_1 is to shield the 317 from high voltages, so is exists happily within a floating voltage 'window' equal to the Zener voltage minus V_{GS}. This type of current source has the advantage of being entirely floating, although the impedance of the current source is effectively in parallel with R_2 which puts an upper limit on what is achievable. Nevertheless, $1M\Omega$ is enough to be considered 'infinite' for most applications. See section 7.7.5 for the series voltage regulator version of this circuit.

The final circuit is the simplest as it uses a DN2540 15-watt depletion MOFET which, like a valve, can be self-biased. The set current is equal to the V_{GS} of the MOSFET divided by R_s but, owing to variation between samples, V_{GS} is not precisely defined. Typically it will be about 1.4V at valve-amp levels of current, or: $i \approx 1.4/R_s$. The 100Ω resistor is simply a gate stopper to discourage oscillation. Again, this is a floating current source.

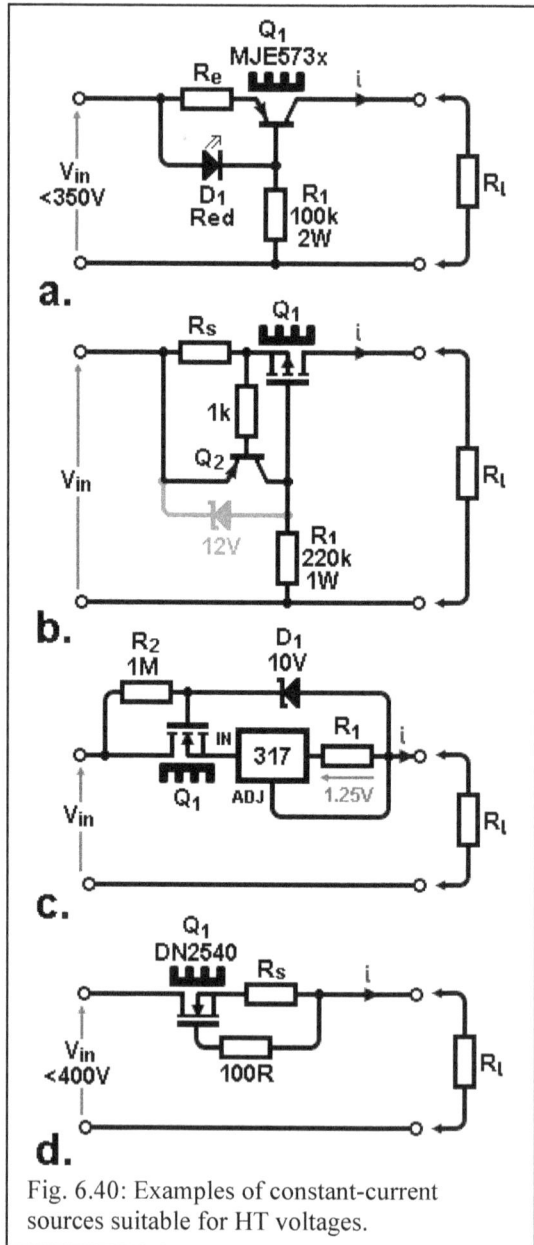

Fig. 6.40: Examples of constant-current sources suitable for HT voltages.

Fig. 6.41 shows the amplified Zener regulator from the previous section, augmented with a constant-current source. A DN2535 is used because that's what the author had in stock (this device is rated for 350V rather than 400V for the DN2540). This circuit has a lot to commend it; the output voltage is easy to define since it is roughly equal

160

to the two Zener
diodes in series, R_s
could be made
variable to allow
the set current to
be adjusted on the
fly, no capacitors
are needed, and
the regulation is
outstanding. The
CCS enhances the
ripple reduction to
the point where it
could no longer be
measured –even
with $25V_{pp}$ input
ripple, the output
was only white noise around 10mV.

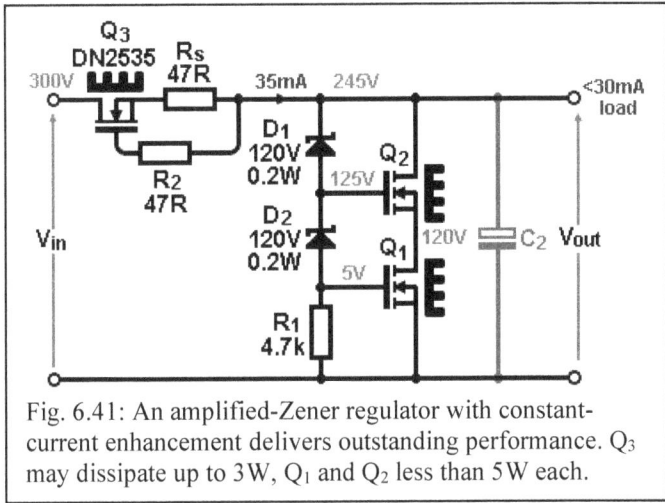

Fig. 6.41: An amplified-Zener regulator with constant-current enhancement delivers outstanding performance. Q_3 may dissipate up to 3W, Q_1 and Q_2 less than 5W each.

Chapter 7: Series Stabilisers and Regulators

Having looked at shunt regulators in the previous chapter, here we will look at the alternative, again with a focus on HT supplies. Many worthy books have been written on the subject of series regulator design, and with enough patience and determination it is possible to devise highly sophisticated 'high-end' regulators with breathtakingly good regulation, transient response, and vanishingly low output impedance.

However, chasing each decimal place of performance tends to lead to an exponential increase in complexity, and regulator design becomes yet another specialism. Moreover, such designs tend to be unstable into certain loads, making them less universally applicable, and a firm understanding of loop compensation is required to optimise them. The author much prefers a robust circuit that is good enough, to an excellent circuit that requires constant parental supervision. Simple circuits have simple shortcomings; designing a power supply regulator that outclasses the amplifier it is expected to mate with does not seem like time well spent

7.1: Series Regulation

Fig. 7.1 shows the essence of shunt and series voltage regulation, reproduced from the previous chapter. With the series regulator, the power control device forms the upper arm in the conceptual potential divider, with the load forming the lower arm. In a practical series regulator circuit, the main regulating component is called the **pass device** since all the load

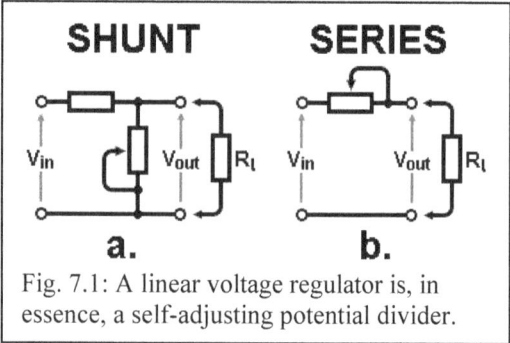

Fig. 7.1: A linear voltage regulator is, in essence, a self-adjusting potential divider.

current passes through it. The pass device needs to adjust the voltage across itself in response to changes in loading or in the raw supply voltage, to keep the output voltage stabilised. Transistors definitely have the advantage over valves here, since transistors can handle large currents while dropping very little voltage, which minimises wasted power and generally makes this type of regulation far more efficient than the shunt type.

7.1.1: Principles of Feedback Regulation

Fig. 7.2 shows the functional parts of a feedback series regulator. It is the same as for shunt regulation (remember fig. 6.2) except for the position of the control device, which is now the series pass device. Again, the error amplifier and the pass device will sometimes be one and the same component,

A fraction of the output voltage –the feedback voltage– is compared with a fixed reference voltage. Any difference between the two is the error, and indicates that the output voltage is not what it should be. This error is amplified and sent to the pass device which does the heavy lifting; conducting less current to let the output voltage fall, or conducting more to pull the output voltage up. As explained in the previous chapter, the greater the loop gain, the lower the output impedance and the better the regulation. The voltage across the pass device itself, which may be varying all the time, is called the **differential voltage** and is the principle cause of loss and inefficiency in a series regulator.

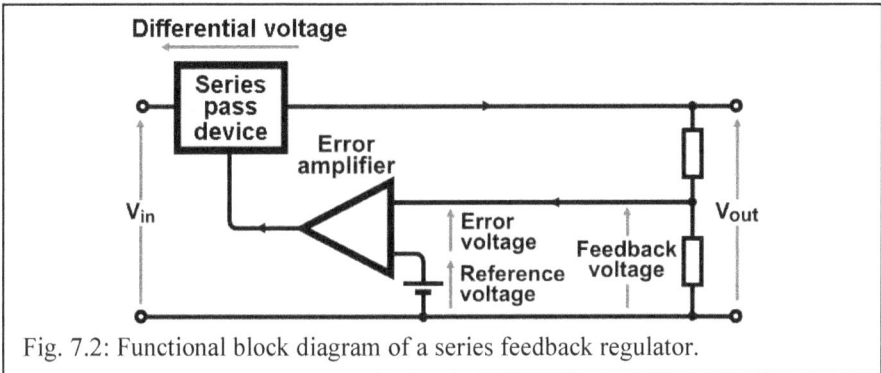

Fig. 7.2: Functional block diagram of a series feedback regulator.

7.1.2: Classic Topologies

Fig. 7.3 shows the most common topologies used for series feedback regulators. All the transistors are depicted as BJTs, but they might alternatively be valves or FETs. In every case Q_1 is the main pass device and R_1 and R_2 form the feedback divider. The upper arm of the divider is normally bypassed with a speed-up capacitor C_1, to increase the loop-gain for AC and also ensure stability, as explained in the previous chapter. D_1 is the voltage reference, shown here as a Zener diode though it might alternatively be a gas tube or some other reference device.

The version in a. is the classic two-transistor design, descended of course from the classic two-valve regulator –see section 7.7.2 later. Q_2 is the error amplifier –a simple common-emitter gain stage– which amplifies the difference between base and emitter, that is, between feedback and reference. The collector resistor R_3 affects the gain of Q_2 which we ordinarily want to be large for better regulation, but it must also supply any base current to Q_1, so there may be a trade-off here.

Fig. 7.3b is the most common next-step in regulator evolution, employing a differential amplifier or 'long-tailed-pair' as the error amplifier. The advantage of this is that it buffers the voltage reference D_1 from the regulator proper. This means the current in D_1 remains substantially constant whatever the regulator happens to be doing, remembering that the cleanliness of a regulator is only as good as the cleanliness of its reference.

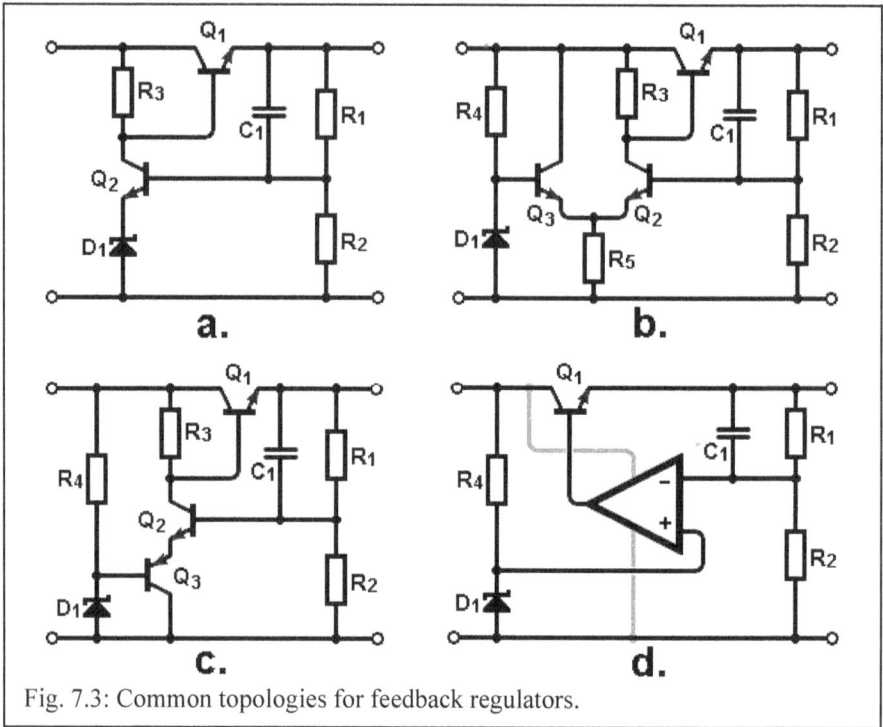

Fig. 7.3: Common topologies for feedback regulators.

Fig. 7.3c works exactly the same as b., but it uses a vertical differential amplifier instead (something which is not possible with valves of course, since there is no such thing as P-type valve). This topology can be useful for high-voltage designs since Q_2 and Q_3 share the total voltage burden, potentially allowing lower-voltage devices to be used –see for example section 10.3.7.

Fig. 7.3d most resembles fig. 7.2 earlier, since it uses an opamp rather than discreet transistors. This topology is commonly used for power supplies up to about 30V since this conveniently allows the opamp to take its power directly from the raw supply (shown faint). Opamps offer extreme loop gain and therefore high regulation figures are possible, and this topology is still often found in commercial bench power supplies despite the development of powerful 3-terminal regulator ICs.

Endless refinements on these ideas are possible, which is how whole books end up being written on the subject. There is not space here to cover all the possibilities, save to mention the most common. In each case R_3 is often substituted with a constant-current source (CCS) to increase loop gain and shield the error amp from input ripple. Similarly, R_4 is often substituted with a CCS to shield the reference from ripple and wall variations. Alternatively, the reference is sometimes fed from the clean *output* of the regulator, though this can create problems with making the circuit switch on politely (or at all). Current-limiting or safe-operating-area protection is often necessary, and this is especially true for high-voltage circuits, and various options are covered later in this chapter.

7.1.3: Voltage Sharing

It is often convenient to use low-power resistors throughout a design, not just because they are inexpensive but because they are the most commonly stocked by hobby suppliers and by surface-mount fabrication houses. But it is easy to forget that resistors have a voltage rating as well as a power rating. 1W-rated resistors are typically rated to withstand 500V across them, whereas the ¼W devices sold by most hobby suppliers are normally rated for 200V, and common 0805 chip resistors are usually rated for 1/8W, 150V. It would be all too easy to exceed such ratings in an HT regulator, especially at switch on.

Fortunately, we can still use such small devices by connecting several in series, so the total voltage and power burden is shared between them. This will often work out cheaper than investing in a single, heavy-duty device, and it can help to spread the heat dissipation over a wider area which reduces hot-spot effects or PCB scorching. The same principle holds for voltage sharing across capacitors too, though electrolytics may additionally need balancing resistors to swamp the effects of leakage, as described in section 2.3.5.

Transistors are also commonly used in tandem to increase power or voltage handling. To increase voltage handing they must be connected in series, i.e. in cascode; examples have already been shown in section 4.4. If only power needs to be shared then they may alternatively be connected in parallel. However, since transistors are never exactly matched, emitter resistors should be included to encourage equal division of current between them. These resistors should be chosen so the maximum average voltage drop across each resistor is significant compared with the V_{be} or V_{GS} of the transistors, e.g. >50mV for BJTs or >500mV for MOSFETs. Fig. 7.4 illustrates how series components would typically be applied to the regulator from fig. 7.3c, along with parallel pass transistors with emitter resistors, R_{e1} and R_{e2}.

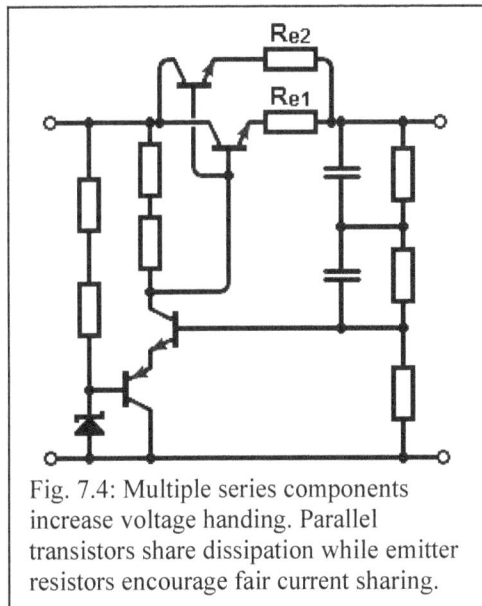

Fig. 7.4: Multiple series components increase voltage handing. Parallel transistors share dissipation while emitter resistors encourage fair current sharing.

7.1.4: BJTs vs. MOSFETs

The main advantage of a BJT pass device is that it has a small and very predictable V_{be} of 0.6 to 0.7V, which minimises differential voltage loss, as well as reducing any need for hand-selection or circuit trimming in certain designs. This is compared with

165

the much larger and more vague V_{GS} of a power MOSFET, which is often between 3 and 5V in the sort of circuits we're interested in. For high-voltage environments, however, MOSFETs are the device of choice.

There was once a cost advantage to BJTs over MOSFETs, but this is no longer true for high-voltage devices; MOSFETs are now just as cheap. There is also a wide selection of MOSFETs to choose from, up to thousands of volts $V_{DS,}$ with high current capability, whereas BJTs are scarce above 400V, or even less for PNP types.

A further disadvantage of high-voltage BJTs is that their h_{FE} is usually very low, typically no more than 30, meaning significant base current may be demanded. We can of course use a Darlington-pair to increase h_{FE} (the BU931 is a 400V-rated Darlington in a single package), but this starts to erode the few reasons for using a BJT in the first place. A MOSFET, on the other hand, has a fully insulated gate and therefore require no 'drive' current at DC or low frequencies. The gate-source junction does need to be protected from voltages larger than 20V, however, as the quartz insulation will break down above this.

BJTs typically have smaller capacitance between collector and base than MOSFETs of similar capability, but whether this is an advantage depends on the application. Since we're not dealing with high-frequency switching circuits, junction capacitance is a minimal concern.

Yet another shortcoming of using a BJT at high voltages is the safe operating area (SOA), which is much more restrictive than for a similar MOSFET. For example, fig. 7.5a shows the SOA of an MJE13700 BJT, lifted from the datasheet. It is advertised as a 400V, 80W device, but it only achieves the 80W figure at low voltages, as indicated by the dashed 'thermal limit' curve on the SOA. When the voltage across the transistor is larger than about 40V, a lower (and variable, and non-intuitive) power limit is imposed by **secondary breakdown**, caused by hotspots appearing in the silicon. This is common to all power BJTs and is a serious handicap for high-voltage regulators, as it makes the transistor unlikely to survive an output

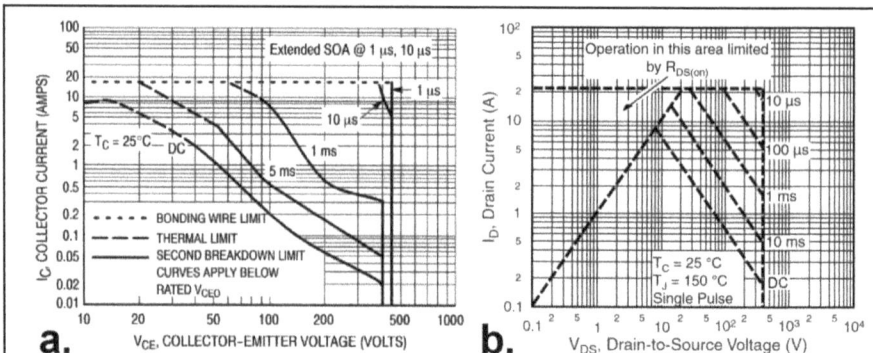

Fig. 7.5: The SOA of high-voltage BJT is more restrictive than for a MOSFET. **a:** BJT (MJE13007, 400V, 80W); **b:** MOSFET (IRF730, 400V, 74W).

short. Even a brief period of inrush current when powering up into a capacitive load might be enough to violate the SOA.

Compare this with the SOA of an IRF730 MOSFET, shown in fig. 7.5b. This is a 400V, 74W device, making it similar in principle to the BJT. However, MOSFETs do not suffer from secondary breakdown, so the full DC dissipation limit is maintained all the way up to the V_{DS} limit. MOSFETs with extremely high dissipation limits are available for extra rugged designs, such as the 600V 540W-rated IXTQ30N60L2; a mere £10 each!

7.2: Current Sensing and Limiting

In a series regulator all the load current flows through the pass device. Therefore, if the load happens to be an uncharged capacitor, or a faulty component, or a screwdriver you have accidentally dropped into the rats nest, the pass device must supply the entirety of the overload current, whether briefly or indefinitely, or die trying. Therefore, most series regulator designs include some form of protection against overloads. This is especially true for bench power supplies, since continuous shorts and general abuse are sure to happen, but even a fixed regulator in a finished amplifier may be subject to stressful conditions at switch-on when things charge up.

Protection could be as simple as a fuse, which might be satisfactory for a valve regulator, but fuses act much too slowly to protect semiconductors, and in either case replacing a fuse is an annoyance. We ought to have some fast-acting, self-resetting solutions in our bag of design ideas. For low-voltage applications there are many options available in IC form, from voltage regulator ICs that already have built-in protection to specialised current and voltage monitoring ICs. These are easy to find in manufacturers' catalogues and need not be explored here. We are naturally more interested in the tricky high-voltage environment where we have to build our own regulators.

Fig. 7.6 shows some textbook approaches for current-limiting a series pass device, represented by Q_1. An enhancement MOSFET is shown but it could equally be a BJT. A gate resistor R_g is shown faint and represents whatever gate-supply circuit is used in practice. All these circuits work by adding a current-sampling resistor R_s to the output of the pass device, and the voltage monitored across this resistor is used to trigger the current-limiting mechanism. Clearly R_s adds to the output impedance of the pass device, but it can usually be made small enough that this is a negligible concern. The current limiting action may fall into three general categories: constant-current, foldback, and trip.

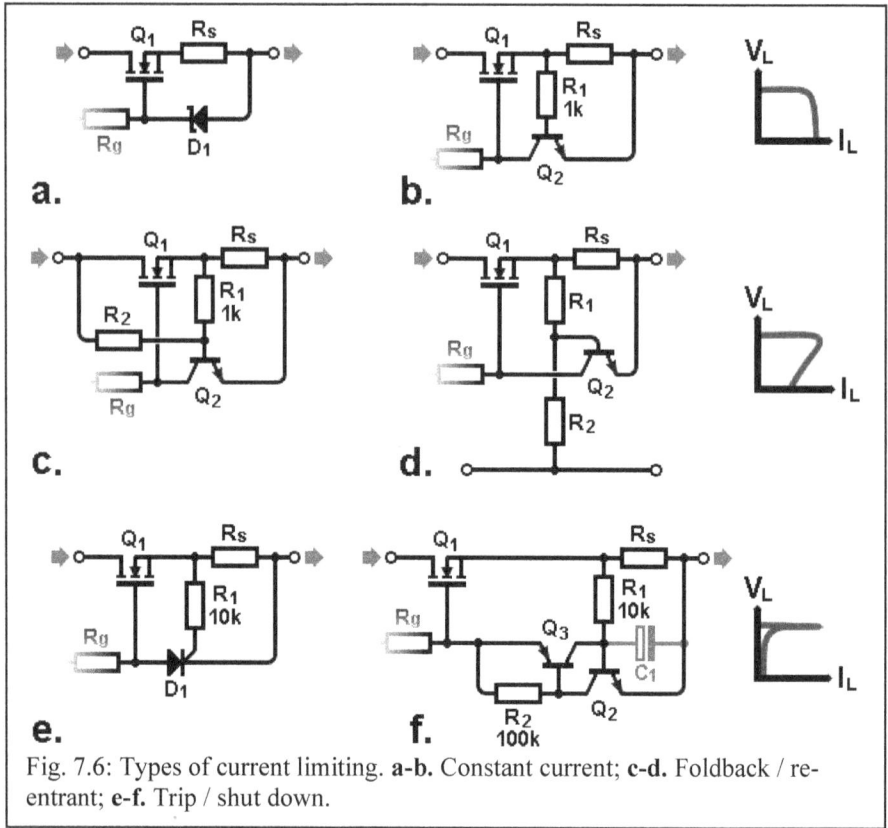

Fig. 7.6: Types of current limiting. **a-b.** Constant current; **c-d.** Foldback / re-entrant; **e-f.** Trip / shut down.

7.2.1: Constant-Current Limiting

With constant-current limiting, the load current is allowed to reach the design maximum but is prevented from climbing any higher. If the load attempts to demand more current, the output voltage will automatically fall to prevent this from happening, holding the current at the design maximum even into a short circuit. This ultimately leaves the full voltage across the pass device, resulting in considerable power dissipation until the short is removed. In an ideal world the protection circuit would be completely invisible until needed, snapping into action and acting like a brick-wall limit. In practice, the threshold where limiting begins will have a somewhat softer 'knee', and the current limit may not be truly constant (depending a great deal on the source impedance of whatever gate-supply circuit is used).

Fig. 7.6a shows the simplest way to add constant-current limiting. As the load current increases, the voltage drop across R_s increases too. When the voltage across R_s, plus the gate-source voltage, becomes equal to the Zener voltage, the Zener will conduct. This steals current from the gate of the pass device, dragging down the gate voltage, and the output voltage therefore follows likewise. The Zener is often needed anyway to protect the gate-source junction from over-voltage, so it does double-duty.

168

The output current limit will be equal to:

$$I_{max} = \frac{V_z - V_{GS}}{R_s} \qquad (7.1)$$

Where:
V_Z = Zener voltage;
V_{GS} = gate-source voltage of Q_1 if it is a MOSFET, or V_{be} if it is a BJT.

The gate-source voltage of a MOSFET is variable between samples and increases with drain current. Nevertheless, for the sort of stabilisers used in this book it will usually be around 4V to 5V. A BJT has a more reliable V_{be} of about 0.7V, but even then we are left with the slightly imprecise threshold voltage of the Zener diode. Fortunately, most amplifier applications don't need high precision, they just need a current limit that is somewhere between 'more than enough' and 'less than catastrophic'.

For more precise constant-current limiting the circuit in fig. 7.6b can be used. Here the voltage across R_s is monitored by a BJT. When the voltage drop across the resistor is equal to V_{be}, Q_2 will turn on and steal current from the gate of the pass device in exactly the same manner as previously. The current limit is therefore:

$$I_{max} = \frac{V_{be}}{R_s} \approx \frac{0.65}{R_s} \qquad (7.2)$$

Since this depends only on V_{be}, which is always 0.65V for a low-power BJT, the current limit can be accurately chosen, regardless of the characteristics of the pass device. Note that Q_2 is safely contained within the V_{GS} window, and it passes only a small current, so it can be a low-voltage, general purpose device. R_1 is a generic value, included to protect the base of Q_1 from excessive transient current, especially when starting up into a capacitive load.

With any form of constant-current limiting, if the output is a dead short the full supply voltage will be dropped across the pass device. Its dissipation will therefore be at a maximum, and in a high-voltage situation this could represent a huge amount of power. Consequently, in high-voltage designs, constant-current limiting is commonly used to protect against *short-term* overloads only, such as the inrush to smoothing capacitors at switch on.

7.2.2: Foldback Current Limiting

With foldback limiting, if the load attempts to demand more current than the design maximum, the output voltage drops at a faster rate than the demanded current. This causes the load current to 'fold back' below the design maximum. This means that while the voltage across the pass device increases during limiting, the current reduces, so the power dissipation in the pass device is not as severe as with constant-current limiting. This may allow the pass device to withstand a continuous short circuit with less heatsinking required. This approach is also called re-entrant limiting or sometimes constant-power limiting.

Fig. 7.6c shows a simple way to accomplish fold-back limiting. The circuit is almost the same as in b. except for one extra resistor, R_2. When the current limiting kicks in and the output voltage starts to fall, the drop across the pass device –and therefore across R_2– increases. This means R_2 supplies even more base current to Q_2, turning it on even harder. In this way the load voltage is forced to drop at a faster rate than the drop in load resistance that caused the change in the first place, so the current limit 'folds back on itself'. The effect will be enhanced even more if the supply voltage un-sags as the load current falls.

Foldback-limiting is an application of positive feedback, which also means the design equations can quickly become unmanageable unless we make some simplifications. Start with formula (7.2) and use it to choose R_s:

$$R_s \approx \frac{0.65}{I_{max}}$$

Then make R_1 at least one hundred times larger than R_s so it does not affect the chosen current limit. Now choose the short-circuit current, which should be between one third to one half of I_{max} (any less and the circuit might not start at all). R_2 can then be found:

$$R_2 = \frac{V_{in}R_1}{0.65 - I_{short}R_s} \tag{7.3}$$

Where:
V_{in} = the unregulated input voltage;
I_{short} = short-circuit current limit, typically one third to one half of I_{max}.

A slightly different foldback configuration is shown in fig. 7.6d. Q_2 is now supplied from a potential divider R_1-R_2, so under light loading its base voltage is below the output load voltage. But as the load current increases, eventually the voltage drop across R_s becomes equal to the drop across R_1 plus V_{be}, causing Q_2 to turn on and start limiting. However, in doing so, the output voltage begins to fall and therefore the voltage across R_1 also falls, pulling the base of Q_2 up and turning it on even harder. With this arrangement the current limit is:[1]

$$I_{max} = \frac{1}{R_s} \cdot \left[\frac{R_1}{R_2} V_{out} + \left(1 + \frac{R_1}{R_2} \right) V_{be} \right] \tag{7.4}$$

Where:
V_{out} = output voltage of the regulator under normal operating conditions;
V_{be} = base-emitter voltage of Q2, which will be about 0.65V.

---

[1] Horowitz, P. & Hill. W. (1989). *The Art of Electronics* (2nd ed.), p317. Cambridge University Press.

170

The short-circuit current is:

$$I_{short} = \frac{1}{R_s} \cdot \left(1 + \frac{R_1}{R_2}\right) V_{be} \qquad (7.5)$$

And the ratio between the two:

$$\frac{I_{max}}{I_{short}} = 1 + \left(\frac{R_1}{R_1 + R_2}\right) \cdot \frac{V_1}{V_{be}} \qquad (7.6)$$

The short circuit current would typically be set between one third to one half of I_{max}.

7.2.3: Overload Trip

With a trip circuit, if the load current exceeds the design maximum the circuit shuts itself off. The load current therefore goes to zero or a negligible value, so there will be practically no dissipation in the pass device. This is really an extreme form of fold-back limiting.

Fig. 7.6e shows an approach using a thyristor, also called a silicon-controlled rectifier (SCR). When the voltage across R_S reaches about 0.6V it triggers the thyristor which instantly snaps on *and stays on,* even if the load relaxes. This effectively shorts the MOSFET gate to its source, turning it off and shutting down the regulator completely, except for the trickle current that continues to flow into the thyristor from the gate supply circuit. The thyristor can only be turned off (reset) again by interrupting this trickle current, e.g. by removing the load completely or by switching the power supply off and on again.

Since thyristors are less likely to be found in the average hobbyist's stock, an alternative is to make your own using a pair of transistors as in fig. 7.6f. When the voltage across R_s reaches about 0.6V the NPN transistor Q_1 turns on, conducting current through R_2. The voltage across this resistor likewise turns on the PNP transistor Q_2, which feeds current into the base of Q_1, keeping it turned on. In other words, the transistors latch permanently on, defeating the MOSFET. Again, the load must be removed before the circuit will reset. Bear in mind that the current threshold needs to be greater than any expected inrush current into the load, or the circuit may trip at switch on. An alternative workaround is to add the capacitor C_1, which takes time to charge up through R_1 and therefore slows down the reaction time of the circuit.

7.2.4: High Side vs. Low Side

For all the current-limiting techniques in fig. 7.6, the current-sense resistor was in series with the high-voltage output of the regulator. This is called high-side monitoring. The current-sensing circuit therefore sits at a similar potential to the pass device, which for simple current limiting is convenient. Dedicated high-side current sense amplifiers are also available for precision applications, such as when monitoring current with a microprocessor. For example, the LTC6101 measures the voltage across a high-side current-sense resistor and produces an output current that is directly proportional to the current being measured. Such devices are normally rated for less than 100V but fig. 7.7 shows how this can be extended by by cascoding with a high-voltage MOSFET. The IC sits within a safe 47V window provided by Zener diode D_1, floating below the HT rail, while Q_1 takes up the excess burden. HT voltages up to the limit of the MOSFET can be tolerated, which is 500V in this example. The output current from the IC is fed down into R_2 which therefore produces a voltage proportional to the HT current, which can be safely read by a low-voltage circuit such as a microprocessor. This output voltage is equal to:

$$V_{out} = i\,R_s\,\frac{R_2}{R_1} \qquad (7.7)$$

With the component values shows this corresponds to 1V output for every 100mA of HT current.

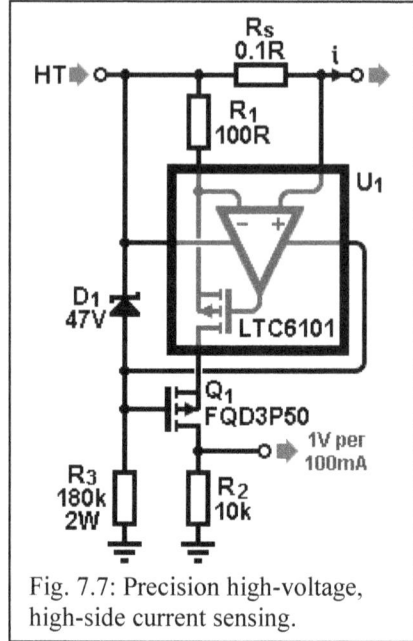

Fig. 7.7: Precision high-voltage, high-side current sensing.

If a direct interface between a high-voltage rail and a low-voltage circuit is inconvenient, the alternative is to place the sense-resistor in the ground return path, which is called low-side monitoring. This approach is often used in bench power supplies. Fig. 7.8 shows two examples of low-side current limiting. A simple Zener-follower regulator (section 7.7.1) is used as the example, but the same approach would apply to any series regulator. In a. when the voltage across R_s reaches the threshold voltage, Q_2 will turn on and pull the gate voltage of the pass device down, shunting the Zener and defeating the regulator. Q_2 must be rated to withstand the full output voltage under normal running conditions which is why a MOSFET is shown rather than a BJT, since we are dealing with HT regulators. Note that R_s is on the unregulated side of the circuit so it does not add to the output impedance of the regulator or affect the regulated output voltage. A ground symbol has also been added to the output for clarity; the input side of the regulator must not be grounded or R_s would be shorted, obviously.

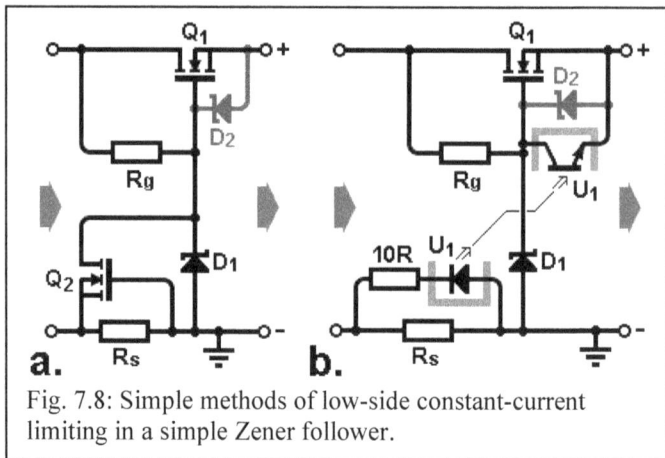

Fig. 7.8: Simple methods of low-side constant-current limiting in a simple Zener follower.

172

In fig. 7.8b an optoisolator is utilised. This contains an infrared LED and phototransistor in a single package. These are very cheap devices that can provide isolation between circuits thousands of volts apart. When the voltage across R_s reaches about 1.1V the LED will begin to turn on, and this in turn will switch on the phototransistor. The 10Ω resistor limits transient current through the LED. In this case the transistor is doing exactly the same job as fig. 7.6b earlier, 'pinching' the gate and source together to choke off current, but R_s no longer adds to the output impedance of Q_1 (if oscillation arises, a 100nF to 1µF capacitor between the gate and source of Q_1 usually suffices). Indeed, the optoisolator brings enormous design freedom –the current monitoring could be done anywhere; high-side or low-side, before or after the regulator. Likewise the phototransistor could be made to do whatever job we want, such as overload trip rather than constant-current limiting.

7.3: Cooling and Heat Sinking:

Heat is the enemy of electronics, valve heaters notwithstanding. There are practical limits to how hot we can allow various components to become. As valve enthusiasts we are spoiled by the fact that we rarely need to use heatsinks, since most valves are designed to handle their anode dissipation limit without assistance, assuming they are sitting in free air. This is one of their few advantages over transistors. It is understandable, therefore, when a beginner buys a transistor boasting 100W dissipation and expects it to do just that, naively thinking that a piece of plastic the size of a thumbnail can do what a valve the size of a cola bottle can do. Some ICs have built-in thermal shut down, but a discrete transistor will simply burn to death. A few dead transistors later and he learns that even for moderate power dissipation, a semiconductor needs a helping hand from a slab of metal. The question is, how big does that slab of metal need to be?

When a semiconductor turns electrical energy into heat, the heat energy has to flow from the silicon junction to the case of the device, then into the heatsink, and finally into the surrounding air. The ease (or rather, difficulty) with which heat can flow through a material object is called its **thermal resistance** and is specified in Celsius per watt (or kelvin per watt, which amounts to the same thing). In other words, it is how much the temperature of the material will rise for every watt of heat flowing through it. The situation is very much like an electrical circuit: heat energy flows like current through the thermal resistance, causing a temperature difference across it. Silver and copper have the lowest thermal resistance; aluminium is about twice as high; steel is ten times worse again. But almost anything is better than silicon.

7.3.1: Thermal Calculations

When a device is fixed to a heatsink, the resistances of each part of the 'thermal circuit' add together in series to form the total thermal resistance between the semiconductor junction and the air:

$$\theta_{ja} = \theta_{jc} + \theta_{cs} + \theta_{sa} \tag{7.8}$$

Where:
θ_{ja} = junction-to-air thermal resistance;
θ_{jc} = junction-to-case thermal resistance;
θ_{cs} = case-to-sink thermal resistance;
θ_{sa} = sink-to-air thermal resistance.
Figures will be provided on the device and heatsink datasheets.

The total thermal resistance we *need* depends on how hot we are happy for the semiconductor junction to run. To stay within the design limit, the heatsink must have a thermal resistance of:

$$\theta_{sa} = \frac{T_j - T_a}{P} - \theta_{jc} - \theta_{cs} \tag{7.9}$$

Where:
T_j = junction temperature;
T_a = ambient temperature;
θ_{jc} = junction-to-case thermal resistance;
θ_{cs} = case-to-sink thermal resistance (e.g. insulation hardware);
P = power dissipated in the device.

The maximum permissible junction temperature will be specified on the datasheet and is usually 125°C or 150°C. Of course, for reliable operation we should aim for something lower, perhaps 100°C at most. Ambient air temperature is often assumed to be 20°C, but inside a stuffy chassis it is better to assume 40°C. We can therefore allow the junction to rise by 60°C, bringing it from 40°C up to our limit of 100°C. If there is too much thermal resistance standing between the junction and the air, i.e. the heatsink is too puny, the junction will rise by more than 60°C because the heat can't get out of the device fast enough.

For a bare metal-on-metal contact the thermal resistance of the case-to-sink interface, θ_{cs}, can be taken as about 1.5°C/W. A thin smearing of thermal compound on the surfaces will reduce this to 0.5°C/W or even less. The compound fills the microscopic air pockets between the surfaces. Only a thin layer is needed; you are not icing a cake! And be warned, it is *extremely* messy stuff that will ruin your shirt as soon as you open the tube. It may also be slightly electrically conductive, so try not to smear it on anything else.

The metal tab of the device is usually connected internally to the junction, meaning it is part of the electrical circuit. If the heatsink is bolted to (or part of) the chassis, we cannot screw the device directly to the sink since we would effectively be shorting the active circuit to the grounded chassis. We must therefore insulate the device from the sink. This can be done with a mica

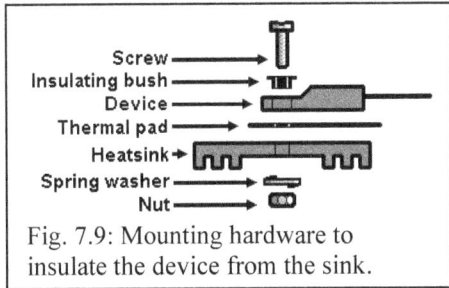

Fig. 7.9: Mounting hardware to insulate the device from the sink.

wafer or a silicone thermal-pad (Sil-Pad®), plus a plastic bushing affectionately known as a 'top hat' to insulate the metal screw from the tab, as illustrated in fig. 7.9. Alternatively, some kind of spring or clamp may press the device firmly against the heatsink, avoiding the need for the top hat and screw. Unfortunately, despite what the advertising says, the mica/pad is a great spoiler of thermal resistance, amounting to 2 to 3°C/W. An all-plastic TO-220FP package with thermal compound sometimes works out better, but surprisingly few devices are sold this way.

7.3.2: Testing Heatsinks

The cost-is-no-object way to test a proposed heatsinking system is simply to build everything and let it run. If the sink stays cold then it's overengineered; if it becomes too hot to hold then it's not big enough! But if a little more finesse is required then the internal thermal-protection feature found in most three-terminal regulators provides a convenient way to test the properties of heatsinks. This is especially useful when planning to employ the chassis or some other bit of scrounged metal as the sink, whose properties we have no clue about.

The LM317, LM350 and LM78xx-type regulators all go into thermal shutdown when the internal junction temperature exceeds 150°C. At this point the output voltage will droop, and if this is not enough to bring the temperature down it will go into a discontinuous mode where the output switches on and off continuously. By attaching the regulator to the proposed heatsink and monitoring the output for these signs of thermal shutdown, we can estimate the thermal resistance. We can also see directly the effects of forced-air cooling, drilled holes, an enclosed chassis, or whatever else may affect heat flow.

The test is carried out as follows. Attach the regulator to the heatsink under test. If you want the effects of mounting hardware to be included in the final figure, then fit that too. Connect the regulator input to a bench power supply. Connect the output of the regulator to a suitable load. If using the LM317, for example, the output voltage will be 1.25V, so a 1Ω 5W resistor could be used to consume 1.25A. The exact figure does not matter; you can use whatever parts you happen to have in your drawer, the aim is simply to make the regulator dissipate enough power to push it to its limit.

Monitor the output with a voltmeter and wait for everything to warm up. You can now adjust the input voltage from the bench supply to increase or decrease the power dissipation in the device, until you find the level at which the output voltage just starts to drop as the IC protects itself. At this point we can assume the junction temperature has reached 150°C. Now measure the current through, and voltage across the regulator, and multiply them to find the power being dissipated; call this P.

Measure the ambient air temperature, T_a, with a thermometer of thermocouple. The datasheet tells us the junction-to-case thermal resistance of the regulator itself, θ_{js}, so we have enough information to estimate the thermal resistance of the heatsink, using equation (7.9):

$$\theta_{sa} \approx \frac{T_j - T_a}{P} - \theta_{js}$$

Entering the figures for an LM317 in a TO-220 package this becomes:

$$\theta_{sa} \approx \frac{150 - T_a}{P} - 1$$

7.3.3: Some Practical Advice

It might be thought that the chassis would make an excellent and basically free heatsink, but while a sheet-metal chassis may have a lot of surface area, most of it is not useful because it is too far away from the device. By experiment the author has settled upon personal rules of thumb of 8°C/W for a typical mild steel chassis, and 2.5°C/W for a 1.5mm-thick aluminium one. Since insulation hardware is usually required too, the chassis is not necessarily very useful as a heatsink, and the hot spot could also be a user hazard. Note that commercial product regulations specify a maximum allowable temperature rise of 40°C above ambient for metal enclosures, or 65°C above ambient for grilles or covers above heatsinks, which must carry the warning "CAUTION – HOT SURFACE".*

Mounting devices on heatsinks while maintaining electrical isolation –and keeping the product easy to assemble and repair– is such a pain that professional designers will go to great lengths to avoid it. This may mean using several devices to share the dissipation so that each one requires only a small PCB-mounted or clip-on heatsink, or maybe nothing at all. Working out how much power a device can dissipate without any heatsink is easy, as the datasheet will quote the thermal resistance of the junction-to-air interface, θ_{ja}. For example, an LM7805 in a TO-220 package has θ_{ja}=65°C/W. Allowing a maximum junction temperature of 100°C in 40°C ambient, the maximum power that can be dissipated with no heatsink is:

$$P = \frac{T_j - T_a}{\theta_{ja}} = \frac{100 - 40}{65} = 0.92\,\text{W} \tag{7.10}$$

---

* See UL 60065 section 7.1.1

Most other TO-220 devices have similar θ_{ja}, so 'one watt for a TO-220' is an easy figure to remember when using such devices.

Although we should always do the maths, the following represent the author's own rules of thumb:

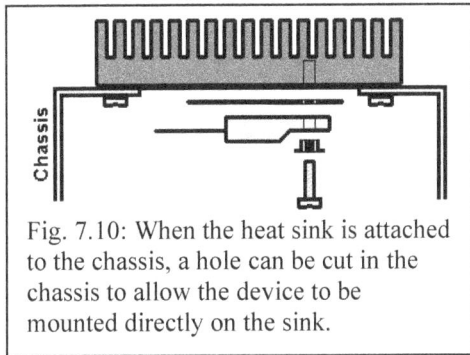

Fig. 7.10: When the heat sink is attached to the chassis, a hole can be cut in the chassis to allow the device to be mounted directly on the sink.

- TO-220 package alone; up to 1W.
- Clip-on heatsink; up to 3W.
- 1mm-thick mild-steel chassis with insulation hardware; up to 6W.
- Finned aluminium heatsink the size of a matchbox; up to 7W.
- Finned aluminium sink the size of a deck of playing cards; up to 10W.
- 1.5mm-thick aluminium chassis with insulation hardware; up to 10W.
- Finned heat sink the size of a housebrick; up to 50W.

Very large heat sinks are typically attached to the outside of the chassis so they are well ventilated. To achieve the smallest thermal resistance the device should be mounted directly to the sink through a cut-out in the chassis, as shown in fig. 7.10. This is absolutely essential if the chassis is made from steel, and is still advisable even if it is aluminium since it avoids a potentially poor interface between the chassis and sink.

Experience shows that if you need to dissipate more than about 15W per device, you will almost certainly need a metal-to-metal contact (no insulating hardware). If this is not possible because insulating hardware is needed, then consider using more devices to share the power burden. Alternatively, bolt the transistor directly to a wide piece of aluminium –then called a 'heat spreader'– which in effect increases the size of the transistor package. The heat spreader can then be mounted on a conventional heat sink with a sheet of Sil-Pad® material for electrical insulation. But be warned, large pieces of Sil-Pad® material are expensive!

Cooling is also influenced by good ventilation and conscientious physical design and layout. Heat sinks should be mounted so their vanes are aligned vertically where possible. Hot components should be given plenty of free air space around themselves, and their leads kept long to encourage heat to wick along them, away from the device itself. Electrolytic capacitors, in particular, should be kept well away from hot areas, and less critical components could also be used as crude heat barriers between areas. Ventilation holes in the top and bottom of the chassis are better than holes in the sides. However, holes in the top might not be allowable if UL certification is sought, owing to liquid spillage tests.

When power levels get serious we may need to add forced-air cooling. The proliferation of computers has made small fans widely available, and they can often be obtained for nothing from scrap equipment. In the author's own experiments

177

using a 3-inch computer fan blowing air over various sinks from 150mm away, the effective thermal resistance was roughly halved in each case. Even with the fan running at half-voltage for lower noise, the effective sink thermal resistance was reduced by as much as a third.

It is more effective to blow cool air (e.g. from outside the chassis) across the devices, than it is to try to suck hot air out of the chassis, or to recirculate the same air. The air current can be optimised by judicious placement of ventilation holes, so the air flows across as many needy components as possible. Typically a fan would be placed near one end of the chassis with ventilation holes in the top of the chassis at the far end, or drilled around top-mounted valve bases so the air flows over the hot glass as it leaves the chassis, as in fig. 7.11.

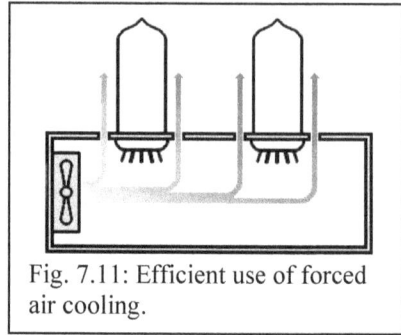

Fig. 7.11: Efficient use of forced air cooling.

A concern with fans is that they can be a strong source of electrical noise which might be picked up by audio circuits. This is particularly true for the DC (brushless) type fans which draw current spikes from the supply at around 300Hz. This noise can be minimised by careful placement away from sensitive circuitry, by running the fan at reduced voltage, or by running it from an isolated power supply such as a wattless dropper (section 5.3.5).

7.3.4: Thermostatic Switches

If the reliability of the heatsink system is in any doubt, consider adding a 'switch of last resort'. A thermostatic switch is engineered to actuate when it reaches a certain trigger temperature. By bolting one to the heatsink near the vulnerable component/s it can be used to switch the whole circuit off, or to limit the power in some other way if the temperature climbs too high. Such switches are available in both normally-open and normally-closed varieties, and many are rated for mains voltage so can be used to switch off the entire appliance if necessary, as illustrated in fig. 7.12. A

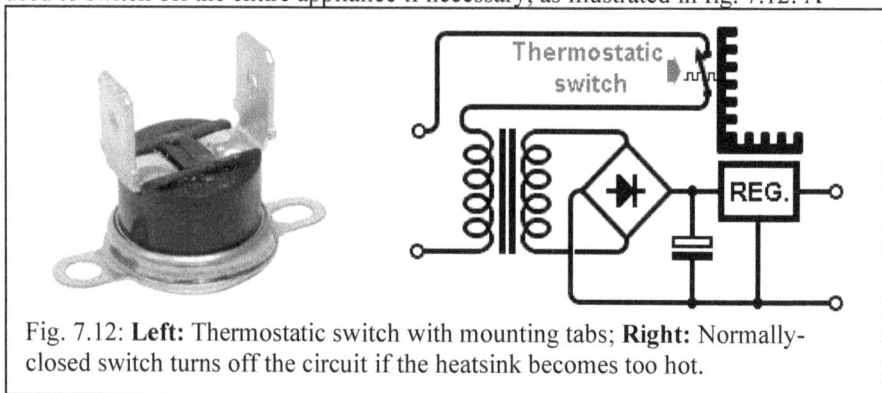

Fig. 7.12: **Left:** Thermostatic switch with mounting tabs; **Right:** Normally-closed switch turns off the circuit if the heatsink becomes too hot.

variety of trigger temperatures are available, and a typical choice for a heatsink is 50 to 60°C.

7.3.5: Dissipation in Bench Power Supplies

In a fixed installation we can usually nudge the overall design in various directions to achieve manageable dissipation in the regulator. Often we are not aiming for a specific HT voltage, meaning we can set it as high as possible to minimise the drop across the pass device. The rest of the amplifier circuit can then be designed to suit whatever HT voltage we are left with. When the design is finished, it is the same every time we turn on the power, so we don't have to guess at any weird or wonderful changes in operating conditions.

A lab bench power supply is different. A bench supply is normally used for experimental work, which could be anything from an open circuit to a short circuit, and any reactance in between. We usually want the voltage to be adjustable over a wide range, and maybe the current limit likewise. It needs to maintain consistent performance at every setting, and it would be nice if overloads were self-resetting, so we don't have to replace a fuse every time we make a mistake. Maybe it needs to be programmable, or data-loggable; the list of demands could go on. This is not a book about commercial instrument design, but it is certainly a book aimed at valve experimenters who might want to build their own bench supply.

Low-voltage bench supplies are easy to buy and not too expensive. Building your own is also quite straightforward thanks to modern technology; an off-the-shelf switch-mode power supply can provide the initial DC voltage very efficiently, while a simple three-terminal regulator can be added to clean up the result and make it adjustable. There is no shortage of choice, from the humble LM317 (1.5A) up to the LT1085 (7.5A), all with built-in current and thermal limiting. Kits and preassembled regulator modules are commonplace. By contrast, a high-voltage bench supply is a specialist item that we will probably have to build from scratch.

This chapter will cover various adjustable HT regulator circuits that could, in theory, be put in a box with a suitable transformer to form a bench power supply. But the handicap is always dissipation in the pass device. The lower the output voltage is set, the more voltage is dropped across the pass device. This forces us to make a design decision. Do we perhaps keep the same maximum DC current limit at all settings, and buy a suitably enormous heatsink to cope with it when the output voltage happens to be set low? Or do we settle for a smaller heatsink and make the current limit shrink as the voltage is lowered? Or perhaps we compromise on voltage, making it variable over a less ambitious range? Or something else?

In a linear bench power supply, the traditional way to ease this problem is to switch the *input* voltage to the regulator, to suit the output voltage. In this way, the voltage across the pass device need not be so extreme when the output voltage is set to a low value. On older power supplies the range switching was done manually on the front panel. Instead of one knob that goes from 0V to 500V, say, it was more manageable to have one knob and a rotary switch or radio buttons that select 0V-100V, 100V-200V… or whatever. The regulator configuration, and corresponding input voltage, are selected at the same time with a ganged switch. In this way there is no need for anything 'smart'; it is all hard wired. The disadvantage with this approach is that an output short-circuit on the highest range will still leave the full voltage across the pass device, so an ordinary fuse would normally be included in series with the output. That is, a fuse that must be replaced whenever a mistake was made.

These days we can do the input-range switching automatically by monitoring the output voltage with some sort of comparator, and switching to the correct input range as necessary. The comparator circuit must include some form of hysteresis, otherwise the switch would chatter back and forth when the output was sitting right on the threshold. Traditionally, the actual switching would be done with a mechanical switch or relay. A relay needs an additional low-voltage power source for the coil, but it is still the simplest way to switch AC and reconfigure transformer windings. If we choose to switch the DC part of the circuit instead, then relays are less ideal, owing to internal arcing, so SCRs or MOSFETs are preferred. Isolation between different parts of a circuit –whether for safety or convenience– is still easily accomplished with optocouplers of one sort or another. Fig. 7.13 shows five common approaches, each with different design tradeoffs.

Fig. 7.13a shows a classic scheme where the rectifier is switched between different voltage tapping points on the transformer. Two settings are shown, but a power supply that has to span a wide voltage range would extend this to several tappings. A possible criticism of this approach (depending on how you look at it) is that the available DC current is the same regardless of the tapping used, being limited by the transformer secondary rating. In other words, switching to the half-voltage tapping does *not* mean we can demand twice the current. This means the transformer VA can be fully exploited at the highest-voltage tap, but will be under-used at the lower voltage tap(s). A similar result could be achieved with an over-wound primary and doing the switching on the wall side. In the US, a standard transformer with a dual primary could be used, switching them between series or parallel. In Europe a special transformer would be needed, or possibly two identical transformers used in tandem.

Fig. 7.13b shows an alternative way to use a centre-tapped transformer. When the switch is closed the whole bridge rectifier is used, the DC voltage being equal to the peak voltage across the entire winding; no current flows in the centre tap at all. Opening the switch effectively removes half the rectifier from the circuit, and current flows from the centre-tap through the extra diode, halving the DC voltage. In either case the full transformer power is available. In other words, in half-voltage mode we can demand twice the current compared to full-voltage mode. We can't expect the user to remember or respect this feature, of course, so it is necessary when switching between ranges to automatically switch in a different current-limiting feature, too.

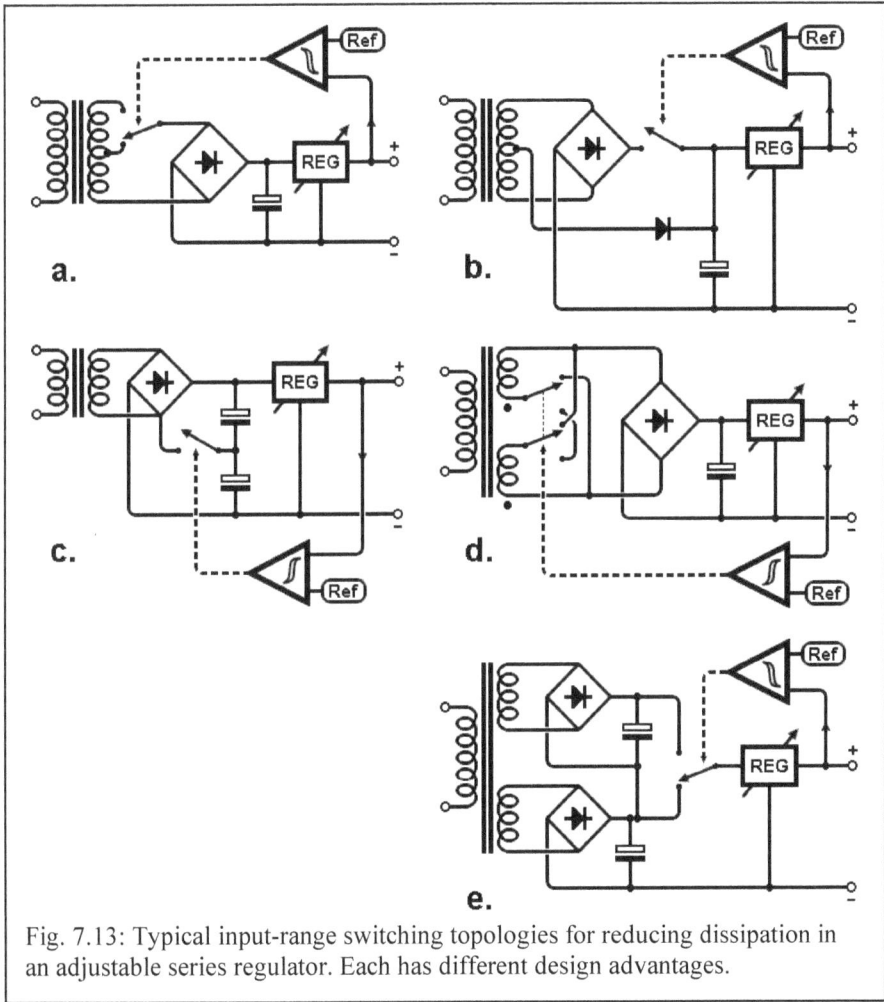

Fig. 7.13: Typical input-range switching topologies for reducing dissipation in an adjustable series regulator. Each has different design advantages.

Fig 7.13c shows a different approach which switches between standard bridge rectifier mode, or full-wave voltage-doubler mode. This economises on the transformer design since it needs only one secondary winding. As previously, the same *power* is available in either mode, which means the available DC current in voltage-doubler mode is half what it is in bridge-rectifier mode. The current-limit should therefore be automatically switched too.

Fig. 7.13d exploits two identical secondary windings, switched between series or parallel. Again, the full transformer power is available in either mode.

Fig. 7.13e uses two (or indeed more) rectifiers, stacked one upon the other. In low-voltage mode only one transformer winding is used while the other sits idle, so the available DC current is the same in either mode. In other words, the transformer VA

181

can be fully exploited in the highest-voltage mode, but will be under-used in the lower-voltage mode(s). A similar topology could be achieved with a single transformer winding, using a capacitor-coupled rectifier stacked upon the main rectifier, but then the available DC current would be halved when switching to the high-voltage mode. Since the switch is now on the pure DC side, MOSFET switches would be a more reliable choice than mechanical relays.

7.4: Three-Terminal Regulator ICs

For regulating low voltages there is a whole world of integrated circuits to enjoy. Among the most common are three-terminal regulators –the bread-and-butter of analog electronics design. Feed a dirty DC voltage to the input and it will give you a nice, well-regulated output voltage, with minimal design-effort required. Nearly all these regulator ICs have built-in short-circuit and thermal shut-down protection, making them virtually fool proof.[*]

Any foray into three-terminal regulators is sure to begin with the old and reliable 78xx (positive voltage) and 79xx (negative voltage) series, produced by many manufacturers.[†] The part number indicates the output voltage, e.g. the 7805 is a +5V regulator, while the 7912 is a −12V regulator. Most versions will handle up to 1A and accept an input voltage up to 35V, but some will handle 1.5A and different (occasionally lower) input voltages. The series quotes about 60 to 70dB of ripple rejection, so 1V of input ripple will emerge as less than 1mV at the output, which is good enough for most audio applications and certainly good enough for heater supplies. Line and load regulation similarly exceed our practical needs. A modern regulator IC can typically hold its output constant within a few millivolts over the full range of expected input voltages and output currents, though this performance does degrade at higher (e.g. audio) frequencies. Application information on this and more is abundant on the internet, so we will cover only the essentials here.

Fig. 7.14 shows a textbook circuit with supporting components. Input capacitor C_1 is a local bypass capacitor, typically about 100nF, mounted close to the regulator. It is only essential if the regulator is more than a few inches away from the reservoir capacitor. Output capacitor C_2 is usually a few tens or hundreds of microfarads. This part is not usually essential for positive regulators, but it does no harm and is thrown in for good measure to improve load transient

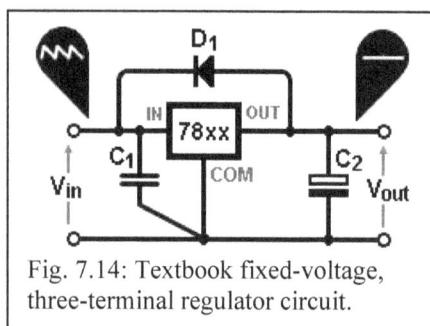

Fig. 7.14: Textbook fixed-voltage, three-terminal regulator circuit.

---

[*] But as Douglas Adams noted, never underestimate the ingenuity of fools.
[†] Various letters are added to the part number to indicate different ratings or origin, such as LM7805, MC7805, KA7805 etc. Usually the differences are subtle and of little significance to the average user, but always check the datasheet.

182

response. However, negative voltage regulators do need this capacitor for stability. An ordinary electrolytic or tantalum should be used here as the ESR is actually beneficial in this respect. D_1 is a rectifier diode which protects the regulator from reverse voltage if, at switch-off, the input voltage happens to fall faster than the voltage stored on C_2.

Like any amplifier, a regulator requires at least some voltage across itself between input and output in order to function properly. If the input voltage drops too low then the device will not be able to maintain the correct output voltage, and it drops out of regulation. The drop-out voltage of the 78xx series it is about 2V. In other words, the input voltage must always be at least 2V higher than the output voltage, and ≥3V is preferred for reliability.

The differential voltage is an unfortunate necessity that accounts for most of the power dissipation in the regulator, i.e. the voltage across the device multiplied by the current through it. Thus if a regulator has a dropout voltage of 2V and we want it to supply 1A, then at the very least it is going to have to dissipate 2W and will need a small heatsink (section 11.7.6). But the chances are that it will be supplied from a simple rectifier, so the average input voltage will actually need to be somewhat higher so the ripple valleys don't swing too low and cause momentary drop out. And we will probably want to supply *even more* voltage to accommodate wall voltage variations. What we anticipated as 2W dissipation is now several times greater, and the necessary heatsink has grown in size and cost. This problem can (sometimes) be minimised by using low drop-out or switching regulators.

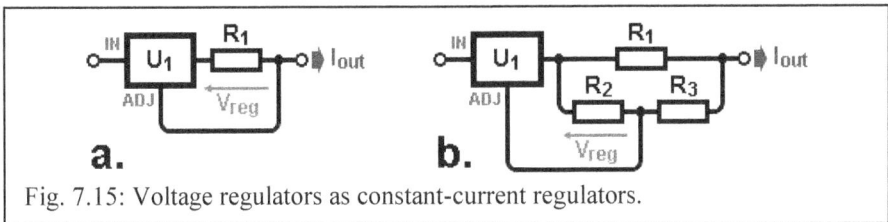

Fig. 7.15: Voltage regulators as constant-current regulators.

A three-terminal voltage regulator can also be configured as a constant-current source, which is useful for series heater supplies among other things. In fig. 7.15a the output current is simply V_{reg}/R_1, and the resistor will have to dissipate I^2R watts. However, at high currents this arrangement can lead to an awkwardly small and precise value for R_1, not easily made using power resistors from the E6 range. A work-around is the arrangement in fig. 7.15b. A convenient power resistor somewhat larger than V_{reg}/I_o can now be chosen for R_1 (e.g. 10Ω), while R_2 and R_3 are 'programming resistors' that set the output current. R_2 can also be chosen to be a convenient value such as 1kΩ, leaving R_3 as:

$$R_3 = \frac{I_{out}R_1R_2}{V_{reg}} - R_1 - R_2 \qquad (7.11)$$

R_3 could be replaced with a pot of course, allowing the output current to be adjusted exactly.

183

7.4.1: Adjustable Regulator ICs

A three-terminal regulator strives to maintain a constant voltage between its output pin and common pin, so if the common pin is raised or lowered by some externally-applied voltage, the output voltage will track the change. For example, an old trick for powering 12.6V heaters is to use a 7812 to get the first twelve volts, then add a diode in series with the common pin, as shown in fig. 7.16. About 5mA of idle current

Fig. 7.16: Increasing the output voltage by the voltage drop across D_2.

flows out of the common pin and into the diode, which is enough to ensure the diode drop is about 0.6V, so the output will be 'jacked up' to 12.6V (but as explained in chapter 9 it is also fine to power 12.6V heaters with 12V). Similarly, a 7805 can be jacked up using two diodes to bring the output close to 6.3V. The common pin may also need to be bypassed with a few tens of microfarads to prevent oscillation with capacitive loads.

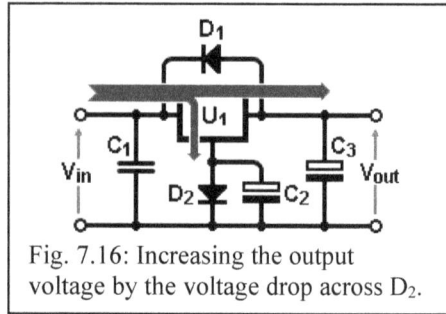

So-called adjustable regulators are really just 1.25V fixed regulators. What makes them 'adjustable' is the fact that the idle current flowing out of the common or 'adjust' pin is only a few microamps and more tightly controlled than in ordinary fixed regulators like the 78xx series. The output voltage can therefore be set accurately anywhere between 1.25V and some maximum limit, using resistors.

Probably the most popular adjustable regulators are the 317 and its negative-voltage twin the 337. These devices are rated for 40V maximum input, 1.5A current, and even better ripple rejection than the 78xx series. Somewhat annoyingly, they have a different pinout from the 78xx series. Close cousins of these devices are the 350 (positive) and 333 (negative) 3A regulators, and the 338 (positive) 5A regulator.

Fig. 7.17 shows the quintessential adjustable regulator circuit. The IC will maintain 1.25V between its output and adjust pins, that is, across R_1. A constant current of $1.25/R_1$ therefore flows down the potential divider, and the default 1.25V output will be jacked up by the resulting voltage across R_2. Assuming the adjust-pin current is negligible, the voltage across R_2 will be $R_2 \times 1.25/R_1$, so the total output voltage will be:

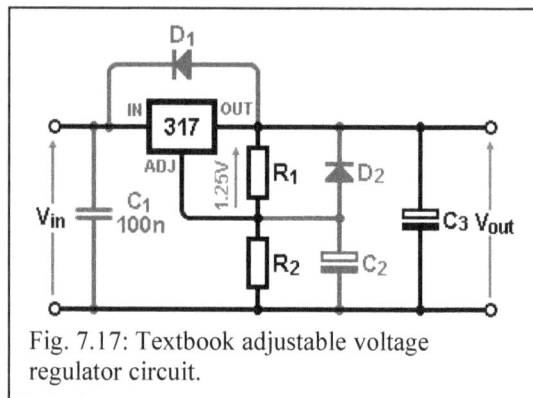

Fig. 7.17: Textbook adjustable voltage regulator circuit.

$$V_{out} = 1.25\left(1 + \frac{R_2}{R_1}\right) \qquad (7.12)$$

Obviously, if R_2 is made adjustable then we get an adjustable-voltage regulator.

Output capacitor C_3 is normally included to improve load transient response. Moreover, for negative-voltage regulators an output capacitor is mandatory for stability, so check the datasheet. D_1 protects the regulator from reverse voltage at switch off. C_2 can also be added to shunt any broadband noise generated by R_2, which will otherwise appear at the regulator output. This will generally be a few tens of microfarads, and D_2 then becomes necessary to protect the out/adj junction from reverse bias when this capacitor discharges at switch off.

The IC requires some load current to flow at all times to maintain good regulation; if the load current is too small then the output voltage will rise above the desired value. For the 317 the minimum requirement is 3.5mA. Therefore, R_1 and R_2 are often chosen so they always drain this minimum current, though this is not essential if an additional load is always present, or if you simply don't care what happens at very small load currents.

7.4.2: Low-Dropout Regulators

Power dissipation in a regulator is the annoyance that turns what should be a cheap and simple circuit into a practical burden, because it demands heatsinking. As their name suggests, low-dropout regulators (LDOs) will function with much less voltage across themselves than ordinary regulators, so the input voltage can be made lower and dissipation thereby reduced. Apart from this they are used in exactly the same way as ordinary regulator ICs, though internally they are rather different.

The most common adjustable LDOs are the 1084 (5A), 1085 (3A), and 1086 (1.5A), which will withstand up to 29V input and have a dropout voltage of 1.5V. All LDOs require an output bypass capacitor for stability, so pay attention to the datasheet recommendations.

7.4.3: High-Voltage Regulator ICs

In the world of regulator ICs, 'high voltage' means anything over about 20V. For example, the ambitiously named 317HV (now obsolete) would accept 50V input, impressing no-one from the world of valves. Slightly more respectable is the TL783 rated for 125V, 700mA. It has the disadvantages that it requires at least 15mA load to maintain good regulation, and has a dropout voltage of about 5V, but it remains eminently useful for series heater chains. Above this the choices all but disappear. The VB408 and HIP5600 would accept $400V_{dc}$ or even $280V_{ac}$, and deliver about 30mA, but they are sadly obsolete. In current production, however, is the LR8.

The LR8 is a three-terminal regulator manufactured by Microchip. It is rated for up to 450V input voltage and has a respectable drop-out voltage of about 10V, meaning the output voltage can be set anywhere from 1.25V up to 440V. Unlike the LM317

185

commonly used in the Maida regulator (section 7.7.5), the LR8 has an adjust-pin current of only 10μA typical, and the minimum output current for stable regulation is only 500μA. This means little or no tweaking of resistance or loading is required to hit a specific output voltage. Like most regulator ICs it has internal thermal and overcurrent shutdown. However, the maximum output current is only 20mA typical, and it's TO-92 package is not easy to heat sink, all of which limits its usefulness.

Fig. 7.18 shows a typical application circuit. It looks like any adjustable regulator but for one subtle feature: since this is a high-voltage circuit, D_1 is needed to prevent the output capacitor (at least 1μF is required for stability) from discharging into the pot if its resistance is suddenly turned down, as this can otherwise burn out an ordinary 500mW carbon track. For the same

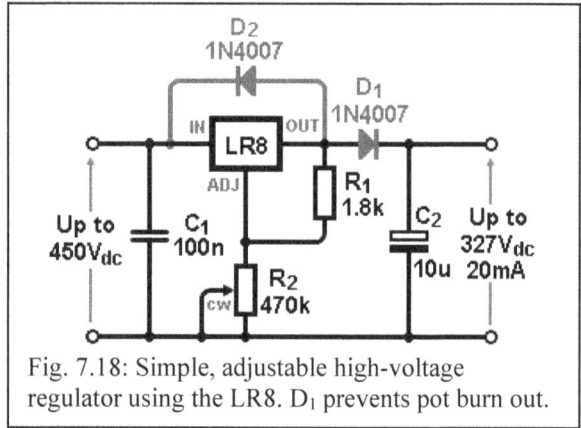

Fig. 7.18: Simple, adjustable high-voltage regulator using the LR8. D_1 prevents pot burn out.

reason, R_2 should not be bypassed with anything larger than about 100nF. Obviously if R_2 is a fixed resistor then neither of these things is an issue.

7.5: Switching Regulators and Converters

The ultimate way to minimise power dissipation is to use a switching regulator circuit. Moreover, switching circuits can convert practically any input voltage into any desired output voltage –whether higher (boost) or lower (buck). The price we pay for this convenience, as with any SMPS, is a more involved design process, and possible switching noise. The former can be avoided by using ready-made circuits and modules, which are getting cheaper and more abundant by the day. The latter may be overcome with plenty of output filtering, shielding, and possibly a linear post-regulator (e.g. section 9.4.5), but be prepared for plenty of manual tweaking if total silence is your aim.

A common example of a low-voltage buck regulator is the LM2596 which, unlike most switching regulators, is available in a non-surface mount package (5-pin TO-220). The adjustable version (LM2596-ADJ) can provide 1.25 to 37V at up to 3A and will accept up to 45V input voltage. Fig. 7.19 shows a practical circuit using this device; all we need to provide is a Schottky diode, inductor, and a couple of resistors and capacitors. It is essential to consult the datasheet for specific application information as the manufacturer goes to great trouble to explain the proper choice of external components and layout. In particular, the inductor must have a DC current rating greater than the load current. In many situations it is more cost-effective to buy a complete, pre-assembled converter module.

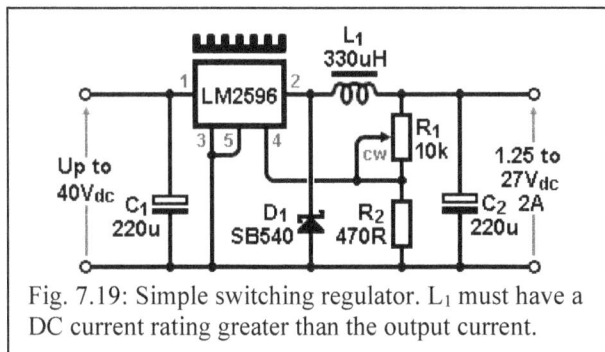

Fig. 7.19: Simple switching regulator. L_1 must have a DC current rating greater than the output current.

The switching frequency of fig. 7.19 varies from about 70kHz under light loading to about 30kHz at full load. This is high enough that further LC filtering is effective using ferrite or even air-core inductors. The input voltage must be at least 1V higher than the output, and the smaller the differential voltage, the better the efficiency. The author's test circuit varied from 80% to over 90% efficient at all load currents. The regulator may therefore have to dissipate up to $P_{out} \times (1-0.8)$ watts, so a heatsink is advisable for outputs greater than about 5W. The near-constant efficiency is a wonderful feature not just because it means less wasted heat but because it makes the input voltage non-critical; we can supply whatever input voltage is convenient.

Simple boost (step-up) regulator ICs are also available of course –such as the LM2577– but are usually limited to about 60V output. Ready-made, high-voltage boost converter modules that may be suitable as HT supplies are also available, and some of the smaller ones are very cheap. This is an area that is still evolving. Popular solutions should emerge as valve enthusiasts explore the territory, perhaps making this book obsolete rather quickly! But a common difficulty with these options is noise coupled into the audio circuit by the switching waveforms, either through stray capacitance or mutual inductance with the switching inductor/transformer. This is normally tackled by adding further filter stages on the output, or a linear post-regulator, and by playing with the physical orientation of inductive components. This can be a nightmarish problem to chase down in a very high-gain environment like a microphone preamp.

7.5.1: 555 Boost Converter

For low-current projects, especially guitar effects pedals and nixie clocks, the boost regulator in fig. 7.20 –and variations on it[*]– has become quite popular. It uses the venerable 555 timer IC configured for pulse-position modulation. Q_1 should be a suitably high-voltage MOSFET with $R_{DS(on)}$ below 1Ω to minimise dissipation; the author used an IPA60R380E6. If only a couple of milliamps of load current is needed, such as when driving one or two 12AX7s, a heatsink is not normally required for the MOSFET. For higher currents a clip-on heatsink is sufficient, or else it can be bolted to the chassis. Q_2 can be any general purpose NPN. D_1 must be a fast-switching diode of course; a junkbox 1N4007 will not work.

---

[*] See also: Lynch S. (2013) HV809 EL Lamp Driver for Battery Powered and Off-line Equipment, *Supertex Inc. Application Note AN-H36.*

This simple circuit tends to work best when the no-load output voltage is adjusted to 300V, but do not expect outstanding performance. The regulation is quite poor, especially with low input voltages, owing to the limited gain in the feedback loop. For example, in the author's test circuit with 9V input and the no-load output adjusted to 300V, the voltage sagged to 274V with 12mA load (an apparent output resistance of over 2kΩ). In other words, the circuit is best suited to applications where the input and output voltages, and load, are fairly constant. This is of course true for most preamp circuits. The switching frequency is in the region of 50kHz with no load, falling to around 25kHz at a practical maximum output current of 15mA. The efficiency is generally between 40% and 60%. A negative bias supply could also be generated from the same circuit using the principle shown earlier in fig. 3.3.

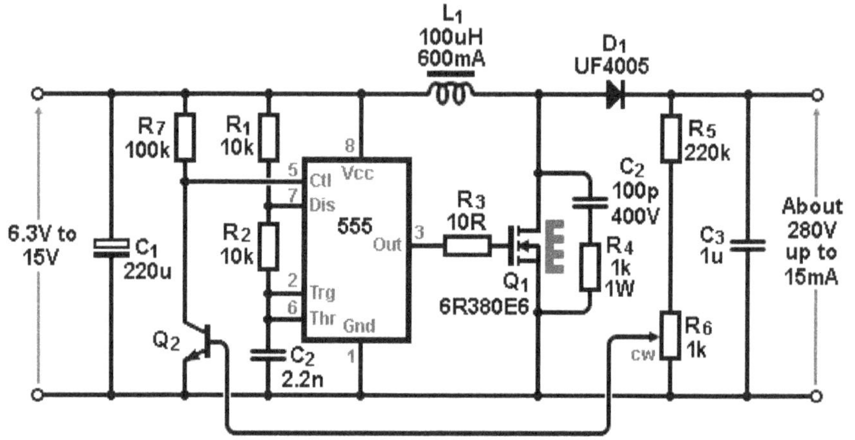

Fig. 7.20: High-voltage boost regulator suitable for small valve projects.

7.5.2: 40106 Voltage Multipliers

Fig. 7.21 shows a circuit[2] which is useful for generating a modest HT voltage for low-current effects circuits and indicator tubes, or for generating microphone phantom power from a low-voltage rail. It may also be useful for beginners who wish to experiment with small valve circuits in a safe way, since it is completely shock free. It is a switched-capacitor voltage multiplier using a common 40106B hex inverter (Schmitt trigger). With ceramic capacitors the circuit can be built very small indeed, even with through-hole components.

The first inverter is a simple square-wave oscillator which runs at about 130kHz. When the inverter output switches low, C_2 is able to charge up to V_{in} through D_1. When it switches high, the whole capacitor is 'jacked up' by an additional V_{in}, and dumps its charge through D_2 into C_3 (since the subsequent inverter output will be low during this time), meaning C_3 charges up to $2 \times V_{in}$. C_3 will in turn be jacked up

---

[2] Holmes, R. (1983). Designer's Notebook, *ETI*, April, pp223-6.

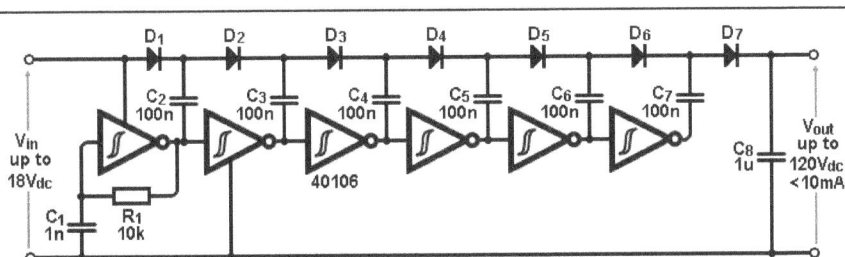

Fig. 7.21: Simple switched-capacitor voltage multiplier using a 40106B IC. All diodes can be 1N4148, C_{2-8} should be rated for the output voltage.

to $3 \times V_{in}$ on the next switching cycle, and the process repeats for each stage, bucket-brigade fashion. The unloaded output voltage will therefore be equal to $(N+1)V_{in}$, where N is the number of inverter stages, ignoring losses. C_8 provides final smoothing.

The switching frequency and output impedance will vary somewhat, depending on the input voltage. Using a HEF40106B with a 15V supply, the idle supply current was 2.3mA and the unloaded output voltage was found to be 102V. With a 10kΩ load the voltage sagged to 79V and the supply current increased to 47mA –an overall efficiency of 88%. The apparent output resistance was 2.9kΩ which might sound large, but it hardly matters for a typical valve effects pedal which needs a couple of milliamps at most. Performance may be slightly better with a CD40106B owing to that IC's larger output current limit, but the author has not tested this.

Fig. 7.22 shows another classic implementation. Again, one inverter is used as a simple square-wave oscillator, but the remaining inverters are now connected in parallel and used to drive a simple ladder multiplier. As drawn, the switching frequency is about 65kHz and the multiplier will approximately triple the supply

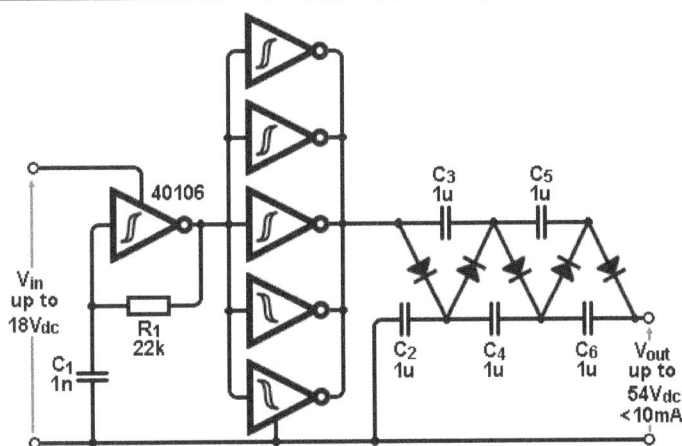

Fig. 7.22: Simple voltage multiplier using a 40106B IC. All diodes can be 1N4148, all capacitors should be rated for the input voltage.

189

voltage, though more stages could be added. Using a HEF40106B with a 15V supply, the idle supply current was 1.4mA and the unloaded output voltage was found to be 47V. The output voltage sagged down to 41V with a 10kΩ load, and the supply current increased to 13mA –an overall efficiency of 84%.

7.6: HT Voltage Reducers

Sometimes we do not need a regulated voltage supply, but we do want to be able to adjust the voltage over a range. This might be for reasons of experimentation, or for power reduction in a guitar amp, as covered in chapter 10.

7.6.1: Voltage Follower

A voltage follower offers a simple way to control a DC voltage, illustrated in fig. 7.23. This concept is of course related to the capacitor multiplier and Zener follower (section 7.7.1). The pot provides an adjustable reference voltage to the base of the follower/pass-device which acts as a simple buffer. It dumbly copies the pot voltage and provides a low output impedance, isolating the pot from the load. The output voltage is simply an attenuated version of the input voltage, including any ripple, which is attenuated by the same proportion. When using a transistor the output voltage will be slightly lower than the pot wiper voltage, whereas in the case of a valve it will be slightly higher, owing to bias voltage. A MOSFET is the logical choice for a high voltage environment. Note that if the pot is turned fully up the base will be shorted to the collector and the pass device will behave as a simple diode; this is a useful feature for power reduction in a guitar amp since it gives us access to normal, unadulterated operation when the pot is at maximum. The pass device may be called upon to dissipate a lot of power depending on how much the voltage has been reduced. Assuming the load resistance is constant, the worst-case dissipation occurs when the output voltage is set to half the maximum, at which point dissipation in the pass device will be equal to dissipation in the load.

The pass device does not necessarily need to work in the pure DC environment as in fig. 7.24a, but can be placed before the reservoir capacitor, effectively creating a

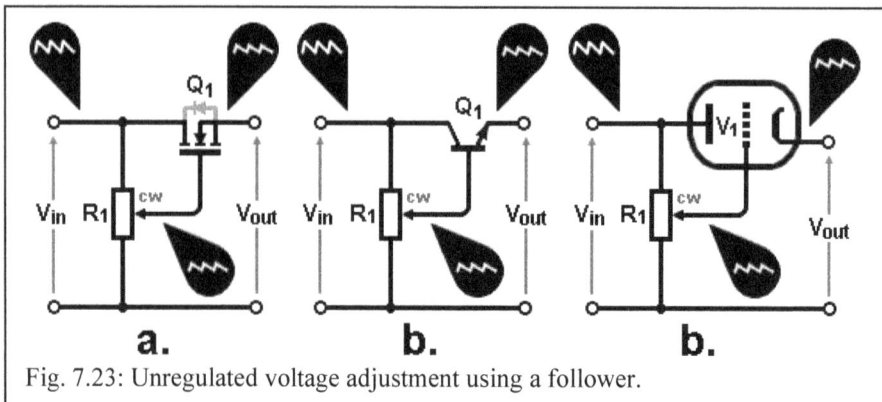

Fig. 7.23: Unregulated voltage adjustment using a follower.

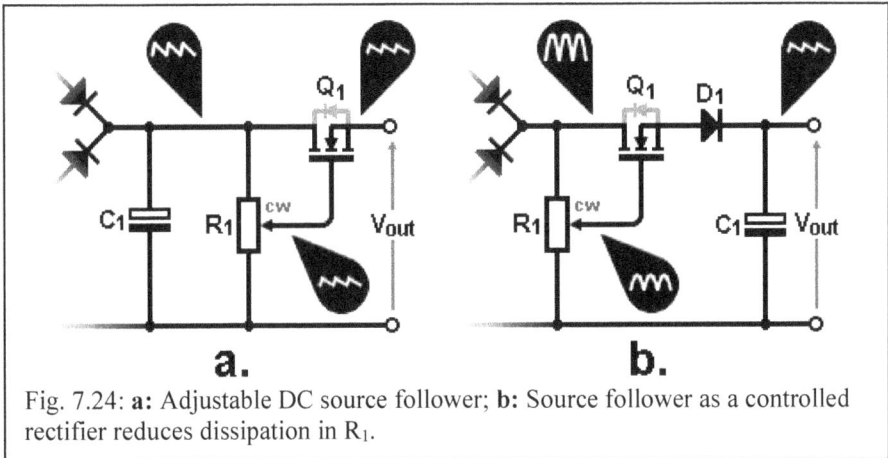

Fig. 7.24: **a:** Adjustable DC source follower; **b:** Source follower as a controlled rectifier reduces dissipation in R_1.

controlled rectifier as in b.[3,4,5] D_1 prevents C_1 from feeding back through the body diode of Q_1, which would otherwise defeat the controlled rectifier action. In this configuration the pass device must handle large peak ripple currents, but less differential voltage, so average dissipation in the pass device is practically the same in either configuration. Some transistors are specifically advertised for pulsed operation, although whether they are really any more reliable this way is not clear. The main advantage of fig. 7.24b is not dissipation in the pass device, but dissipation in the potentiometer. Instead of being exposed to the full DC voltage, the pot only sees rectified pulses which have an average value of $2V_{pk}/\pi$. From $P = V^2/R$ this means dissipation in the pot is 0.63 times lower than in fig. 7.24a. Additionally, any current limiting circuit added to fig. 7.24b will not add significantly to the overall power-supply output impedance.

Whichever arrangement is used, gate-source voltage protection and current limiting are needed. The latter should be set so that it protects the MOSFET from inrush current at switch on, but not so low that it clamps normal ripple current, or dissipation in the MOSFET will increase considerably. Setting the limit to be ten times greater than the

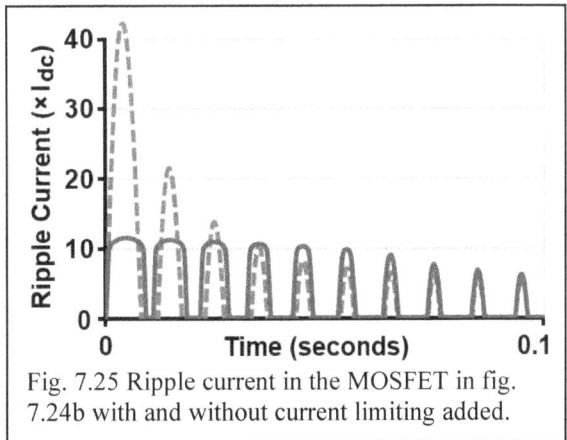

Fig. 7.25 Ripple current in the MOSFET in fig. 7.24b with and without current limiting added.

[3] Walker, A. (1952). Variable H.T. Power Pack, *Wirelsss World*, Sept, pp374-6.
[4] Banthorpe, C. H. (1956). Simple Adjustable Voltage Supplies, *Practical Wireless*, August, p380.
[5] Brandon, D. A. (1962). A Versatile Low-Current Variable Supply, *Practical Wireless*, July, pp237-8, 241.

maximum expected DC load current should cover nearly all situations; choose a MOSFET with a pulsed drain current rating that exceeds this. Fig. 7.25 shows the expected current waveforms at switch-on for the circuit arrangement in fig. 7.24b. Without any current limiting (dashed) there is a very large inrush, limited only by the transformer winding resistance. The solid trace shows the effect of including a simple current limit set to ten times the expected DC load current. This quashes the inrush current pulses but has no further effect during normal running.

Fig. 7.26 shows a practical circuit suitable for supplies up to 1kV. Standard 0.5W carbon pots are normally restricted to situations where the voltage across the pot is less than 300V, so for higher voltage supplies it will be necessary to use a high-power pot, even though actual dissipation may not be an issue. All other resistors only see brief dissipation spikes at

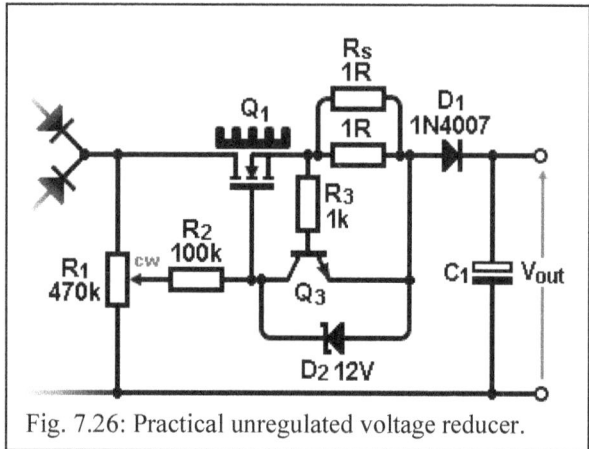

Fig. 7.26: Practical unregulated voltage reducer.

switch on. Note that the maximum available load current from this circuit is the same at any voltage setting; whatever the transformer can safely provide at the maximum setting, we can draw the same current at any other setting too, without overloading the transformer.

7.6.2: Conduction-Angle Regulator

A different way to control a DC voltage is to go back to the source and control the AC supply instead, or the murky world between pure AC and pure DC. With some sort of controlled-switch we can chop-up the AC so that current flows into the reservoir capacitor for a shorter slice of time, so it charges up to some less-than-maximum voltage. In other words, we control the conduction angle. This can be a highly efficient way to control the DC output voltage, since the control device need not burn off heat all the time; it can alternate between being fully off (no dissipation) or fully saturated (minimal voltage drop so minimal dissipation). There are many ways such control can be envisaged. In times past, triodes were used as controlled rectifiers. Today we have the solid-state alternatives of opto-couplers, triacs, silicon-controlled rectifiers (SCRs), and of course our most useful servant the MOSFET.

Fig. 7.27 shows an example; the output voltage is fed through R_3 to the gate of the MOSFET Q_1. As long as it is above the MOSFETs' threshold voltage, Q_1 will turn on and allow the reservoir capacitor C_1 to charge. At the same time, the output voltage is supplied through a potential divider R_1-R_2 to an ordinary NPN transistor

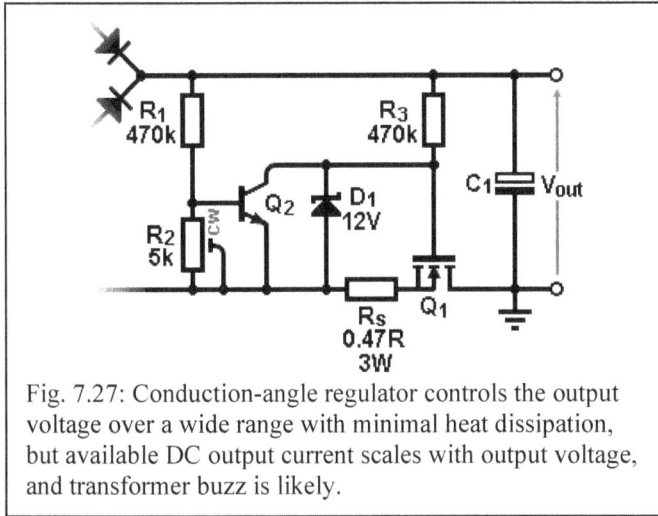

Fig. 7.27: Conduction-angle regulator controls the output voltage over a wide range with minimal heat dissipation, but available DC output current scales with output voltage, and transformer buzz is likely.

Q_2. When the voltage across R_2 reaches about 0.6V, Q_2 will turn on and shunt the MOSFET's gate to source, turning it off. This halts the charging of C_1, stopping the output voltage from rising any further. As C_1 discharges, eventually Q_2 will turn off, allowing Q_2 to turn on and top-up C_1 again, and so on. Use a MOSFET whose $R_{DS(on)}$ is less than 0.5Ω to minimise dissipation. D_1 protects both the MOSFET gate and Q_2 from excessive voltage, and in combination with R_s also limits peak inrush current. By making R_2 adjustable, the output voltage is adjustable. When R_2 is turned down to 0Ω, the circuit is completely unregulated and delivers full output, like any simple rectified supply. As the resistance is turned up, the peak output voltage falls to:

$$V_o = V_{be} \frac{R_1 + R_2}{R_2} \tag{7.13}$$

In this case it is variable down to about 60V. In a typical HT supply where the load current is a good deal less than 500mA, the dissipation in the MOSFET will normally be less than 1W, meaning it will be happy with a clip-on heatsink or maybe no heatsink at all.

Note that this circuit provides line regulation but not load regulation. The output still has ripple, and it is the maximum peak of this ripple that is being regulated. The *average* output voltage will still sag under load much like an ordinary unregulated supply. But if all you need is an HT supply that can be 'turned up and down' then this circuit has appealing economy. If a cleaner DC supply is needed, we could bolt a capacitor-multiplier on the output, which also needs little or no heatsinking since it 'tracks' its input voltage.

So why do we not see this solution used more often? There are two main reasons. The first is that with this approach the available output current is roughly proportional to the output voltage. This is because, as we reduce the conduction angle, we degrade the power factor. Ultimately, the safe DC current limit is imposed by the power transformer. In other words, when the output voltage is set to maximum (unregulated), the DC output current can be whatever the transformer can safely provide, in the same way as for any unregulated power supply. But when the

193

voltage is adjusted down to half the maximum, say, then the DC output current can likewise only be half the maximum value; trying to draw more DC current will overload the transformer. This is unlike the follower-type voltage reducer which permitted the same output current at any setting. The second reason is noise. This circuit will radiate sharp EM noise from the ugly ripple current and may cause the transformer to buzz loudly at lower output settings, due to the narrow conduction angle. Nevertheless, the circuit may still be useful as a clamp or pre-regulator, to make the job of a subsequent regulator easier. But always remember, as you reduce the voltage, the safe available DC current is reduced too.

7.6.3: Optical HT Control

Fig. 7.28 shows how a pass device can be controlled using a phototransistor optocoupler. R_g pulls the gate of the MOSFET up and therefore fully on, but if the transistor is allowed to conduct it pinches the gate and source together, choking off the MOSFET's conduction. Controlling the output voltage is therefore simple matter of lighting the LED inside the optocupler. Since the high-voltage part of the circuit is galvanically

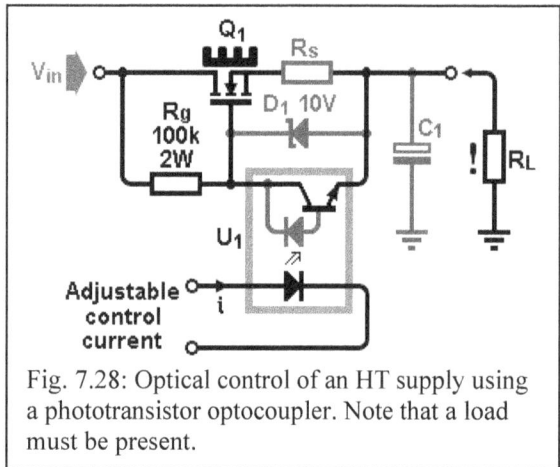

Fig. 7.28: Optical control of an HT supply using a phototransistor optocoupler. Note that a load must be present.

isolated from the control section, a low-voltage circuit such as a microprocessor can safely control the high voltage. The output HT voltage is directly proportional to the LED current, i, so if voltage-control is wanted then a suitable V-to-I converter must be devised. In theory, this topology could be used to build a feedback regulator,[6] but the author's attempts met with stability problems so it is recommended here only for open-loop control. Since the circuit works by pulling current though R_g, that current must have somewhere to go. In other words, a DC load must always be present or the output voltage will remain stuck at maximum. The circuit works like a potential divider where the upper arm is optically controlled, and the lower arm is formed by the load, just like fig. 7.1a. This means the output voltage cannot be reduced below:

$$V_{o(min)} = V_{in} \frac{R_L}{R_L + R_g} \qquad (7.14)$$

Some optocouplers such as the 6N136 contain a photodiode which should be connected as shown, but otherwise leave the transistor base unconnected. R_s can be selected to provide a current limit in the usual way. A typical output smoothing

---

[6]https://www.edn.com/use-an-optocoupler-to-makea-simple-low-dropout-regulator/

capacitor, C_1, is recommended to prevent oscillation (without it the circuit can become quite a good longwave transmitter). Note that the current fed into the LED will be greater than the current needed to flow in R_g, because optocouplers have a current gain less than unity. This is normally quoted on data sheets as the current-transfer ratio (CTR), and is normally expressed as a percentage. This might typically be 30%, meaning whatever current we shove into the LED, the phototransistor will allow up to 30% as much to flow in itself.

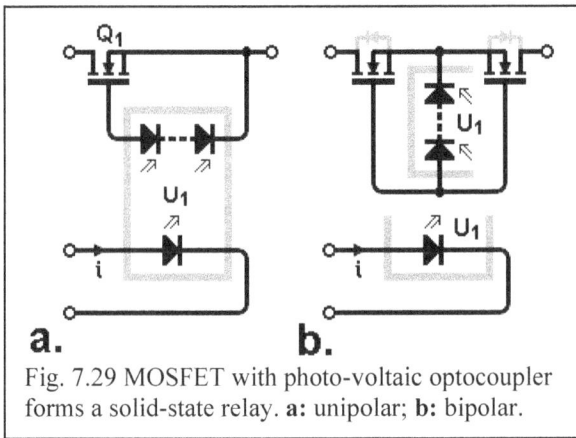

a.

b.

Fig. 7.29 MOSFET with photo-voltaic optocoupler forms a solid-state relay. a: unipolar; b: bipolar.

Fig. 7.29 shows the ultimate way to control a MOSFET optically, using a particular type of optocoupler called a Photo-Voltaic Isolator, or PVI. The LED inside the PVI shines light onto an array of photodiodes which generate a (high impedance) DC voltage across themselves – enough to turn any MOSFET on. The MOSFET combined with the PVI creates a solid-state relay which can be switched from completely off, to fully saturated, just by lighting the LED. The fact that full saturation is achievable means high voltages or currents can be switched on and off with negligible loss; only the $R_{DS(on)}$ of the MOSFET introduces any voltage drop. With two MOSFETs as in fig. 7.25b, AC signals can be switched, with current flowing alternately through the channel of one MOSFET and the body diode of the other.

PVIs are not a common item, and through-hole versions can be fairly expensive, but they do make high-voltage switching very easy. Example devices include the VOM127, TLP190B TLP590B, TLP591B, TLP3906, and PVI1050.

PVIs can also be used to turn the HT voltage up and down, not just on and off, but the conduction of the MOSFET is very sensitive to the LED current, which might typically range between 1.5mA (MOSFET basically off) to 1.6mA (MOSFET fully on), a mere 0.1mA difference. Therefore, linear control normally requires some form of overall negative feedback to prevent wild drifts with temperature –section 7.7.7.

7.7: HT Series Regulators

Having looked at some unregulated methods of voltage control we can move on to proper HT stabiliser and regulators, which attempt to maintain a more constant output voltage in the face of changing wall voltage and load current.

7.7.1: Zener Follower

The Zener follower is a sibling of the capacitor multiplier and voltage-follower reducer. As shown in fig. 7.30, it consists of a follower whose base voltage is stabilised by a Zener diode or gas reference tube, or any other fixed voltage reference. The follower is simply a buffer, dumbly copying the reference voltage and isolating the reference from the load. Unlike its simpler siblings, the Zener follower delivers a fixed output voltage, inasmuch as the reference voltage is fixed and the pass device is able to follow it. The output voltage is not exactly equal to the reference of course, since the V_{GS}, V_{be}, or V_{gk} of the pass device is subtracted from it. In the case of a valve this means the output voltage will be a little *higher* than the reference voltage.

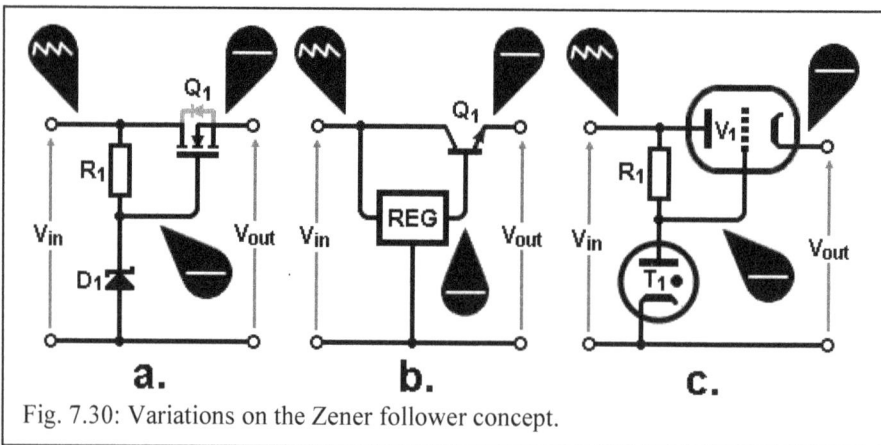

Fig. 7.30: Variations on the Zener follower concept.

The line regulation of the Zener follower is dependent on two things; the stability of the reference device, and the natural PSRR of the follower. Variations in input voltage which cause the current flowing down through R_1 and D_1 to vary, will in turn vary the reference voltage owing to the internal dynamic resistance of the Zener or gas tube. The pass device will of course slavishly follow any such variation. This source of error can be eliminated by replacing R_1 with a constant-current source, or by using a superior reference device like an LR8 regulator IC. However, we cannot change the intrinsic PSRR of the follower. This is not likely a concern when using a MOSFET, but may be rather poor for a power triode, since it is roughly equal to its μ. In other words, if the triode has a μ of 10, a 10V variation input voltage will cause the output voltage to vary by about 1V, in addition to any variation in the reference voltage.

196

Load regulation depends on the output impedance of the circuit. This is potentially very small for a MOSFET, but a current-limiting resistance will often be required, which will add to the figure. The output impedance of a valve cathode follower is of course larger than for a transistor.

A final source of output voltage drift is the tempco of the reference device. The voltage across a Zener diode (strictly an avalanche diode) will increase by perhaps 2 to 3% for a 30°C rise in temperature, though this is a minimal concern once everything has reached a stable running temperature.

The Zener follower may lack the precision of a 'proper' feedback regulator, but it has the advantage of being conceptually simple, well-behaved, versatile and cheap. Its broadband output noise can also be made unusually low. If you only need the HT supply to be 'fairly' well regulated but not 'extremely' well regulated then the Zener follower is the topology of choice.

7.7.2: Classic Valve Regulator

Whole books were written on the design of valve regulators, each presenting ever more advanced and complicated refinements. Academic shots were fired over whether a cascode, pentode, or differential-pair error amplifier was a better choice to achieve the best regulation. Everything was tried to minimise voltage drop across the pass device, from floating screen-grid supplies to cathode followers for driving the grid into the positive grid-current region. It became common to find half-a-dozen different valves in a modest regulator design, with just as many transformer secondary windings needed (look up the Heathkit IP-32 for one of the simpler designs). There isn't enough space to devote to these things here, nor is there much point when valve regulators are so wasteful. We will cover only the classic, two-triode topology illustrated in fig. 7.31. The pass device is a cathode follower, and the error amplifier is an ordinary gain stage. The reference is a gas stabiliser tube, though a Zener would also work of course.

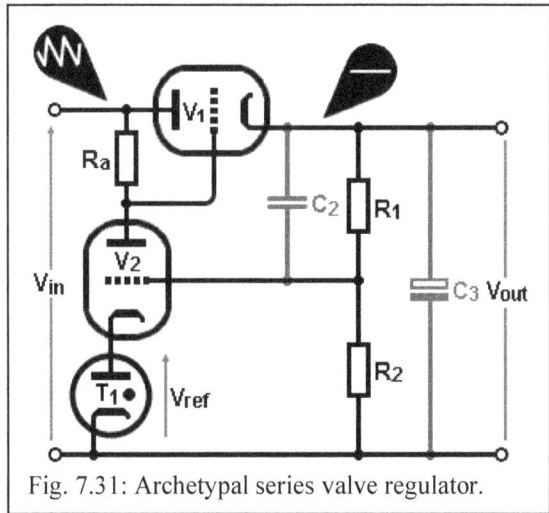

Fig. 7.31: Archetypal series valve regulator.

The classic pass-device for a valve regulator is the 6AS7/6080 –one of the last valves to be specially designed for the task. It contains two 13 W triodes with an extraordinarily small internal anode resistance of 300Ω, and a $V_{hk(max)}$ rating of

300V. However, these properties make this valve far more valuable today for use in output-transformerless amplifiers, so we will use something else. The ECL86/6GW8 is a convenient candidate for a small regulator, as it contains a high-gain triode (error amplifier) and a 9W power pentode (pass device). The PCL86/14GW8 is the 14V-heater version of the same device, which is usually cheaper. In any case, the design procedure can be followed cook-book fashion for any combination of valves. For simplicity we will make the pass device a triode by connecting the screen-grid to the anode.

An early decision we must make is, how much input voltage variation do we expect to cope with? If the nominal rectified voltage is $400V_{dc}$, say, then a 10% wall-voltage variation would cause 40V change. On top of this is transformer regulation which might add yet another 10%. This is getting out of hand. We will assume the wall voltage is fairly reliable and/or the load current is fairly constant and say the input voltage is 400V nominal, ±40V. Now we can examine the grid curves of the pass device to see what it is capable of, as shown in fig. 7.32.

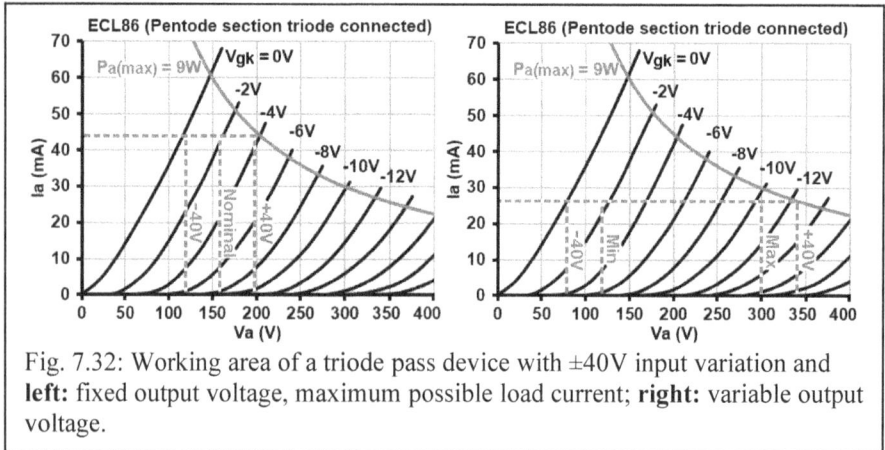

Fig. 7.32: Working area of a triode pass device with ±40V input variation and **left:** fixed output voltage, maximum possible load current; **right:** variable output voltage.

For a triode, the Safe Operating Area (SOA) is roughly triangular, with the $V_{gk} = 0V$ grid curve forming one side, and the anode dissipation limit forming the other (in this figure the maximum anode voltage limit is beyond the right-hand side of the graph). Strictly speaking, since the anode and screen grid are connected together in this scenario, we could add the 1.5W screen dissipation limit to the 9W anode dissipation limit and call it a 10.5W triode overall, but let's not split hairs. The peak of this triangle represents the maximum current we can safely or realistically pull through the device. If our main design goal is to get as much current as possible from this particular valve choice, then we must centre our operation here, as illustrated by the left-hand diagram in fig. 7.32. Allowing for the input voltage to vary by ±40V we are forced to accept a nominal 160V drop across the pass device. Since the nominal input voltage is assumed to be 400V the regulator would be designed for a fixed output voltage of 400−160 = 240V. Under these conditions the available current is up to 44mA. The only way to get more current would be to use a different device, or several in parallel.

On the other hand, maybe we don't need so much load current. If we relaxed our requirement to 28mA, say, then the right-hand graph shows the nominal drop across the pass device could be anywhere from 120V (min) to 300V (max) while still leaving margin for ±40V input variation. With 400V nominal input we could therefore make the output adjustable from 100V to 280V. When the output voltage is set to an intermediate value then more than 28mA is technically available, up to 44mA if the output happens to be set exactly as in the left-hand scenario. We'll proceed with this design.

The voltage across the error-amplifier triode will be equal to the output voltage minus the grid bias of the pass device, minus the reference voltage. If we use a neon lamp as a roughly 65V reference voltage, then with the regulator output at 100V this would leave less than 35V across the error amplifier. This probably won't be practicable using the ECC83-type triode contained in the PCL86. We could use a lower-voltage Zener

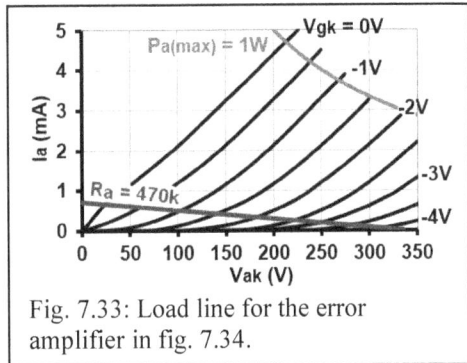

Fig. 7.33: Load line for the error amplifier in fig. 7.34.

diode, but that hardly seems in the spirit of a valve regulator, so we'll see how close we can get with the neon. Assuming a raw input voltage of 400V and a reference voltage of 65V, the effective HT seen by the error amplifier is $400 - 65 = 335$V. Selecting a large 470kΩ anode resistor –both to achieve high gain and to access low anode voltages– produces the load line in fig. 7.33. Ordinarily we might be worried about Miller capacitance or slew rate when using such a large resistance, but that hardly matters in this low-frequency application. All that is left is to add in the feedback divider, a feed resistor for the neon to ensure it starts reliably, good-will grid stoppers, and a grid-cathode protection diode. The completed circuit is shown in fig. 7.34.

The practical output range of this circuit turned out to be 120V to 260V. Any less and there is too little voltage across the error amplifier, any more and the pass device enters the region of grid-current. The regulation performance varies with setting. At the higher end of the range, the output voltage varied by less than 1V over the full range of input voltage and load current, and the peak-to-peak ripple reduction was better than 56dB. At 120V output, where the error amplifier operates at low current and therefore less gain, the output varied by as much as 4V, with about 41dB ripple reduction.

The maximum heater-cathode rating of the ECL86 is 100V. If we elevate the heater supply so that it floats mid-way between the two cathodes, the output can be as high as 260V and $V_{hk(max)}$ will not be exceeded –the heater supply will be almost 100V

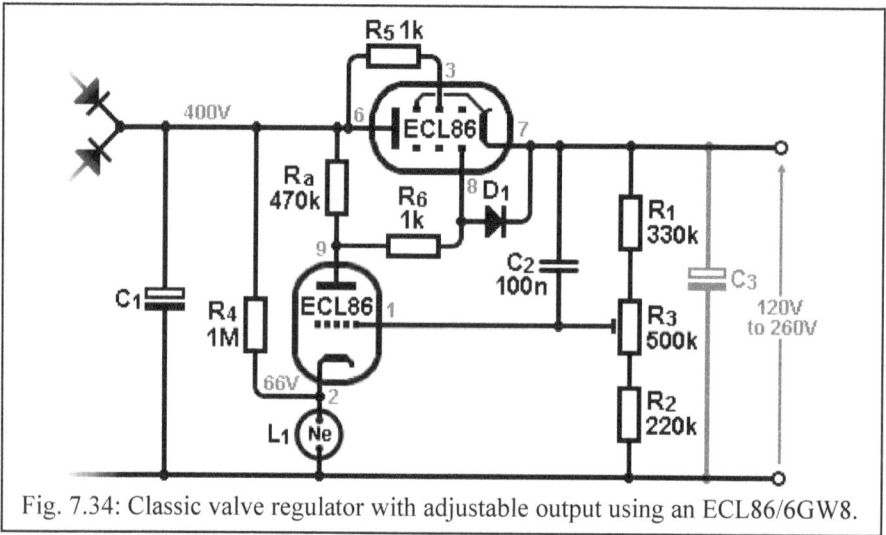

Fig. 7.34: Classic valve regulator with adjustable output using an ECL86/6GW8.

above the lower cathode, and almost 100V below the upper cathode. Fig. 7.35 shows how the heater supply can be elevated so that it always sits mid-way between the two cathodes. With higher-voltage valve regulators, however, the pass device will need to be a separate bottle with its own dedicated heater supply, referenced to its cathode. If the transformer has a spare 5V rectifier winding then this can often be used –if it is under-run then the voltage will usually come out closer to 6V.

Implementing clever short-circuit protection for a valve regulator usually means adding a negative voltage supply. This is because valves are depletion devices, so if the output is shorted to ground, the grid of the pass device must be pulled *negative* to choke off the current.[*] Since a negative supply may be inconvenient, it is common to rely on a simple fuse for

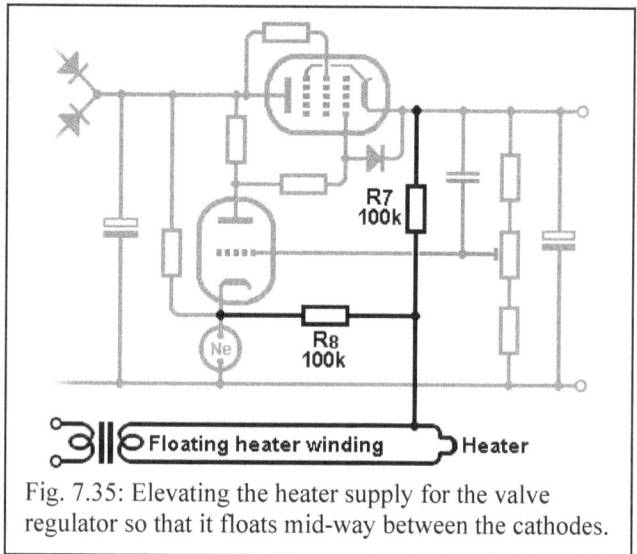

Fig. 7.35: Elevating the heater supply for the valve regulator so that it floats mid-way between the cathodes.

---

[*] For example, see: Michaelis, M. L. (1962). Experimenter's Power Pack, *Practical Wireless*, January, pp820-24.

protection, or a cut-out relay.[†] Unlike solid-state devices, valves are robust enough to withstand an overload for the short time it takes for the fuse or relay to react. Another consideration is that the output voltage could rise higher than the design value at start up, owing to difference in heater warm up and the presence of the grid-cathode protection diode. Still worse if the error valve is faulty or pulled from its socket. Critical applications might therefore demand preheating or crowbar features to be added. Putting it another way, don't use valve regulators for critical applications in the twenty-first century.

7.7.3: Classic Transistor Regulator

Designing a high-voltage regulator using MOSFETs is far more straightforward than with valves. Power MOSFETs are built for much more demanding applications than we are likely to see in a valve amplifier, so it is easy to buy devices with far more voltage and current capability than we need. The main pass device must be capable of handling the total power that must be dissipated: $P = (V_o - V_{in})I_{dc}$, but most of the actual design thought will go into the heatsink and mounting considerations, rather than which device to buy.

Fig. 7.36 shows the quintessential, cheap-and-cheerful two-transistor (Q_3 doesn't count!) HT regulator using MOSFETs. The similarity to the previous valve regulator should be obvious. For this example the author used a pair of STP4NK60ZFPs rated up to 600V, 70W, but the same circuit can be adapted for most applications up to several hundred volts and several hundred milliamps if necessary, using suitably heavy-duty devices.

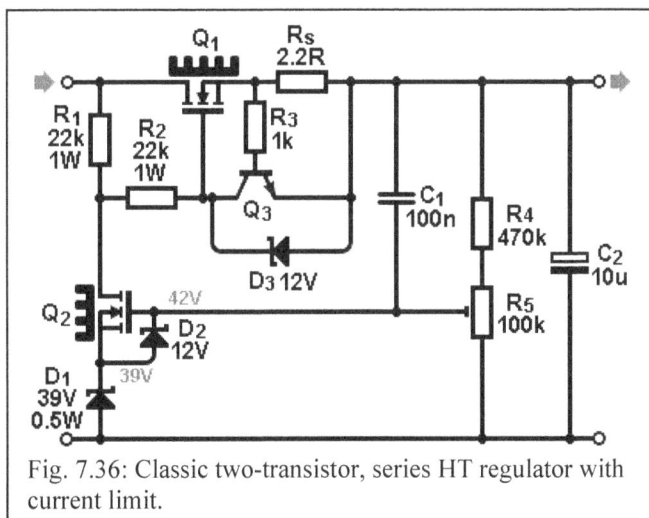

Fig. 7.36: Classic two-transistor, series HT regulator with current limit.

Zener diode D_1 provides the reference voltage; the higher it is the greater the loop gain and potential regulation performance. The author happened to have a 39V ½W device on hand. The voltage across Q_2 will be roughly equal to $V_o - V_{ref}$ and must pass enough current to bias the Zener reliably (unless extra current is provided through a boosting resistor). Depending on how ambitious we are, this might lead to

---

[†] For example, see the Hewlett-Packard *711A* or: Newell, N. S. (1967). General Purpose PSU with Auto Cut-out, *Practical Wireless*, December, pp570-2.

a couple of watts dissipation in Q_2, remembering that a single TO-220 device can only handle 1W with no heatsink. Since we already have a high-voltage, high-power MOSFET for Q_1, it is often convenient to use the same type of device for Q_2, which could be mounted on the same heatsink, or have its own small sink.

It is essential that Q_1 has some form of current limit to protect it from capacitor inrush at start up. In this case a simple 270mA constant-current mechanism has been added, utilising Q_3, as described in section 7.2 earlier. D_3 protects Q_3 and the gate-source junction of Q_1 from any possible over voltage, so Q_3 can be any low-voltage NPN. Similarly, D_2 protects the gate of Q_2 from the transient that would otherwise result from charging C_1 at switch on. These Zeners could be any convenient value under 20V.

The feedback divider could be made from any convenient values that do not drain excessive current from the output, remembering that it must divide the desired output voltage down to the reference voltage, plus the gate-source voltage of Q_2. The latter is likely to be a couple of volts for a power MOSFET operating at low current. Fixed resistances could be used, but a trimpot takes the headache out of exact calculation. In this case the output can be trimmed from no regulation at all, down to about 240V.

Performance of this simple topology is not spectacular, but better than a simple valve regulator, which means it is certainly good enough for most valve circuits. With 320V input and the output set to 250V, the peak-to-peak ripple reduction was 35dB with no capacitors, rising to 51dB with C_1 and C_2 included. Strictly, C_2 adds very little to the ripple reduction, but it is needed for stability with no load. The drop-out voltage is around 15V –lower than any valve regulator could hope to manage– though if the differential voltage gets close to this then performance inevitably degrades as less current is available to flow in Q_2 (reducing its g_m) and D_1 (raising its internal resistance). For example, with the output voltage set to 300V, and the input voltage changed from 360V down to 320V, the output voltage dropped by 3.3V and ripple reduction degraded to 23dB. Of course, from the point of view of a typical valve amp circuit, this still qualifies as a tightly regulated and sparkling clean HT.

7.7.4: Hybrid Maida Regulator

When it comes to solid-state HT regulation there are few if any off-the-shelf IC solutions, so we must be creative. Fig. 7.37 shows how an ordinary low-voltage regulator IC can be shielded from high voltages by placing it in the bias 'window' of a valve. The 317 is arranged as a standard three-terminal regulator set to whatever output voltage we desire (whether fixed or adjustable), which will vary the voltage across itself –that is, the grid-cathode voltage of the valve– to control its conduction and so maintain regulation. The valve dumbly does as it is told, dissipating the bulk of the heat and withstanding the high input voltage.

This circuit has the advantage of relative simplicity, but the SOA of the valve inevitably imposes a large differential voltage drop. Moreover, the regulator IC cannot withstand more than 40V across itself, so we must be cautious about using a valve like the 6AS7 or KT88 that might demand V_{gk} to be more than $-40V$ at small load currents. The IC also cannot operate with less than about 2V across itself, so we cannot access the triode's SOA above the $-2V$ grid curve. But at least the valve is naturally current limited, meaning the IC is protected from momentary output shorts (a fuse or other protection measure may be needed for *sustained* output shorts, as with any series valve regulator). This topology has been on the books for a long time,[7] but since it satisfies neither valve purists nor those trying to improve on classic valve regulation, it has rarely (if ever?) been used in practice. It is shown here only because it is a nice stepping-stone between the classic valve regulator and the all-silicon Maida regulator.

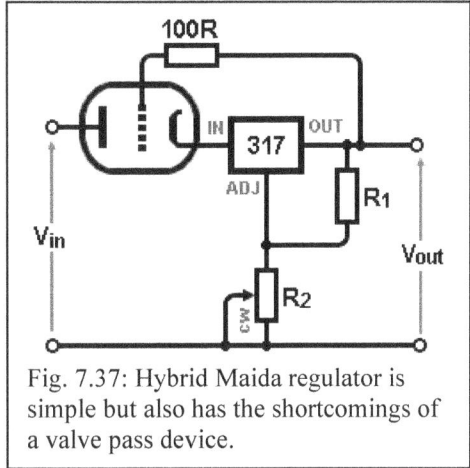

Fig. 7.37: Hybrid Maida regulator is simple but also has the shortcomings of a valve pass device.

7.7.5: Maida Regulator

Fig. 7.38 shows the simplest form of the all-silicon Maida regulator, so-named after an application note written in 1980.[8] Its relatively low-cost, high-performance and versatility makes it an ideal solution for valve amp designers looking for a stiffer regulator than the Zener follower, so it is worth covering in detail.

Looking at the BJT version in fig. 7.38a, U_1, R_3 and R_4 form a standard 317 voltage regulator set up to produce the desired output voltage (whether fixed or adjustable). To set a high voltage, R_4 will be relatively large, meaning the adjust pin current – which might be up to $100\mu A$– may significantly affect the output voltage. R_4 may therefore require some adjustment-on-test to hit a specific design voltage.

Meanwhile, Q_1 does the hard work of dissipating heat and withstanding the high input voltage. R_1 feeds a small current into D_1 which is a low voltage Zener diode. The voltage across U_1 is equal to the Zener voltage minus V_{be}, so U_1 is protected at all times, floating within a 'window' of a few volts. R_2 may be included to provide a crude current-limiting function; as load current increases, the voltage across R_2 increases, eventually leaving too little voltage for U_1 and forcing it to drop out. D_2 provides protection against reverse bias. The drop-out voltage of the circuit overall is roughly equal to the Zener voltage. Since this only needs to be large enough to

---

[7] National Semiconductor (1980). *Voltage Regulator Handbook, pp10-22.*
[8] Maida, M. (1980). High Voltage Adjustable Power Supplies. *National Semiconductor Linear Brief 47.*

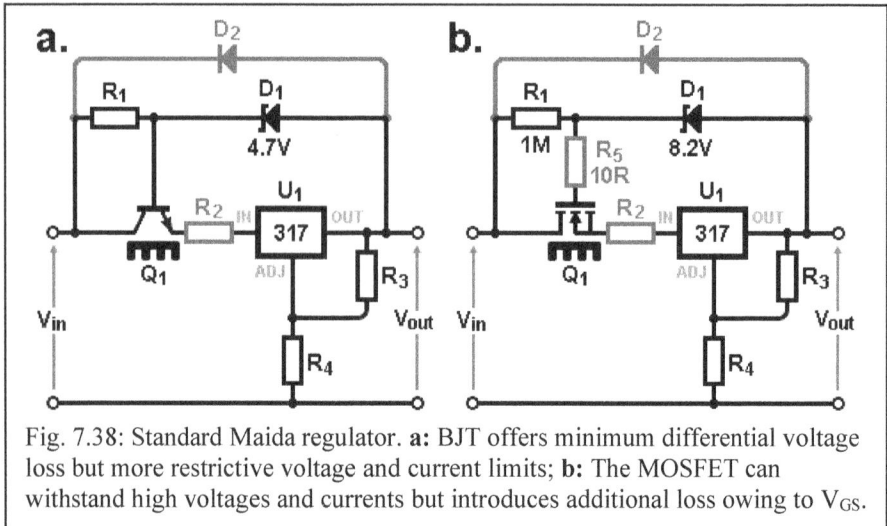

Fig. 7.38: Standard Maida regulator. **a:** BJT offers minimum differential voltage loss but more restrictive voltage and current limits; **b:** The MOSFET can withstand high voltages and currents but introduces additional loss owing to V_{GS}.

accommodate V_{be} (typically 0.7V) plus about 3V for U_1 to function normally, the Zener could be a 4.7V device.

The minimal drop-out voltage is the one advantage of using a BJT for Q_1. But as discussed in section 7.1, high-voltage BJTs have low h_{FE}, meaning significant base current may flow in R_1 which might therefore need to be a relatively small value, high-power resistor. This in turn introduces a subtle failure mode because U_1 cannot sink current into its output. Under no-load conditions, the current flowing through R_1, D_1, and down through R_{3-4}, might be enough to forcibly pull the output voltage up above the design value, destroying U_1. A Darlington could be used instead (as in the original application note) so that R_1 can be made large, but this does not evade the other big problem with using a BJT: secondary breakdown. BJTs cannot withstand large currents when the voltage across them is also large, which will be the case if the output of the regulator is shorted. Even the inrush to a large capacitive load could prove destructive. The current limit we choose to build in must therefore be very conservative and possibly onerous. In short, the BJT version of this circuit is best avoided.

The MOSFET version of the regulator –fig. 7.38b– works in exactly the same way as already described. The main difference is that V_{GS} will usually be 4 to 5V, with an additional 3V or so needed across U_1, so the Zener voltage must therefore be a little larger, and with it the drop-out voltage of the whole circuit. R_5 is a generic gate-stopper added to discourage oscillation, which should be mounted physically close to Q_1. Since the MOSFET demands no gate current, R_1 can now be very large.

The 317 needs over 3.5mA load current to maintain regulation, otherwise the output voltage will climb. This is not a problem in a fixed installation where there is always a load, but it might be unwelcome behaviour in a general-purpose power supply. We could perhaps allow the full 3.5mA to flow down R_{3-4} at all times, but in a high-

204

voltage situation this means R_4 will need to be a high-power resistor; yet more inconvenient if it is a pot. An alternative is to substitute a more sophisticated regulator IC such as the LT3080 which has only 10μA adjust-pin current and 0.5mA minimum load requirement.[9]

It is tempting to add a capacitor in parallel with R_4 to shunt broadband noise, as is standard practice in a low-voltage situation. But be warned; every capacitor added to a high-voltage, solid-state regulator hangs like a sword of Damocles; it introduces the possibility of destroying something due to large transient charging or discharging currents. It is all too easy to create a regulator that appears to work superbly for a while, but which dies for unclear reasons. Robustness is generally the reward for going without capacitor 'enhancements'.

Speaking of robustness, U_1 has built-in current and thermal limiting, neither of which are actually used in the Maida regulator, which seems like an opportunity squandered. The author therefore developed the circuit variation in fig. 7.39 which is robust enough to withstand continuous shorts (with suitable heatsinking for Q_1 of course), so it can be used as a bench supply. The secret to this circuit is R_2 and U_1, the latter being an LM317L. The 317L is the little brother of the 317 and comes in a TO-92 package. Consequently, it has a much smaller built-in current limit of 200mA (still enough juice for most valve projects) which is more precise than the simple current-limiting method in fig. 7.38. R_7 is now included only to discourage oscillation rather than to provide current limiting.

R_3 allows $1.25V/1.5k\Omega = 0.83mA$ to flow down the voltage-setting divider, and another 100μA may flow from the adjust pin. This means dissipation in the pot is safely below 500mW. $R_4 \| R_5$ have a combined resistance of 427kΩ and set the maximum output voltage to 400V. However, the minimum load current specified in the 317L datasheet is 1.5mA, so the output voltage may be higher than expected if no load is attached.

The main reason for using the 317L, however, is its built-in 2W thermal limit which, by judicious choice of R_2, is exploited to protect the MOSFET too. Ignoring R_7 which is small, the voltage across U_1 is equal to:

$$(V_{in} - V_{out} - V_Z)\frac{R_2}{R_1 + R_2} + V_Z - V_{GS}$$

Thus when the output voltage is set to the maximum value of 400V there is only about 6.8V across U_1, so the full 200mA limit is available. But as the output voltage is turned down, the drop across R_2 –and therefore across U_1– increases, causing U_1 to dissipate more heat (U_1 should *not* be attached to a heatsink). R_2 is selected so that if the output is a dead short, the voltage across U_1 reaches almost 40V –the maximum allowable– so it will limit the current to approximately $2W/40V = 50mA$. The

---

[9] Horowitz, P. & Hill. W. (2015). *The Art of Electronics* (3rd ed.), p698. Cambridge University Press.

remaining voltage across the MOSFET will be about 420−40 = 380V so it will dissipate 380×50mA = 19W, which is manageable with a decent heatsink (<1.5°C/W). What's more, the thermal lag in the 317L means it will in fact deliver the full 200mA *briefly* even when the output voltage is set low. In other words, the circuit can deliver high

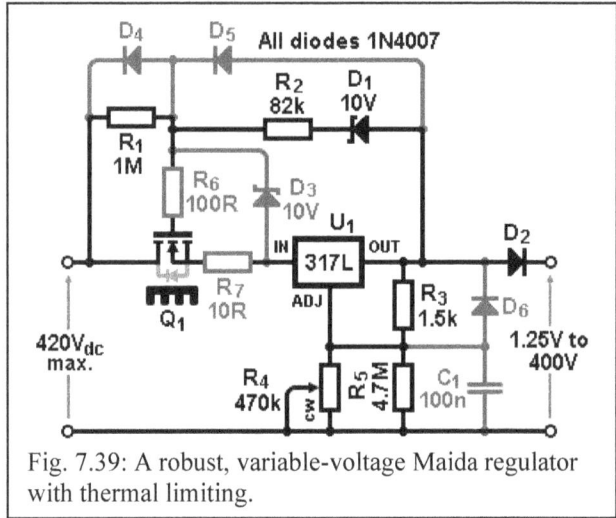

Fig. 7.39: A robust, variable-voltage Maida regulator with thermal limiting.

voltage *and* high power under many practical conditions but is smart enough not to overheat.

Various diodes have also been added for protection. Some of them are probably unnecessary but the author does not have the fortitude to find out which, having blown up several earlier circuit variations to reach this point. Diodes are cheap, and so far the circuit in fig. 7.39 has proved indestructible. It can of course be adapted for lower input voltages by increasing R_2 accordingly.

Fig. 7.40 shows another variation on the Maida regulator which the author has used as a bench supply. The shell of the circuit is a conventional Maida, but an adjustable constant-current limit is inserted between Q_1 and U_2. This is a textbook application

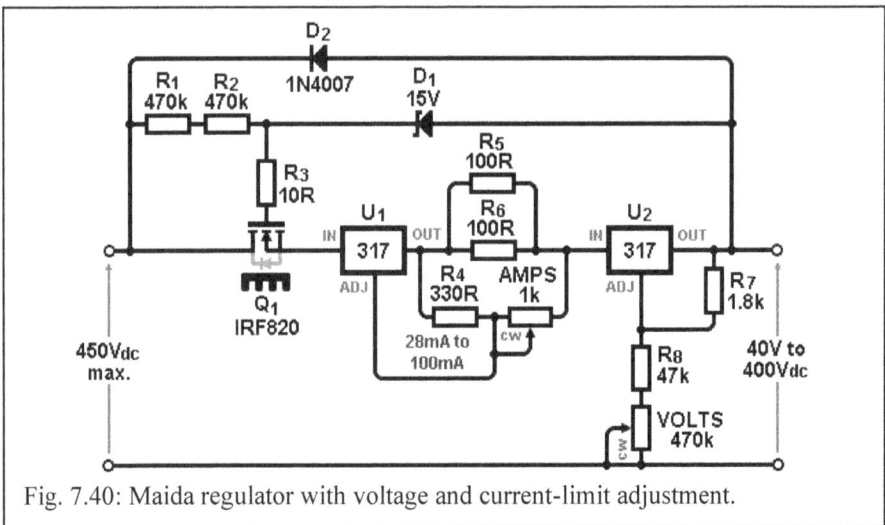

Fig. 7.40: Maida regulator with voltage and current-limit adjustment.

of a pair of ordinary 317 regulators; the functioning of the current-limiter U_1 was explained in section 7.4 earlier. If the load current is less than the limit, U_1 drops out and the circuit delivers a constant voltage output as determined by U_2, but if the load demands more than the limit, U_1 will kick in and increase the voltage across itself, forcing U_2 to drop out instead.

R_1 and R_2 share the total voltage during an output short, and $R_5\|R_6$ share the total current, which allows ¼W devices to be used throughout. No bypass capacitors are used, to avoid any damaging transient currents, as explained earlier. The maximum available current-limit is 100mA which is within the capability of Q_1 even with a large voltage across itself, assuming it has a suitable heatsink, but this circuit lacks the 'smart' power limiting of the previous circuit.

7.7.6: Floating Error Amp (T-Reg)

One of the shortcomings of a classic two-transistor series regulator (e.g. fig. 7.3a) is that in a high-voltage environment, a large voltage is imposed across the error amplifier which brings various design difficulties, and in an adjustable regulator this voltage may vary over a wide range, which brings more. A way around this is to provide the error amp with its own separate, low-voltage power supply, and merely reference this to the high voltage output. This allows the error amplifier to use low-voltage parts, floating safely on top of the high voltage.

An example of this floating-error-amp approach is the 'T-Reg' published by Didden,[10,11] shown in fig. 7.41. This circuit uses the existing heater supply of the pass valve to provide the floating power supply for the error amplifier. Since the heater supply is referenced to the output voltage this also neatly avoids violation of $V_{hk(max)}$, in the same manner as in a valve rectifier (on the other hand, it means the same heater supply cannot be used for other valves in an amplifier). In fig. 7.41, the lower dashed box contains an ordinary high-voltage power supply, with a KT88 as the series pass device. The upper dashed box contains the low-voltage supply and error amplifier, stacked on top of the high voltage section. The job of the error section, as usual, is to control the pass device by suitably squeezing the grid and cathode together, or indeed apart.

The voltage drop across the LED, D_4, provides the reference voltage for the entire regulator. An LM334 ensures constant current through the LED for maximum stability. Q_1 and R_1 together form another constant current source that feeds 1mA down through R_3, which therefore creates an amplified version of the reference voltage across R_3. Q_2 and Q_3 form a vertical differential amplifier that compares this reference voltage with the final output voltage (R_4 limits the current through the

---

[10] Didden, D. (2009). T-Reg: A High-Voltage Regulator for Tube Amps, *Elektor*, (March) pp22-8.
[11] https://audioxpress.com/article/t-reg-a-high-voltage-regulator-for-tube-amps. The original design also included a hot-switching automatic time-delay, not shown here.

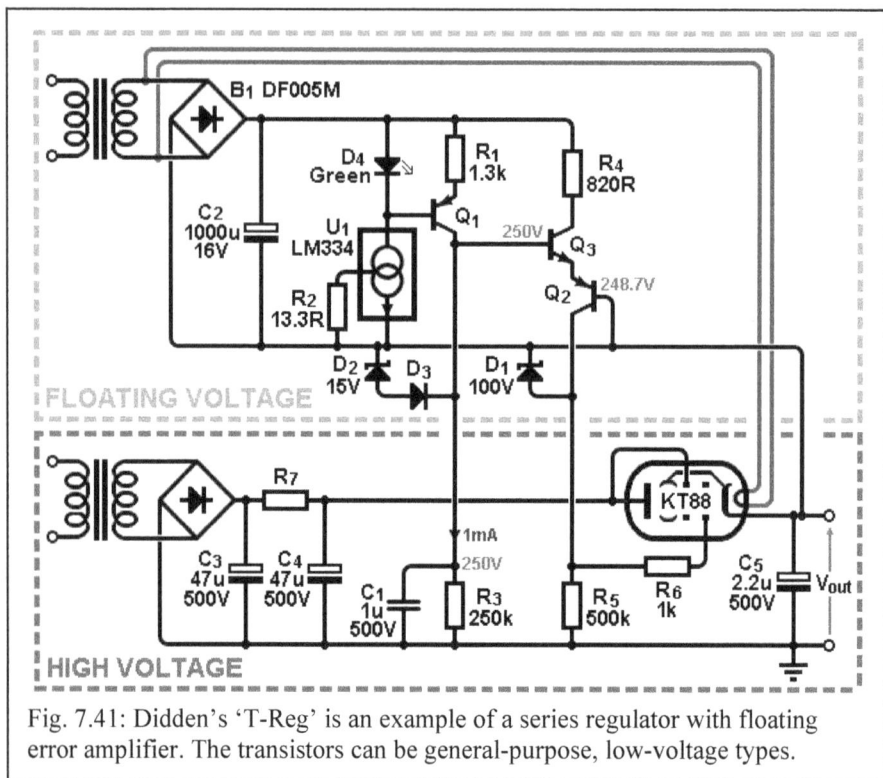

Fig. 7.41: Didden's 'T-Reg' is an example of a series regulator with floating error amplifier. The transistors can be general-purpose, low-voltage types.

differential amp at start up). Collector current from this pair flows down through R_5, causing a voltage drop across it which is supplied to the valve grid. The valve then dumbly follows this voltage. If the output voltage tries to rise, for example, the whole floating supply will rise with it, which pulls up the base of Q_2. Therefore, Q_2 will conduct less current, meaning less voltage drop across R_5, which pulls the grid of the valve down and counteracts the original rise. Because Q_2 has high g_m, R_5 is very large, and the whole of the output voltage is fed back, the loop gain is measured in the thousands. Regulation is therefore extremely good, with the reported output impedance below 0.1Ω.

The output voltage is equal to the voltage across R_3 minus two V_{be} drops, or about 348V with the values shown. This can be adjusted (within reasonable limits) by changing the value of R_3. Different valves or even a MOSFET can be used instead of the KT88, depending on current requirements, and the original author goes into considerable detail with performance data and design variations –see the original references.

7.7.7: Opto-Coupler Regulator

A different way to avoid needing a high-voltage error amplifier is to use optical coupling between the high- and low-voltage portions of the regulator,[12,13] as already noted in section 7.6.3. Fig. 7.42 shows a simple example of this using a PVI. Here the reference voltage is a 12V supply (which might be a pre-existing DC heater supply) and is presumed to be already regulated. A PNP transistor then serves as the error amplifier and drives the LED inside the opto-coupler. If the output voltage attempts to rise too high, it pulls up on the base of Q2, reducing its conduction and reducing the current into the LED. This reduced the drive voltage generated inside the PVI, which reduces the conduction of the pass device. D_2 provides reverse voltage protection for Q_2.

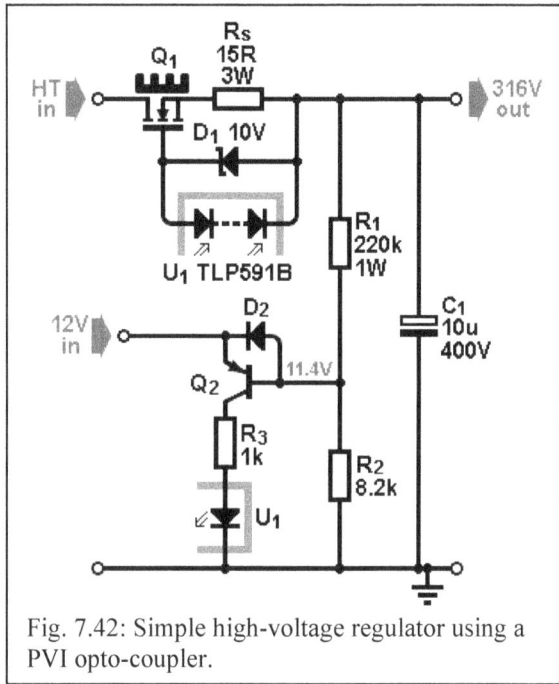

Fig. 7.42: Simple high-voltage regulator using a PVI opto-coupler.

Using an STP4NK60ZFP for Q_1 and a BC327 for Q_2, the author measured a peak ripple reduction factor of 46dB for this circuit. No doubt superior performance can be obtained with greater sophistication.

---

[12] https://www.edn.com/use-an-optocoupler-to-makea-simple-low-dropout-regulator/
[13] https://www.edn.com/regulate-a-0-to-500v-10-ma-power-supply-in-a-different-way/

Chapter 8: Valve Bias Supplies

In addition to the high voltage HT supply and the heater supply, many valve amps also require a negative supply for biasing the output valves. Since a bias supply is usually regarded as a special section of the power supply (as opposed to a fully bipolar power supply where the negative rail can be treated just like the positive rail) it deserves some separate discussion. There are various features of bias supplies which are common to many valve amplifiers, plus some special considerations that are not usually an issue for other parts of the power supply, so the reader should be made aware of all these things. As usual, there are some improvements which can be made to traditional, minimalist circuits. It should perhaps be pointed out that this chapter will adhere mainly to the convention of drawing positive voltage rails above the zero-volt ground rail, and negative voltage rails below it. If this is at all confusing, try turning the book upside-down.

8.1: Bias Voltage Stabilisation and Regulation

The bias voltage keeps the quiescent current in the power output valves at the safe design value. In most amplifiers the HT voltage is unregulated and is free to rise and sag in accordance with load current and wall voltage fluctuations. If the HT –or more particularly the screen voltage in the case of pentodes/tetrodes– increases, the quiescent current in the power valves will tend to increase as a result. To maintain roughly the same operating point and dissipation in the power valves, the bias voltage must therefore go more negative as the anode/screen voltage goes more positive, in almost direct proportion. Conversely, if the HT or screen voltage falls then the quiescent current will also tend to fall (colder bias) unless the bias voltage goes proportionately more positive at the same time.

In general then, the HT and the bias voltage are best derived from the same power transformer so all the voltages track one another naturally (a separate bias winding will still track the HT winding if it is on the same transformer). If the bias is supplied from an entirely separate transformer then it will track with variations in wall voltage, but not with load current. Following the same line of thinking, if the HT uses passive smoothing then the bias supply should too; if the HT is firmly regulated then the bias supply should be too, i.e. the two supplies should complement one another. Simple Zener stabilisation is usually adequate for a bias supply. The alternative is to control the bias supply actively with a bias tracking servo circuit. This is a more involved approach, of course, and some solutions to it are presented in chapter 10.

8.2: Bias Adjustment and Balancing

In many fixed-biased amps, provision is made for adjusting the bias voltage over a modest range, to accommodate differences between valves and changes due to aging (as valves age their g_m falls and a less negative voltage is needed for a given operating point). Sometimes the bias is user accessible, but generally it will be a trimpot inside the amplifier which must be adjusted by someone with at least a rudimentary understanding of what they are doing. If the bias voltage is too small then the valves will run hot (under biased)[*] and the resulting dissipation may damage them. The familiar sign of insufficient bias is **red-plating**, when the anode begins to glow red hot. Conversely, if the bias voltage is too great (over biased) then the valves will run cold or may even be cut off. This does no physical damage but will instead spoil the audio performance by reducing the available output power and increasing harmonic- and crossover distortion.

The appropriate range of bias voltage adjustment will vary depending on the types of power valves being used, the topology, the desired quiescent dissipation and so on. However, there are some simple principles which will aid new designs. A theoretically perfect triode would be biased at cut-off when the magnitude of the bias voltage is equal to the anode-to-cathode voltage divided by the valve's amplification factor μ. In other words, its grid base would be equal to $-V_{ak}/\mu$. A real triode is not perfect of course, but neither do we actually need to bias at true cut-off, so it is safe to assume the largest bias voltage we will ever need in a triode amp is equal to $-V_a/\mu$. A theoretically perfect pentode/tetrode would be biased at cut-off when the magnitude of the bias voltage is equal to the *screen-to-cathode* voltage divided by the *triode* amplification factor μ_{triode} i.e., the μ of the valve when connected as a triode. Again, even when using real pentodes/tetrodes it should be safe to assume the largest bias voltage we will ever need is equal to $-V_{g2}/\mu_{triode}$. These simple rules give us a maximum bias voltage to aim for in a new power supply design –after all, it is easier to throw voltage away after a circuit is built than to try to get more.

Table 8.1 gives the μ of some popular power valves. Immediately it can be seen that the power triodes have very low μ, so they will need fairly large bias voltages. Most of the pentodes/tetrodes have amplification factors of around 10, so we can usually aim for a raw negative voltage that is about 10% of the HT voltage in magnitude, e.g., if the HT voltage is 400V then a maximum negative voltage of -40V will accommodate most of the pentodes and beam tetrodes we are ever likely to use. The notable exceptions are the EL84 which needs comparatively little bias voltage, and the 6550 which needs a fair bit more.

Valve	Type	μ_{triode}
EL34	Pentode	10.5
EL84	Pentode	20.0
KT66	Beam tetrode	7.5
KT77	Beam tetrode	11.0
KT88	Beam tetrode	7.5
2A3	Triode	4.2
300B	Triode	3.9
5881	Beam tetrode	8.0
6550	Beam tetrode	6.8
6L6GC	Beam tetrode	8.0
6V6GT	Beam tetrode	9.8

Table 8.1: Some popular power valves and their triode amplification factors.

[*] Rather annoyingly, in transistor terminology the terms 'under biased' and 'over biased' are used the other way around.

8.2.1: Bias Voltage Adjustment

Once the raw negative voltage is available, we can devise a way to tweak it down to the value needed by the power valves (the exact value will depend on the specific audio circuit and is beyond the scope of this book). The simplest option is a single bias-adjust pot, but there seem to be as many bias adjustment circuits found in schematics as there are different amplifiers! This is somewhat surprising considering how simple the task is. The bias is applied to the valve grids through their grid-leak resistors, and this demands virtually no current since the grids are high impedance. The only significant current flow is therefore in the bias-adjustment network itself, which is usually just a few milliamps at most. If the bias supply should fail it may lead to destruction of the power valves due to over dissipation, so a couple of fail-safe measures are (usually) found in bias circuits, and fuses can be added to the cathode circuit of the power valves too, if necessary.

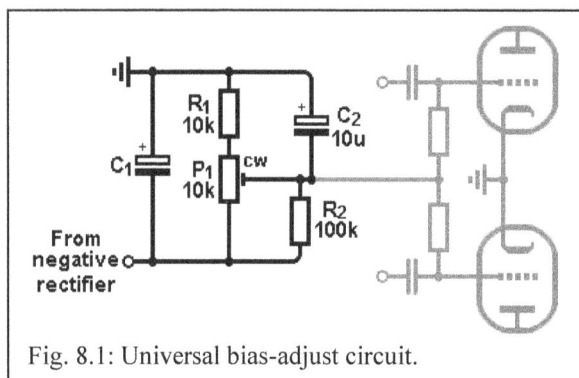

Fig. 8.1: Universal bias-adjust circuit.

A universal bias-adjust circuit is shown in fig. 8.1. Connection to a typical push-pull pair of valves is shown faint, though it could also be single ended. C_1 represents the main bias reservoir capacitor. It is easy to become accustomed to dealing with positive voltages all the time, so positive signs have been added to the capacitors as a visual aid. It is good design practice to limit the range of voltage adjustment so it is impossible to reduce the bias to nothing (0V) as this gives a modicum of protection against accidental or unskilled adjustment. This limit is provided by the end-stop resistor R_1 which forms a potential divider with P_1. There needs to be enough range to accommodate the usual spread in valve characteristics, especially with age, so the ratio of R_1 and P_1 will vary somewhat depending on the type of valves being used and the raw bias voltage available, so a little experimentation will be necessary. Typically though, R_1 might end up roughly equal to P_1 in value.

The total resistance of R_1+P_1 forms part of the power valves' grid leak resistance, that is, the total resistance between grid and cathode. There will be a maximum allowable grid leak resistance given in the data sheet, and note that this is a DC limit; we cannot cheat it by bypassing the bias network with capacitors. Since we will usually want to make the grid leak resistors as large as possible to avoid loading down the driver stage, the resistances used in the bias supply will necessarily be small, perhaps less than 10% of the total grid leak resistance. A 10kΩ bias pot is a nice, common value that should be suitable for most designs.

Further protection is provided by fail-safe resistor R_2 which ensures that if the pot wiper fails to make good contact with the track (because it is dirty or worn out, say) the grid leaks will effectively be connected directly to the raw bias supply, biasing the valves safely cold. R_2 can be about ten times the value of P_1 so it has negligible effect on the rest of the design.

C_2 forms an RC smoothing filter with P_1 and also decouples the grids from one another. Since the bias supply draws very little load current, one stage of smoothing in addition to the reservoir capacitor is usually quite enough to reduce any ripple to negligible levels, without resorting to enormous values of capacitance which would extend the charging time.

8.2.2: Bias Balance Adjustment

In a hi-fi amp or any amp with multiple power valves in parallel, it is often desirable to provide each symmetrical pair –or even each individual valve– with its own bias pot. This allows the quiescent anode currents to be individually adjusted if the valves are not well matched, which reduces DC offset current in the output transformer and can improve the power output stage's PSRR, which reduces hum.

One way to do this is to add more bias-adjust networks in parallel, though a small saving can be made by letting all the bias pots share the same end-stop resistor R_1, as shown in fig. 8.2. A similar circuit was used in the venerable Radford ST-70, for example. Notice that R_1 is half the value used earlier since twice the current now flows in it.

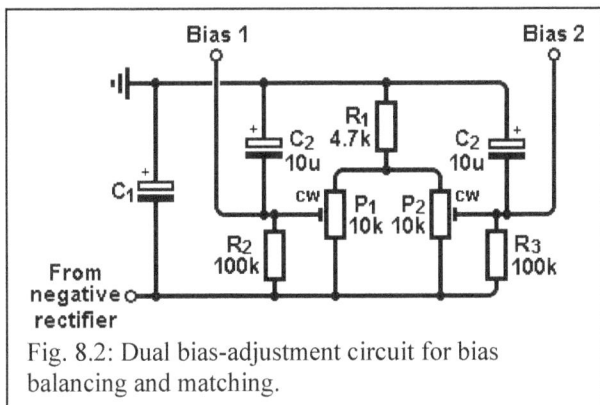

Fig. 8.2: Dual bias-adjustment circuit for bias balancing and matching.

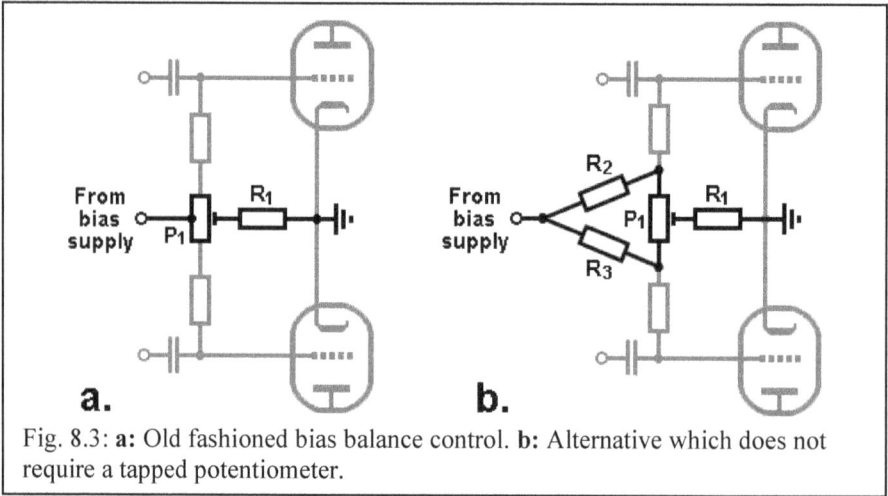

Fig. 8.3: **a**: Old fashioned bias balance control. **b**: Alternative which does not require a tapped potentiometer.

Another method of bias balancing, popular in vintage amplifiers, is to use a tapped potentiometer as shown in fig. 8.3a. With P_1 at the centre setting both valves receive the same bias voltage, but when P_1 is rotated either way the bias voltage for one valve rises while the other simultaneously falls, so we get a see-saw adjustment facility. This makes things intuitive when searching for a hum null. The main problem with this

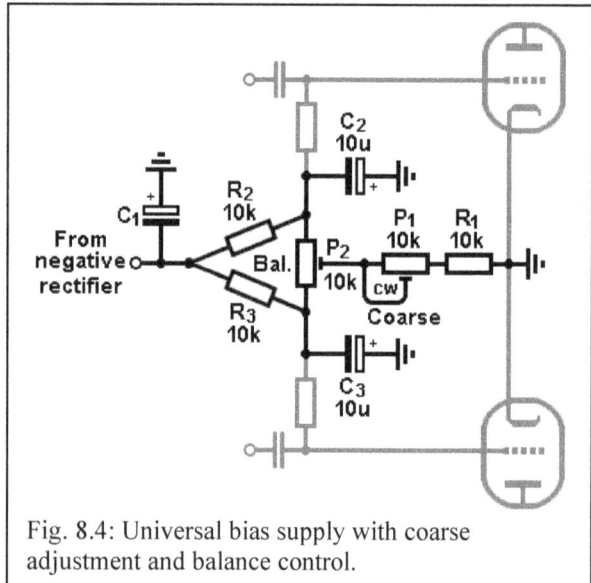

Fig. 8.4: Universal bias supply with coarse adjustment and balance control.

approach is that tapped potentiometers are now a rarity, but this is easily sidestepped by making a star-delta transformation and using the variation in fig. 8.3b instead, which uses an ordinary pot. Fig. 8.4 expands this into a practical version with smoothing capacitors and a coarse bias adjustment pot. Notice that if either wiper breaks contact with its track the bias voltage will go safely more negative.

8.3: Nuisance Fuse Blowing

One minor consideration when designing a bias supply is the time it takes for the bias voltage to reach its normal working value. Often the bias supply is derived from the from the high-voltage

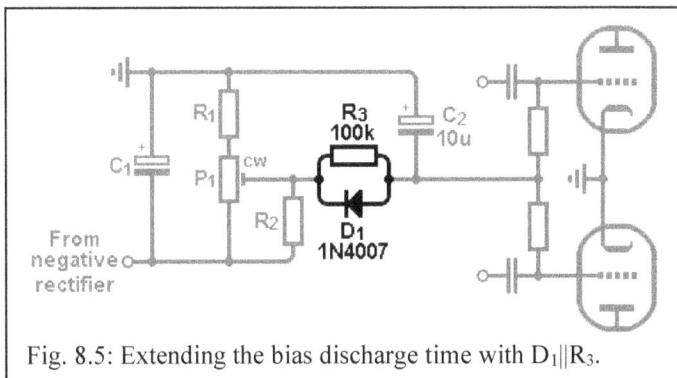

Fig. 8.5: Extending the bias discharge time with $D_1 \| R_3$.

transformer winding through a large dropping resistance plus RC smoothing, so it may take several seconds for the bias supply to charge up. Depending on how quickly the cathodes warm up, the lack of bias may cause a current surge in the power valves. This is rarely a problem for the power valves themselves as they can withstand short lived current overloads quite harmlessly, especially while the anodes are still cold. The problem is instead nuisance blowing of the primary or (more likely) secondary fuse. This is most likely to happen if the power is briefly switched off and the bias supply discharges, but then power is restored while the cathodes are still hot, leading to an abnormally large surge which blows the fuse. This may only be a user annoyance rather than a serious problem, but nonetheless it is worth making the bias charging time as quick as possible and/or the discharging time as slow as possible. The latter can be accomplished using the modification in fig. 8.5. At start-up C_2 can charge quickly through D_1, but it can only *dis*charge through R_3, which is large. D_1 carries almost no current but for convenience it will probably be the same type of diode used for the high-voltage rectifier, such as 1N4007. However, an unfortunate side effect of the added diode is that it is liable to worsen any blocking distortion in the output valves, so it may be unwise to include it in a guitar amplifier unless anti-blocking measures are also added elsewhere.

8.4: Sourcing Bias Voltage

Before solid-state rectifiers were available, valve rectifiers were the only option. Most power rectifiers contained two diodes with a shared cathode, suiting the familiar two-phase rectifier configuration. Most valve equipment required only a positive supply voltage, but if *only* a negative voltage was needed then the same arrangement could be used simply by swapping the output terminals. Producing proper bipolar rails from one transformer winding would have required four diodes implying at least three bottles,[*] so individual transformer windings were used instead, since copper and iron was cheaper than rectifiers in those days. However, if the negative voltage was needed only for biasing –which requires little to no current–

---

[*] A rare exception is the 6JU8 quadruple-diode, but it is only rated for small currents.

215

then it was practical to use half-wave rectification. Additionally, since the negative voltage did not have to be very great, a dropping resistor could be added in series with the diode as in fig. 8.6a, and this allowed the use of low voltage capacitors and a cheaper, low-power diode (often a dual signal-diode was used with the sections in parallel). Alternatively, a lower-voltage tapping point on the transformer could be used as in fig. 8.6b, resulting in less wasted power, faster charging time, and more available current.

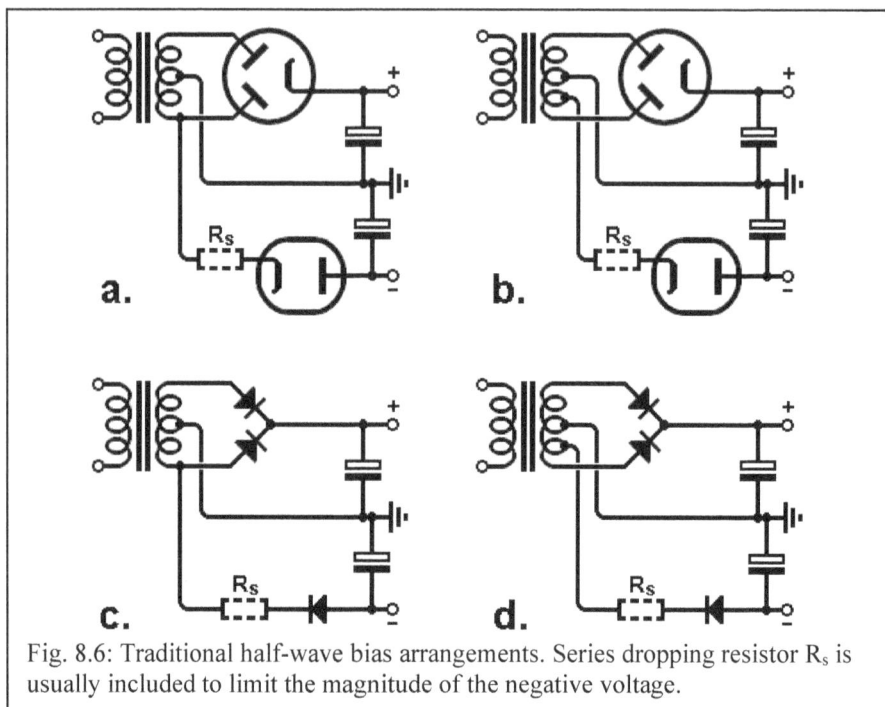

Fig. 8.6: Traditional half-wave bias arrangements. Series dropping resistor R_s is usually included to limit the magnitude of the negative voltage.

Note that in fig. 8.6 the dropping resistor (shown dashed) is placed on the cathode-side of the diode; this reduces the voltage swing at the cathode and therefore minimises the stress on its V_{hk} limit. Nevertheless, it does swing all the way from the negative rail to the HT, so a dedicated heater supply – referenced to the cathode– was often necessary. When this was an annoyance the alternative in fig. 8.7 was sometimes used.[1] Here C_s drops excess voltage without dissipating heat (a 'Wattless dropper') and since the cathode of the bias rectifier is connected to ground it could share the same heater supply as the other valves in the set. Note that R_s is also required to isolate the small capacitance C_1 from the large bias reservoir –this makes the arrangement unsuitable for bias supplies that need significant current.

---

[1] Woodville, G. R. (1948). Economical 50-Watt Amplifier, *Wireless World*, December, pp457-8.

When silicon rectifiers finally appeared they were simply substituted for the old valve rectifiers in many circuits, so the old fashioned half-wave bias rectifier remained a common feature of valve amps. Many readers will be familiar with the arrangements in fig. 8.6c and d. There is really no need to stick to this convention, however, as full-wave rectification brings several advantages –section 8.4.2.

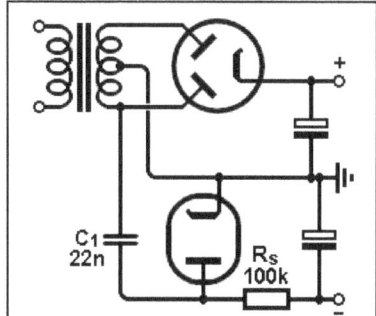

Fig. 8.7: Half-wave bias supply which eliminates stress on the bias rectifier's V_{hk} limit.

8.4.1: Half-Wave Bias Supply

It must be appreciated that there is nothing special about the bias rectifier; it works just like the positive part of the power supply, but it is often drawn differently on circuit diagrams which makes it look like it has special status. Fig. 8.8 attempts to break this illusion by showing more clearly how the traditional half-wave bias rectifier is derived from the more familiar bipolar power supply in fig. 8.8a. Remember, this may look like a bridge rectifier but really it is a pair of two-phase rectifiers with opposite polarity. However, we don't need as much negative voltage for the bias supply, so we can add a series resistor R_s to drop some voltage, taking us to b. Historically it was expensive to use two diodes, so let's throw one of them away, leaving us with the half-wave negative supply in c. Finally, the

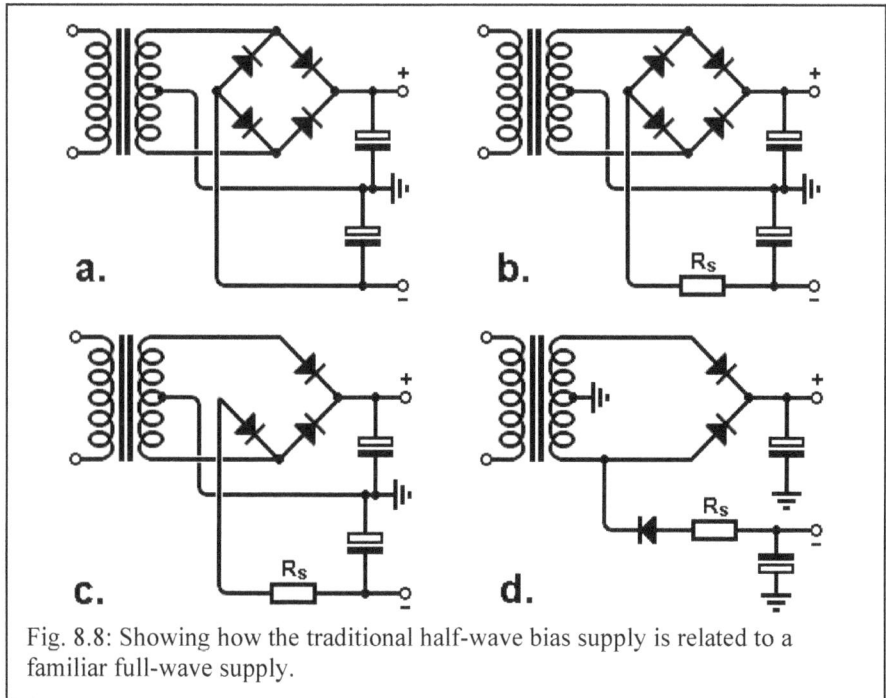

Fig. 8.8: Showing how the traditional half-wave bias supply is related to a familiar full-wave supply.

circuit in d. is functionally identical to the one in c. but has been redrawn the way it is frequently seen on schematics; R_s can be positioned on either side of the diode.

As explained in section 2.3, a large series resistance before the reservoir capacitor defeats the sample-and-hold effect of the rectifier, making it behave instead like an averaging circuit. The DC output voltage is then simply the average value of the rectified sine wave, multiplied by the potential divider formed by R_s and the load resistance R_l, which in this case will be formed by the bias adjust network:

$$V_{dc} = \frac{V_{pk}}{\pi} \cdot \frac{R_l}{R_s + R_l} \tag{8.1}$$

This observation allows us to make some further mathematical simplifications. In a typical amplifier the HT voltage for the output valves will be roughly equal to the peak AC transformer voltage, and we know that we want a raw negative DC voltage to be equal to this divided by the triode μ of the valves:

$$\frac{V_{pk}}{\mu} = \frac{V_{pk}}{\pi} \cdot \frac{R_l}{R_s + R_l}$$

If we have a bias adjust network in mind then we know what R_l is, so we can solve for R_s:

$$R_s = R_l \frac{\mu - \pi}{\pi} \tag{8.2}$$

For example, if we were to use the bias adjust circuit in fig. 8.1 then the load resistance would amount to 20kΩ. Referring to table 8.1 we might select $\mu = 10$ to work with most pentodes, so we would calculate R_s as:

$$R_s = 20\text{k}\Omega \times \frac{10 - \pi}{\pi} = 43.6\text{k}\Omega$$

A close standard is 47kΩ. This will produce a raw negative voltage which is close to one tenth of the HT voltage in magnitude. Fig. 8.9 shows the full circuit for completeness. Notice that we did not have to specify the actual transformer voltage; this universal approach is possible because μ is relatively constant and therefore the required bias voltage always tracks the HT (or screen-grid) voltage, whatever it happens to be. Although this supply is half-wave and therefore leads to net DC in the transformer, R_s provides so much current limiting that the RMS and average ripple current only amount to a couple of milliamps, which is not a burden we need to worry about.

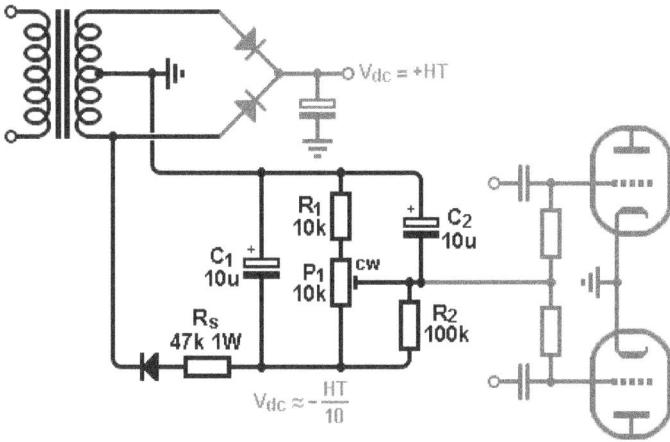

Fig. 8.9: Practical, traditional half-wave bias supply.

8.4.2: Full-Wave Bias Supply

The use of half-wave rectification for a bias supply is a relic from the days of valve rectification; there is no good reason to persist with it in a new design. Full-wave rectification is easier to filter, allows faster charging time, and avoids net DC in the transformer completely. A half-wave bias supply is easily converted into fig. 8.10 with the addition of a single diode, so there can be no excuse for being old fashioned. Again, if a large dropping resistor is assumed, the DC voltage can be calculated from the average value of a rectified sine wave (full-wave this time):

$$V_{dc} = \frac{2V_{pk}}{\pi} \cdot \frac{R_1}{R_s + R_1}$$

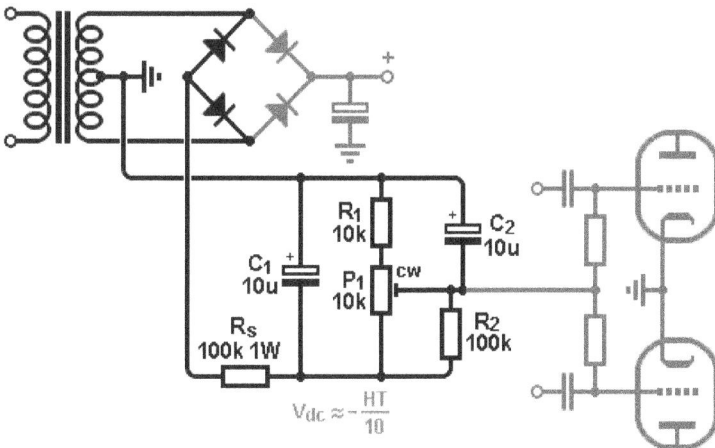

Fig. 8.10: Practical, full-wave bias supply.

So the formula for R_s becomes:

$$R_s = R_1 \frac{2\mu - \pi}{\pi} \tag{8.3}$$

Using the same figures as previously R_s would be:

$$R_s = 20k\Omega \times \frac{2 \times 10 - \pi}{\pi} = 107.3k\Omega$$

Fig. 8.10 shows the full circuit for completeness.

8.4.3: Capacitor-Coupled Bias Supply (No Centre Tap)

A bias supply can also be derived from a transformer winding with no centre tap, but it must be capacitor coupled. Fig. 8.11 shows the half-wave approach, used in the Marshall *JCM900*, for example. The coupling capacitor C_1 initially charges up positively through R_1 (beginners often forget this resistor; it is essential). On the next half cycle it discharges through D_1, effectively transferring its charge to C_2 and thereby generating a negative voltage across it –this is a charge pump. The maximum voltage which could ever appear across C_1 is equal to the peak transformer voltage, so it should be suitably rated. Similarly, D_1 should be rated for at least the peak transformer voltage.

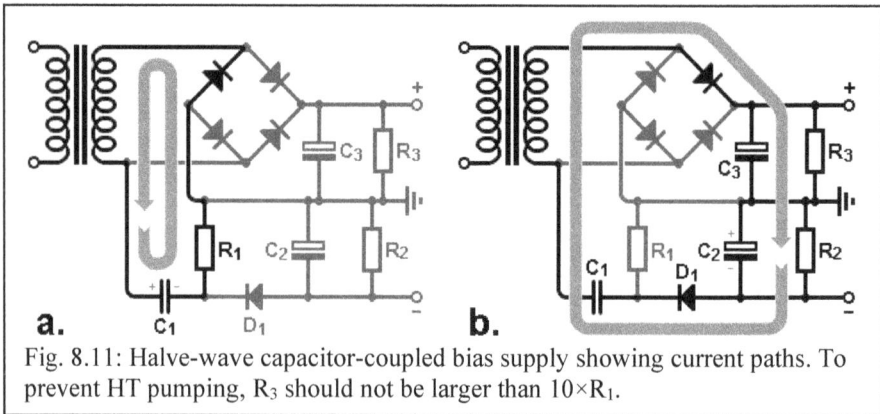

Fig. 8.11: Halve-wave capacitor-coupled bias supply showing current paths. To prevent HT pumping, R_3 should not be larger than $10 \times R_1$.

Notice that this circuit relies on there being a DC path from the HT rectifier output to ground. If no load is attached then no negative voltage will be developed –the circuit will instead pump C_3 up to as much as twice the peak transformer voltage, with consequent danger to its integrity! To prevent this from happening, the load resistance represented by R_3 must not exceed ten times R_1. Under normal operation the amplifier itself will satisfy this requirement, but at start up when the valves are cold, or if a standby switch is used, there may be negligible load on the HT. Therefore, a permanent bleeder resistor smaller than $10 \times R_1$ is typically needed. A further requirement of this circuit is that the total load resistance on the bias supply – represented by R_2– must not be smaller than R_1. If R_2 is too small then it will drain

220

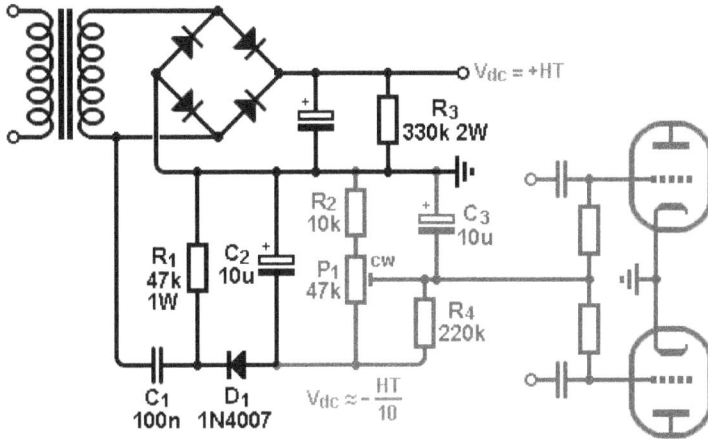

Fig. 8.12: Universal capacitor-coupled, half-wave bias supply.

charge out of C_2 too quickly, resulting in not enough negative voltage pumping. Since R_2 must be larger than R_1, which works best in the 33kΩ to 47kΩ range, this has knock-on implications for the grid-leak path of the power valves, meaning this sort of bias supply is not really suitable for big-bottle amplifiers.

Fig. 8.12 shows a practical version of the circuit which generates a raw negative voltage that is at least one tenth of the HT in magnitude –assuming the load resistance on the bias supply is greater than R_1– so it should be suitable for many amplifier designs where the HT is anything up to 600V, without changes. C_2 takes only two seconds to become fully charged. Increasing C_1 to 470nF will increase the magnitude of the negative voltage to about one fifth of the HT voltage. Reducing it has the opposite effect, of course. A typical bias adjust circuit is shown faint, and note that the total load resistance is greater than R_1. Although this is notionally a half-wave rectifier, both halves of the AC cycle are used, so there is no net DC in the transformer. Ripple current due to the bias supply is less than 5mA which is so small it can be ignored.

8.4.4: Capacitor-Coupled, Full-Wave Bias Supply

The previous bias supply can be improved upon by using a capacitor-coupled bridge rectifier. This provides more current, lower impedance, and better smoothing. Indeed, it can be used to achieve a fully bipolar power supply as was explained in section 3.4, but for biasing purposes we can relax the design rules. Fig. 8.13 shows a practical example where the coupling capacitors C_1 and C_2 are small, meaning they can be non-polarised (they must be rated for more than the peak transformer voltage). What's more they behave as wattless droppers, limiting the raw bias voltage to about one tenth of the HT, and limiting the ripple current to just a few milliamps, without dissipating any heat. A typical bias adjust network is also shown, and notice that it has a total resistance of less than 10kΩ making this supply perfectly suitable for big bottles that need a low-impedance grid leak path. Doubling

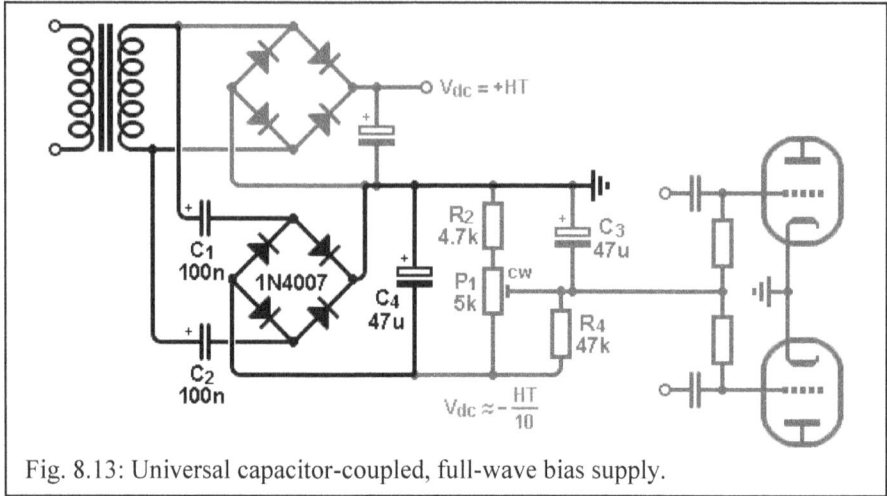

Fig. 8.13: Universal capacitor-coupled, full-wave bias supply.

the size of C_1 and C_2 will almost double the voltage into the same load, so even power triodes could be provided for.

8.4.5: Auxiliary Winding Bias Supply

If all else fails, the bias voltage can always be sourced from its own dedicated transformer winding or even from a small auxiliary transformer (remarkably, half-wave rectification is *still* often used in commercial amps even for this kind of bias supply; such an approach is indefensible these days). Fig. 8.14 shows two examples of simple auxiliary bias supplies using a bridge rectifier or voltage doubler. These should not need any special explanation since they are really just small power supplies in their own right.

When less voltage is needed, such as for biasing EL84s, it could be similarly derived from a heater winding. For example, a $6.3V_{ac}$ heater supply in

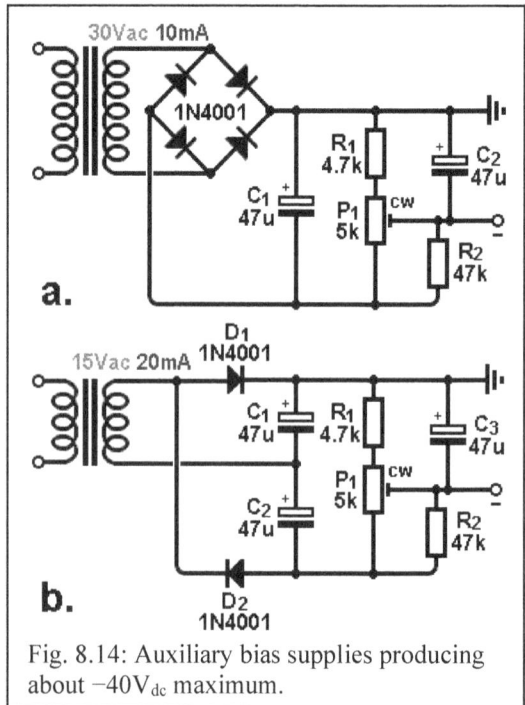

Fig. 8.14: Auxiliary bias supplies producing about $-40V_{dc}$ maximum.

combination with a voltage doubler should yield about $-15V_{dc}$.

8.4.6: Back Biasing

Another way to bias valves is to use back-biasing, illustrated in fig. 8.15. Here the whole of the amplifier HT current flows through a back-biasing resistor R_1 which is connected in the ground return path of the HT supply; bridge rectifier and two phase rectifier are illustrated for clarity. If R_1 is placed before the reservoir capacitor, as shown, then C_1 should be included to bypass ripple current (typically C_1 will be about ten times the HT reservoir capacitor C_2) and maintain a relatively a smooth DC drop across R_1, which can be used for biasing the power valve/s. This is similar in principle to ordinary cathode biasing except the current variations in the bias components will be somewhat more constant since part of the total is provided by the preamp, which is usually class-A.

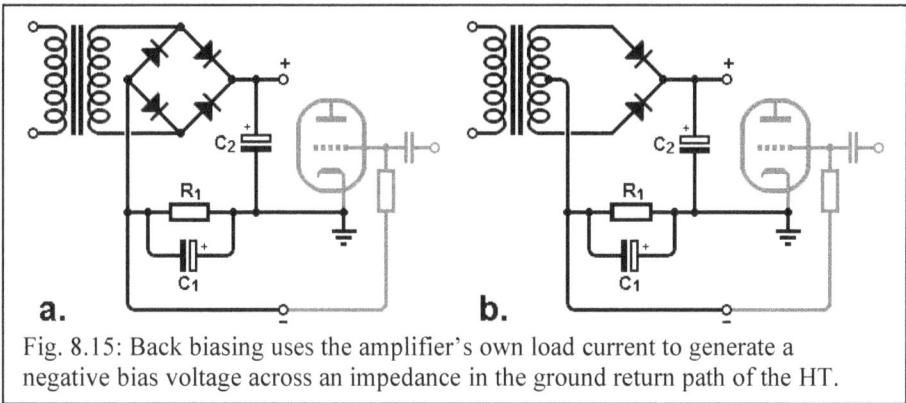

Fig. 8.15: Back biasing uses the amplifier's own load current to generate a negative bias voltage across an impedance in the ground return path of the HT.

What's more, any increase in the total amplifier current not only causes an increase in back-bias voltage, but simultaneously subtracts the same amount from the effective HT voltage too, both of which counteract the increase in current. This enhanced self-regulation may be an attractive feature for a Class-A amplifier or shunt regulator that draws constant average current, but for other amplifier classes it may be a disadvantage. Ultimately, since the negative voltage is produced at the expense of the HT voltage, back-biasing is perhaps most useful when the power transformer delivers more voltage than we really want for the HT (in the author's experience this is frustratingly often). After all, if we need to throw some HT voltage

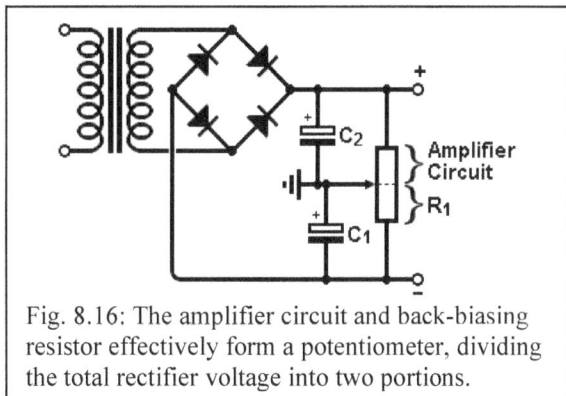

Fig. 8.16: The amplifier circuit and back-biasing resistor effectively form a potentiometer, dividing the total rectifier voltage into two portions.

away, why not put it to good use?

Back-bias may look peculiar to those who are used to seeing a dedicated bias rectifier, but it is really very simple. The main rectifier produces a DC voltage approximately equal to the peak AC transformer voltage, quite normally, but this total DC voltage is measured *between the HT and the bias rail*. In other words, back-biasing takes the (roughly constant) total DC voltage and divides it up into two parts: HT and bias. Hence whatever raw bias voltage we choose to generate is subtracted from the effective HT voltage. You can think of the amplifier circuit and back-biasing resistor as forming a potentiometer, as illustrated in fig. 8.16. For example, if we want a raw negative voltage that is one tenth of the HT, R_1 must be equal to one tenth of the load resistance presented by the amplifier circuit itself. The same principle applies to ripple voltage; if the two capacitors are equal then both rails will have equal ripple, while if C_1 is ten times larger than C_2 then ripple on the bias supply will be one tenth of that on the HT. Moreover, the ripple on either rail is 180° out of phase with its partner, which poses the intriguing possibility of tweaking the value of C_1 to null out any power supply hum in the audio output stage (since a conventional output stage is inverting), giving it superior PSRR. Again, this will be case specific. Alternatively, we may prefer to make the bias supply as smooth as possible, or adjustable in the usual way, as illustrated in fig. 8.17.

The raw negative voltage developed is simply $I_{dc} \times R_1$, where I_{dc} is the mean average load current of the whole amplifier (ripple current is diverted through C_1). R_1 is therefore likely to need a fairly large power rating since it will dissipate I^2R watts. A further cautionary note is that during start-up a heavy inrush current must charge up the capacitors, and this can cause the voltage across R_1 to be greater (more negative) than normal. The maximum possible value this voltage might reach is determined by the divider formed by C_1 and C_2:

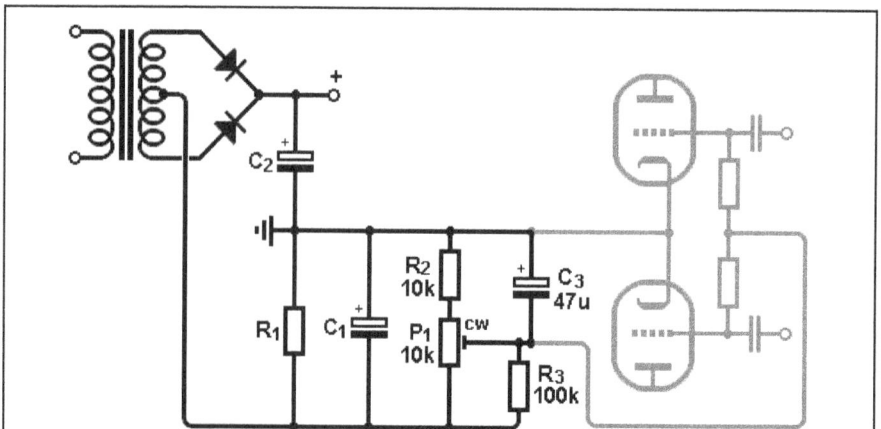

Fig. 8.17: Back-biasing with bias adjust. The values of R_1 and C_1 will depend on the amplifier circuit itself.

$$-V_{max} = -V_{pk} \frac{C_1}{C_1 + C_2}$$ (8.4)

Therefore, C_1 and any other smoothing capacitors used in the bias supply ought to be rated to withstand this maximum.

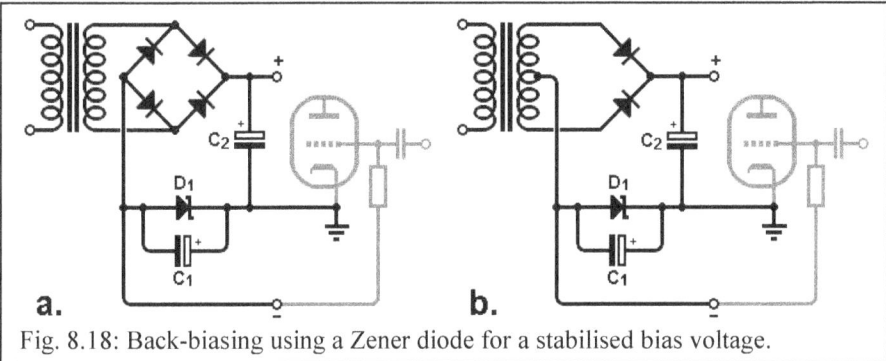

a. **b.**

Fig. 8.18: Back-biasing using a Zener diode for a stabilised bias voltage.

Alternatively, the bias voltage can be stabilised against variations in amplifier current and start-up inrush by using a Zener diode instead of a resistance, as in fig. 8.18, just as cathode biasing can alternatively be done with LEDs or Zeners. Provided D_1 is bypassed by a large capacitor C_1 (e.g. ten times C_2) the Zener will dissipate the same amount of power as an equivalent back-biasing resistor. A bias-adjust circuit can also be added, as before. Having a stabilising bias voltage makes this form of back biasing (somewhat) more suitable for amplifiers beyond Class-A.

Back biasing was exploited in many early radio sets by placing the conventional HT smoothing choke or field coil in the ground return path (*after* the reservoir capacitor). This is illustrated in fig. 8.19. R_1 should be several kilohms to avoid shunting the choke impedance, while C_1 provides additional bias smoothing. The DC voltage drop across the wire resistance of the choke, although relatively small, was sufficient to back-bias the various receiving

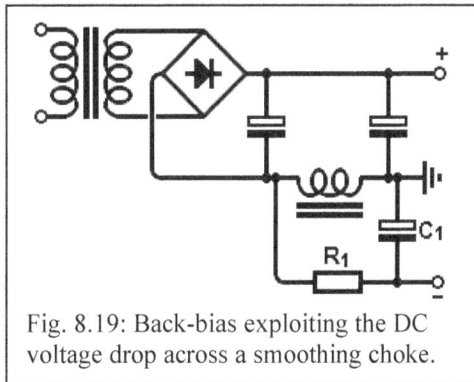

Fig. 8.19: Back-bias exploiting the DC voltage drop across a smoothing choke.

valves; a nice example of efficient design in the days when every penny counted. The same principle could of course be used for audio preamp valves.

Chapter 9: Valve Heater Supplies

Every valve amplifier has a heater supply. In vintage equipment this is invariably a plain AC supply from the transformer, but in modern equipment, especially hi-fi amps, it may be DC. As usual, it may be as simple or as complicated as our patience permits.

9.1: Valve Heaters

The heater or filament in a valve is a very simple element that can be likened to an ordinary incandescent lamp. The earliest valves were directly heated, meaning the heater and cathode were one and the same thing. Indeed, the very earliest types were essentially light bulbs with ambitions, where the cathode –more properly called the **filament** in such devices– was a simple tungsten wire which had to be heated to around 2500K for satisfactory electron emission.

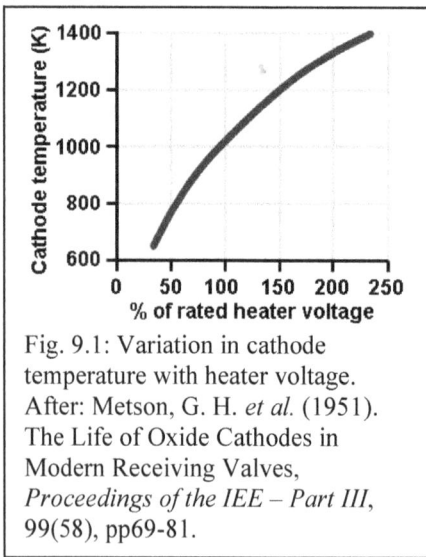

Fig. 9.1: Variation in cathode temperature with heater voltage. After: Metson, G. H. *et al.* (1951). The Life of Oxide Cathodes in Modern Receiving Valves, *Proceedings of the IEE – Part III*, 99(58), pp69-81.

Later directly-heated valves used **thoriated-tungsten** filaments. These mixed thorium oxide with the tungsten to reduce the work function of the metal, so it only had to be heated to about 2000K. Thoriated-tungsten cathodes are quite fragile and such valves must be handled with care; they also benefit from soft-start heating circuits. Power triodes such as the 2A3, 300B, 845 and many transmitting valves fall into this category.

The third stage of development was the indirectly heated cathode in which the cathode is heated by an entirely separate heating element (often omitted from the valve symbol used on circuit diagrams). This has many advantages over direct heating, including reduction of hum and the freedom to operate many valves from the same heater supply. Moreover, indirectly-heated cathodes operate at a tepid 1000 to 1100K and the heater only needs to run a little hotter –about 1500K– to achieve this. For interest, fig. 9.1 shows how the cathode temperature varies with applied heater voltage.

The heater is made from a tungsten wire which is folded or coiled, and coated with aluminium oxide to insulate it from the nickel cathode tube. The heater may further be twisted to create a double helix which improves the cancellation of its magnetic field and so reduces hum.[1] The thickness of the insulation determines how much

---

[1] Hasset, W. A. (1961). The Materials and Shapes of Vacuum Tube Heaters, *Electronic Industries*, December, pp118-22.

voltage can be tolerated between heater and cathode before too much current leaks between them. Not surprisingly, indirectly-heated valves became the standard receiving type from the 1930s onwards, except for some rectifiers, transmitting valves, and notably for small battery-operated valves such as the DF96 whose filament needs just 1.4V/25mA.

9.1.1: Heater Voltage and Current Rating

Heater power scales approximately with the size of the bottle; power valves must handle large currents and therefore have large cathodes that need commensurately higher heater power than small preamp valves. In fact, the heater voltage/current is often quite unique to certain valve types, and when faced with an anonymous valve from which the printing has rubbed off, measuring the heater current can provide a useful clue about what type it may be.

Valve characteristics are normally expected to be within tolerance when the heater voltage is within ±10% of its nominal value, which accommodates mains voltage variations (Mullard specified ±7% for heater voltage, and some special-quality valves expect ±5%). Operating the heater *above* its allowed range will lead to premature ageing without bringing many significant advantages and is therefore to be avoided. Conversely, many designers prefer to operate at the lower end of the allowed voltage range as it encourages longevity while still meeting the datasheet specifications –voltage regulators also come in round numbers like 12V rather than 12.6V. The lower temperature reduces cathode evaporation and interface-resistance growth, and improves heater reliability.[*] Potential lifetime therefore increases, and under certain conditions noise is reduced too.[2]

It is sometimes argued that reduced heater power will lead to reduced lifetime, but this is only half true. When the heater power is reduced, the cathode emission reduces, and therefore its saturation current is also reduced. If the heater voltage is *excessively* low then it is possible that an anode current that would normally be considered quite safe will in fact result in saturation, and this will indeed lead to reduced lifetime. Also, reducing the heater voltage too much causes the anode characteristics to 'slump', increasing r_a and reducing g_m (μ remains largely unchanged and may even increase slightly[3]), thereby reducing the valve's electrical usefulness. However, operating the heater only 10%-low results in negligible slumping of the characteristics and tends to maintain them for longer too, i.e. it

---

[*] According to RCA Electron Tube Division, (1962), *Electron Tube Design,* p91, heater failure rate is approximately proportional to the 12.5[th] power of heater voltage. Reducing the heater voltage by only 10% may therefore improve heater reliability by almost four times.
[2] Blencowe, M. (2013). Noise in Triodes with Particular Reference to Phono Preamplifiers, *JAES*, 61 (11), November, pp911-16.
[3] Winter, A. J. (1953). The Effect of Filament Voltage Upon Vacuum Tube Characteristics, *Trans. IRE* (January), pp47-59.

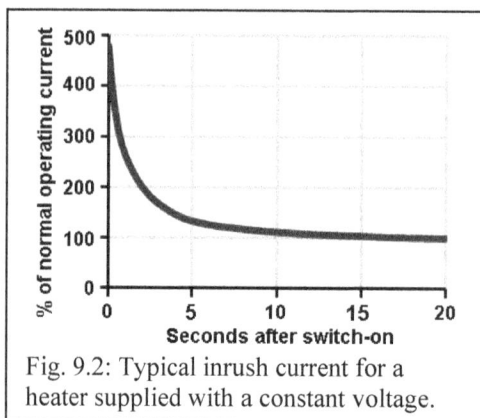

Fig. 9.2: Typical inrush current for a heater supplied with a constant voltage.

increases the useful lifetime.[†] For a 6.3V heater this means operating at not less than 5.7V.

When the heater is cold its resistance is very low, typically about one fifth of its normal hot resistance. When a voltage is first applied there will be an inrush current that decays exponentially as the wire heats up and the resistance rises, until the steady state is reached. Fig. 9.2 shows the typical current variation after the rated voltage is applied. This inrush current may require the use of unexpectedly large fuses in the power supply, as mentioned in chapter 5. It also causes some heaters to flash bright white at power on –a frequent cause for alarm among beginners– but this is normal and is simply a matter of the specific valve design. Heater failure is extremely rare, and soft-start circuits are quite unnecessary for indirectly heated valves (but they don't hurt, either).

The datasheet will sometimes state that the heaters are intended to be supplied from a constant voltage, i.e. parallel connected, the current rating being only approximate. Alternatively, the specification may be for a constant-current supply, i.e. series connection, the voltage rating being only approximate. Valves advertised for series connection were often part of a manufacturer's family of devices such as the Philips/Mullard 'U' and 'P' series, which all have 100mA and 300mA heaters respectively, and matching warm-up characteristics so no individual heater is over stressed after switch on. The Philips/Mullard 'E' series (and in some documents also the 'P' series) were advertised as being suitable for both series and parallel connection. This included the popular ECC83/12AX7 dual triode and its brethren, which have three available heater pins so the two triodes' heaters can be connected in parallel for 6.3V/300mA[*] or in series for 12.6V/150mA.

However, these differences in usage appear to be mainly historical, and even the manufacturers' databooks are inconsistent with their advice.[††] Heaters are trivially simple devices and are so consistent and predictable that they can nearly always be

---

[†] According to Martin, A. V. J. (1967), Factors Determining Tube Life, *Radio & TV News* (July), p111, cathode life is inversely proportional to the 9th power of heater voltage. A 10% reduction may therefore more than double the cathode lifetime.

[*] The odd figure of 6.3V is borrowed from old standard battery voltages. If valve datasheets had been written today they would almost certainly have specified a round number instead, to suit modern electronics design trends.

[††] For example, manufacturers couldn't always agree about what the heater voltage was supposed to be! RCA quoted the 7F7 as having a 6.3V/300mA heater, whereas Sylvania quoted it as a 7V/320mA, but then only provided application data for 6.3V.

operated in series or parallel as desired, whatever the datasheet implies. The so-called 'approximate' voltage or current ratings are usually quite accurate, and it is unlikely that the values in practice will come out more than 10% off the quoted values, whether constant voltage (parallel) or constant current (series) operation is used. With series operation it is only necessary to ensure that no heater is over stressed during the warm-up phase, which is easy enough with modern electronics, so there is no longer a pressing need for heaters with matching warm-up times. However, it is worth pointing out that if valves have previously been tested and carefully matched with the heaters connected in parallel, then connecting them in series may cause them to become unmatched, and vice versa, due to the small differences in heater power in each case

9.1.2: Heater-Cathode Voltage Rating

The datasheet will normally quote a maximum permissible voltage allowed between heater and cathode, $V_{hk(max)}$, below which heater-to-cathode leakage current is guaranteed not to exceed a certain value. Most valves have a guaranteed heater-cathode insulation resistance of $>10M\Omega$, and under normal conditions it actually reaches hundreds of megohms, so leakage current can be expected to be less than a microamp in either direction.

Heater-cathode leakage is mainly due to the emission of electrons from the cathode or the exposed ends of the heater, which are then attracted to whichever electrode is more positive. Some leakage is also due to migration of ions directly through the heater insulation. Leakage is usually worse when the heater is positive with respect to the cathode than when it is negative,[4] and a few valves quote different V_{hk} limits for positive and negative polarities.

However, $V_{hk(max)}$ is more of an advised limit than a true maximum, and exceeding it may not cause immediate problems; leakage will instead increase gradually over time, leading to excessive hum or intermittent pops and crackles. Admittedly, the heater insulation *will* break down completely if enough voltage appears across it, but this limit is well beyond the numbers quoted on datasheets. For many valves the stated maximum is ±90V (when in doubt this is a safe value to assume) but some go much higher, although whether this is always because of a genuine difference in heater insulation or simply a relaxing of the specification remains to be seen.

Leakage between heater and cathode is most annoying if the heater is AC powered since an AC current leaking into the cathode circuit will develop an AC hum voltage across any impedance it encounters, which will be amplified like any audio signal. Leakage is not a problem if the heater is powered with regulated DC, unless it becomes so excessive that the valve develops intermitted crackle; this is a known problem in certain amplifier circuits using high voltages and cathode followers with no heater elevation. However, if a valve does develop this problem then it may still

---

[4] Dingwall, A. G. F. (1962) Heaters. In: RCA Electron Tube Division, (1962). *Electron Tube Design*, pp232-43.

be perfectly usable if it is substituted into in a different circuit location where it is no longer stressed.

Obviously, the heater-cathode voltage rating is of particular importance in circuits where the cathode is at a very different voltage from the heater, such as cathode followers, SRPPs, μ-followers and so on. In such cases it may be necessary to elevate or 'float' the heater supply on a voltage that is close to the cathode voltage – see section 9.3.6. When many valves share the same heater supply it will be necessary to elevate it to a voltage that accommodates the $V_{hk(max)}$ ratings of all the devices. This may require a compromise value that satisfies all the valves at idle, but may be exceeded when signal voltages are present, e.g. when the cathode voltage of a cathode follower swings up to its peak value.[*] If heaters are connected in series then they can be arranged so the valves with the highest cathode voltages are supplied from the higher-voltage end of the heater chain. In more extreme cases it becomes necessary to use more than one heater supply for different parts of a circuit, each elevated to a different voltage.

At the points where the heater actually touches the inside of the cathode tube a kind of semiconductor is created. At these spots the leakage becomes quite non-linear with applied voltage. It is not necessarily zero when V_{hk} is zero, and may level off or saturate if V_{hk} is large enough. Consequently, it is often found that elevating the heater by a few tens of volts (up to ±50V, say) will reduce hum –either by minimising leakage or by saturating it– even if other factors don't require it. Usually, though, we have little choice about the elevation voltage because it is dictated by a cathode follower or some other amplifier stage with a high cathode voltage.

9.1.3: Heater-Cathode Resistance Rating

Many datasheets quote a maximum allowable resistance between cathode and heater, $R_{hk(max)}$, which often takes the form of a bias resistor; a quoted value of 150kΩ is typical. This limit is determined by AC leakage current between heater and cathode; the larger the resistance between the two, the greater the hum voltage that will develop across it due to the leakage current. Again, this is more of an advised limit than an absolute one; exceeding it is not likely to cause a reliability problem (provided $V_{hk(max)}$ is not also exceeded) but rather the device may not meet hum guarantees.

---

[*] Curiously, the presence of signal voltage on top of the average heater-cathode voltage may actually improve matters. See: Gentry, C. H. R *et al.* (1965). Cathode/Heater Insulation Failure in Oxide-Cathode Valves. *Proc. IEE*, 112 (8), pp1501-8.

9.2: AC Heater Supplies

The simplest heater supply is an AC supply, straight from the power transformer. Most valve amplifiers use the popular 6.3V valves with all the heaters connected in parallel, supplied from a dedicated winding as in fig. 9.3.[†] The transformer must supply the current demanded by each individual heater, so the total is simply the sum of them all, as indicated in the figure

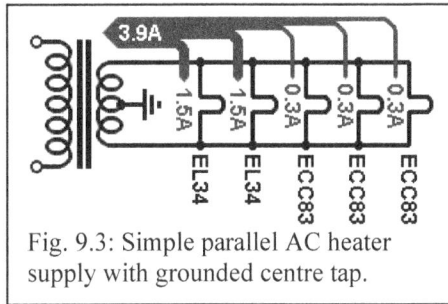

Fig. 9.3: Simple parallel AC heater supply with grounded centre tap.

$$I_{total} = I_{h1} + I_{h2} + I_{h3}...$$

In this example the total is 3.9A, so the transformer must be rated for at least this much, in practice probably a round number like 4A or 5A. It can of course be rated for even more current, but the voltage will rise above the nominal value if it is not loaded to its full rated capacity. Of course, the small excess could be dropped with resistors, diodes, or by some other means if necessary, so this is not a big problem.

If a separate, dedicated transformer is used for the heaters it will typically be an off-the-shelf 6V type, rather than exactly 6.3V, but this is within 5% which is close enough. Besides, it will probably have a regulation figure of around 12% so will deliver 6.3V when loaded to only 63% of its full rated capacity. In other words, if we need 6.3V at 2.5A then we could select a 6V transformer rated for 2.5A/0.63 = 4A, and it should produce almost exactly 6.3V, assuming the wall voltage is ideal. On the other hand, as explained earlier, it can be beneficial to run heaters slightly below nominal anyway.

Using a separate transformer for the heaters has other benefits too:

- Rectifier noise on one transformer is not easy coupled to another transformer, since it is shunted by the low impedance of the wall supply. This is significant, because borrowed rectifier noise is often a cause of obnoxious buzzing (rather than low hum) occurring on heater supplies using standard valve transformers (see section 9.3.3). By using separate transformers, each is magnetically isolated from the other.
- It gives greater design freedom over the voltages used, rather than being endlessly coerced into using the popular 6.3V valve types. Valves with other heater voltages are often much cheaper.
- It allows greater flexibility of construction. Low-profile, toroidal, or special transformer packages can more easily be fitted into the available space.

[†] Parallel connection means each heater receives the same voltage, ignoring any losses in the connecting wires, which is why textbooks and datasheets may refer to it as 'constant voltage' operation. This is not meant to imply that the voltage source is necessarily stabilised or regulated, though of course it may be.

When more than one heater voltage is possible –such as 6.3V or 12.6V for an ECC83/12AX7– then the lower voltage is preferable from the point of view of hum and buzz because it minimises the electric field around the heaters and associated wiring (coupling of the magnetic field is usually a lesser influence than that due to the electric field). On the other hand, 12V operation of such valves makes the socket wiring simpler since pin-9 is left unconnected, and 12V is sometimes a more 'useful' voltage with regard to powering other circuit elements such as relays and opamps.

9.2.1: Series and Parallel Combinations

Although some valve datasheets list certain valve heaters as specifically for parallel operation or series operation, heaters are so consistent and predictable that it is possible to arrange them in almost any desired series or parallel combination. However, series heaters on AC are more likely to induce hum, owing to the larger voltage swings occurring along the heater chain. The higher the AC voltage, the greater the problem is likely to be, so long AC heater chains running off a high-voltage supply (traditionally the wall supply) have more or less fallen out of use. Today we are more likely to put just two or three heaters in series, often because there is a suitable voltage supply already available for some bit of solid-state circuitry, e.g. in a hybrid amplifier.

If heaters of different current ratings are to be connected in series then we need to supply each one with the correct current. This is easily achieved by adding resistors in parallel with the lower-current heaters, to divert any excess.[5] For example, if an ECC83 and EL84 were to be supplied from a 15V source, the circuit might look as in fig. 9.4. The ECC83 is connected for 6.3V, so the two valves together will need 12.6V. This leaves 2.4V to be dropped by R_1. The EL84 needs the most current at 760mA, which decides the total, so R_1 must be 2.4V/0.76A = 3.2Ω. The ECC83 needs only 300mA, so 460mA must be diverted

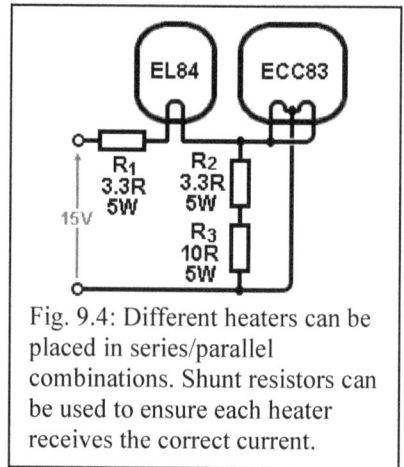

Fig. 9.4: Different heaters can be placed in series/parallel combinations. Shunt resistors can be used to ensure each heater receives the correct current.

around it by R_2 and R_3, which in total must be 6.3V/0.46A = 13.7Ω. In practice we use standard values which get us close enough. This approach is of course very wasteful of power –in this case burning off 4.7W uselessly– but in some applications this may still be cheaper than any alternative.

A further concern with series operation is that some heaters may take longer to reach normal (hot) operating resistance than others, which in turn means the quicker

---

[5] Palmer, G. (1959). Alternations to Series Heater Circuits, *Practical Wireless*, July, pp379-80.

heaters may be subjected to excessive current during the warm up period, which invites premature failure. The obvious solution is to add an NTC thermistor as in fig. 9.5. Its cold resistance should be selected to be around one quarter to one half of the total cold-resistance of the heater chain.

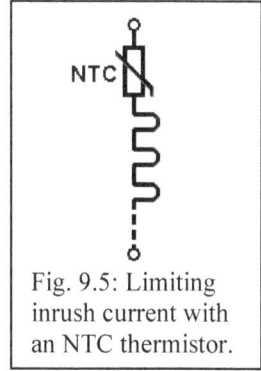

Fig. 9.5: Limiting inrush current with an NTC thermistor.

9.2.2: Dropping Heater Voltage

Depending on the mains voltage and the power transformer being used, it is not unusual to find the AC heater voltage is somewhat higher than we would like it to be. This is not a great concern if it is within 10% of the nominal voltage e.g. not more than 6.9V when aiming for 6.3V. However, if it is uncomfortably high then it can be reduced by adding some resistance immediately in series with the transformer, as illustrated in fig. 9.6. The resistance is easily calculated from the desired V_{drop} divided by the total (nominal) heater current, and the power dissipation will be I^2R.

If the heater winding uses a centre tap for balancing (section 9.3.5) then the total dropping resistance should be split into two equal parts to maintain symmetry, as shown in a. This is not essential if an artificial centre tap or humdinger is used, since the dropping resistance can be inserted before the balancing resistors as in b. The dropping resistance can also do double-duty by creating a hash filter if we also add a 100nF to 1µF capacitor, shown faint.

Another simple way to drop a little heater voltage is to insert a pair of suitably-rated rectifier diodes into the heater feed, as shown in fig. 9.6c. This will reduce the RMS heater voltage by about 0.8V. This is a dirty trick, however, since it cores-out the zero-crossing of the heater waveform, and an ugly waveform potentially leads to buzz. On the other hand, this effect may turn out to be negligible compared with other sources of hum in a vintage or high-gain guitar amplifier.

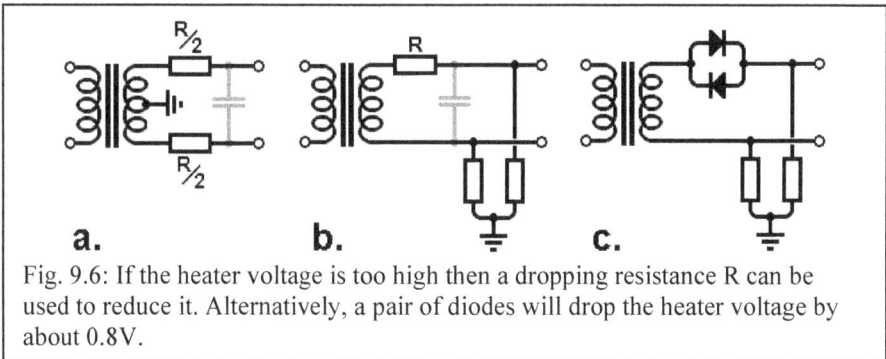

a.   **b.**   **c.**

Fig. 9.6: If the heater voltage is too high then a dropping resistance R can be used to reduce it. Alternatively, a pair of diodes will drop the heater voltage by about 0.8V.

9.2.3: Wattless Dropper for Series Heaters

On an AC heater supply it is also possible to drop excessive voltage with a capacitor, as in fig. 9.7, which therefore dissipates no heat. This is most useful when the voltage to be dropped is very large, such as a heater chain supplied directly from the wall supply in a vintage radio. If the dropped voltage is fairly small, the required amount of capacitance may be impractically large, given that a quality non-polarised capacitor is needed. Moreover, when the dropped voltage is large, i.e. when the reactance of the capacitor is significantly larger than the heater resistance, the capacitor dominates the total impedance, making it behave more like a constant-current supply.

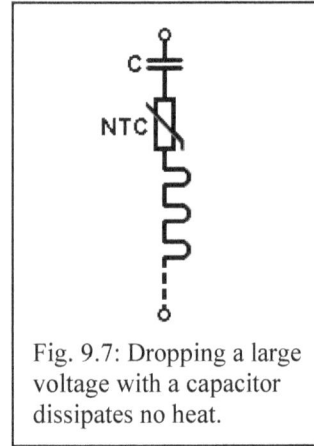

Fig. 9.7: Dropping a large voltage with a capacitor dissipates no heat.

Choosing the capacitor is not as simple as making its reactance equal to the dropping resistance we might otherwise have used, because reactance and resistance do not add directly. Remember, the voltage across the capacitor and heater chain will be 90° out of phase. The reactance of the capacitor must instead be equal to:[6]

$$X_c = \frac{1}{2\pi f C} = \frac{V_c}{I_h} = \frac{\sqrt{V_s^2 - V_h^2}}{I_h}$$

Rearranging the above to find the capacitance:

$$C = \frac{I_h}{2\pi f \sqrt{V_s^2 - V_h^2}} \tag{9.1}$$

Where:
V_s = supply voltage;
V_h = desired heater voltage;
I_h = nominal heater current.

Since the capacitor must run continuously at fairly high current it needs to be a hard-wearing device, which means a plastic capacitor (e.g. polyester/MKT), rated for the full supply voltage. If supplied directly from the mains then it should additionally be a Class-X or -Y safety capacitor. Since capacitors have notoriously poor tolerance, it may be worth using several in parallel to improve precision, and this often works out cheaper and easier than finding one big plastic capacitor anyway.

Finally it is worth noting that, depending on where in the AC waveform the supply is switched on, a very large current transient may rush in, potentially stressing the heaters. An additional NTC thermistor is therefore recommended to encourage a more controlled start. Choosing this part is not an exact science, but as noted earlier,

---

[6] Bradley E. N. (1950). Practical Series Condenser Heater Circuits 2, *Practical Wireless*, February, pp58-9.

try one with a resistance equal to one quarter to one half the total cold-resistance of the heater chain

9.3: Hum

Hum (and buzz) is frequently caused by interference from electromagnetic fields at mains frequency, which may originate inside or outside the amplifier. Outside sources include the electric fields produced by fluorescent lights, televisions, computers and so forth, and the magnetic fields produced by transformers and high-current carrying cables. Inside the chassis, hum comes mainly from the electromagnetic fields originating in the power transformer, mains wiring, and AC heaters. It may also be caused by residual ripple on the power supply, including DC heater supplies.

9.3.1: Electric Fields

The electric field strength –also called the **electrostatic field intensity**–around a conductor is related to the voltage on the conductor, and for simple parallel plates or wires it is:

$$E = \frac{V}{d} \tag{9.2}$$

Where:
E = electric field strength, in volts per metre;
V = potential difference between the conductors, in volts;
d = distance between conductors, in metres.

If the electric field is varying then it will induce a noise *current* into any nearby circuit. The more rapidly the electric field changes, the greater the induced current. This is really just a physical way of describing how voltage signals couple from one wire to another through the stray capacitance between them, and since stray capacitance forms a CR filter with the resistances in each circuit, high frequency signals are coupled more easily than low frequency ones.

Shielding against electric fields is very easy; any bit of metal will do, it just needs a low-impedance connection to ground. The chassis provides the ultimate shield against external electric noise, but shielding may still be required internally, either to protect sensitive wiring against noise coming from other areas, or to enclose noisy conductors and prevent their electric fields from permeating the air. In particular, the wires leading from the input connector to the first valve will sometimes be shielded to discourage parasitic feedback.

Noisy wiring should run close to the chassis wall or, better still, be pushed into the corners to maximise the capacitance to earth. The earthed chassis will then have a similar effect to a Faraday cage and will tend to draw the electric field towards itself. This often avoids the need to shield wires completely, especially heater wires.

Large power supply smoothing capacitors can sometimes be judiciously placed to provide some screening between different parts of a circuit too, since the metal can is normally connected to the negative terminal, which is often grounded. Likewise, other grounded metal objects like smoothing chokes or reinforcing brackets can be used to advantage. Many valve sockets have a central spigot that can be grounded to provide a modicum of shielding between opposing pins. In extreme cases the chassis may be partitioned into sections with bulkheads to create isolated chambers that are shielded from one another. In some sensitive amplifiers the heater wiring was even run on the outside of the chassis, feeding each valve socket by entering through nearby hole to minimise the length of noisy conductors actually inside the chassis.

9.3.2: Magnetic Fields

A wire carrying current generates a magnetic field around itself according to Ampere's law:

$$B = \frac{\mu I}{2\pi r} \qquad (9.3)$$

Where:
B = magnetic flux density in tesla, or webers per square metre
μ = permeability of the material around the conductor, which for a simple wire can be taken to be the same as air: $4\pi \times 10^{-7}$ H/m
I = current in the wire, in amps
r = radial distance from the wire, in metres

If there is any fluctuation in the current, there will be a corresponding fluctuation in the magnetic field which, in accordance with Faraday's law, will induce a fluctuating *voltage* (EMF) into any wire loops which happen to be within the magnetic field at the time. The larger the loop area, the greater the induced voltage. Wires carrying AC mains and AC heater wires are the main offenders here since they carry the largest AC currents in the appliance. Transformers also generate strong magnetic fields, and steps are often taken to minimise their effect on other circuitry.

Shielding against magnetic fields is much more difficult than shielding electric fields, for various reasons. At low frequencies the metal used for shielding must have high magnetic permeability; the usual choice is mu-metal. This has one hundred times higher permeability than steel, and considerably higher cost. Its permeability degrades when the metal is machined, so it must be annealed once the final shape has been formed.

Furthermore, magnetic shielding is most effective when it forms a completely unbroken band or box, since the idea is to encourage any interfering flux to flow in the shield rather than in the thing being screened. Any gaps or seams will prevent the flux from flowing in a complete loop, so magnetic screens are often stamped out, rather than being cut and folded. This is an expensive process. Explicit magnetic shielding is therefore rarely used except on phono/microphone signal transformers. A mild-steel shroud may have a slight magnetic advantage over a stainless steel or aluminium one, but it is hardly noteworthy unless the steel is unfeasibly thick. Instead, the most practical techniques for minimising the effects of magnetic fields are to adopt a tight audio layout with good lead dress and twisted wires to minimise loop areas; to orientate transformers favourably; and to keep transformers and high-current carrying wires well away from sensitive circuitry.

9.3.3: Lead Dress

Lead dress refers to the physical arrangement and layout of conductors in a project, and it is important with regards to hum. Electromagnetic fields decay with distance from the source, so physical separation is our friend. If wires that may interfere with one another must cross then they should do so in a perpendicular fashion, to minimise the area of wire 'seen' by each other. Allowing a noisy wire to run in parallel with a sensitive signal wire is a cardinal sin.

Signal wires leading to valve pins should be kept as short as possible (indeed, signal wiring in general should be kept as short as possible). This obviously requires mounting the valves physically close to the circuit area they serve, i.e. form should follow function.

Pairs of noisy AC wires such as mains and heater feeds should be neatly twisted so the opposing magnetic fields around each wire are forced to occupy the same space, causing them to cancel each other out. Note that a loose twist is useless, only a tight twist will do! But neither should the twists be *too* tight as this may lead to fatigue and internal breakage; the old rule of three twists per inch is usually fine. Twisting is easily done by anchoring the wires at one end and holding the other end in the chuck of a drill, keeping reasonable tension on the wire while twisting. Stretching it gently before releasing will discourage it from twirling back on itself. It is better to use stranded wire for this, as solid-core wire will quickly develop fatigue and break internally.

Heater supplies normally daisy-chain from one valve socket to the next, supplying many heaters in parallel, so current flowing in the *supply* end of the chain is greater than the current flowing near the last valve in the chain. Power valves require the most current and are the least sensitive to heater hum, so they should be at the supply end. The heater chain should then progress logically through the amplifier with the input valve last in the chain where the current –and therefore field strength– is smallest. Heavier gauge wire may be needed at the supply end of the chain where more current is being carried, but a common error is to use wire which is *too* heavy, making it difficult to manipulate. Ordinary 16×0.2mm (~20 AWG) equipment wire has a resistance of about 0.04Ω per metre and is sufficient for up to about ten amps.

When heater wires approach a valve socket the twisting must be kept tight right up to the socket since the other valve pins are in close proximity here, so we must suppress the magnetic fields as much as possible. Allowing the twists to become loose near the socket will spoil a lot of hard work. It is also important not to create a loop of heater wiring around a valve socket since any other wires or valve pins inside the

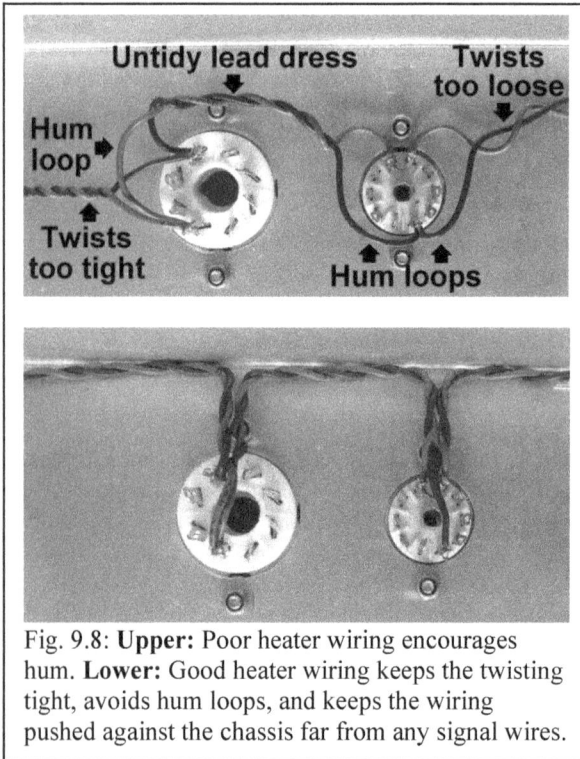

Fig. 9.8: **Upper:** Poor heater wiring encourages hum. **Lower:** Good heater wiring keeps the twisting tight, avoids hum loops, and keeps the wiring pushed against the chassis far from any signal wires.

loop will be subject to increased EM interference. The heater wiring should approach from one side of the socket and, if it must cross it, should jump directly across the socket and straight back. It helps to orientate valve sockets so the heater feeds can all approach from the same direction, usually from a trunk feed pushed into the corner of the chassis.

An example of good heater wiring is shown in the lower image in fig. 9.8. This takes care and patience, and it will usually be obscured by other wiring once the amp is complete, so it is worth getting it right first time. The upper image shows some typical mistakes. On a PCB layout, heater traces cannot be twisted of course, but they should at least run side by side very close to one another. On a double-sided board a sort of quasi-twisting can be achieved by 'saddle stitching' the traces back and forth between layers.

9.3.4: Rectifier-Induced Hum

Since AC heaters operate at mains frequency, which is quite low, hum due to electric-field coupling *ought* to be negligible if sensible lead dress has been observed;. After all, even a whopping 10pF of stray capacitance has a reactance of over 300MΩ at 50Hz. However, if the same transformer supplies both the HT (or some other DC supply), *and* the heater supply, the clipped waveform produced by rectifier action will be reflected through the

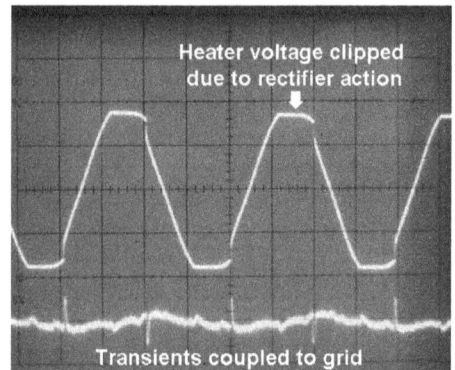

Fig. 9.9: Oscillogram showing how the heater voltage waveform has been clipped by a rectifier on another winding of the same power transformer. The transients picked up on the grid correspond exactly with the cliff-edges of the heater waveform.

transformer to the heater winding too. The result is that the heater voltage will not be pure 50/60Hz but will contain all the high-frequency hash that rectifiers produce, plus any noise coming from the wall supply itself, of course.

Fig. 9.9 shows an oscillogram of the heater-voltage waveform from a transformer which also provides the HT in a small valve amp, and it is obviously not a very good sine wave. The top of the wave is clipped and shows the classic 'hangover' explained earlier in section 2.4.3. The high dV/dt of this hangover, plus any rectifier switching noise that may exist too, will easily couple into the audio circuit via valve interelectrode capacitances. The lower trace shows the signal picked up on the grid of the input valve (which had a 1MΩ grid leak) and transients coinciding with each hangover are clearly visible. It is these transients which often cause heater interference to sound buzzy, rather than the low hum we might otherwise expect. Heater balancing can suppress this differential-mode noise.

9.3.5: Electrical Heater Balancing

The heater supply must always have a reference to audio ground, which may be a direct connection or an elevating circuit (next section). This is equally true for AC or DC supplies. Leaving the heater supply floating will result in almighty hum due to primary-to-secondary transformer leakage current, and is a common beginner's error. AC heater supplies should also be balanced to suppress the EM field. Not only does this reduce the magnitude of the voltage on each wire (e.g., each wire handles ±3.15V rather than one wire handling 6.3V and the other zero), but the opposing fields will tend to couple equal-but-opposite hum signals into the audio circuitry, which should cancel each other out.

If the transformer heater winding has a centre tap it can be grounded to create a balanced heater supply as in fig. 9.10a, and this is by far the most common approach found in vintage equipment. However, if a short develops between the anode and heater pin of a valve socket (on popular octal valves like the EL34 these pins are unfortunately adjacent) then it will short the anode directly to ground through the heater supply, which is the worst possible fault condition. It is therefore preferable to insert a small resistor (ideally fusible / flameproof) into the heater centre tap to serve as an emergency fuse, as in fig. 9.10b. This is also a good habit for beginners to adopt, as they often become fixated with grounding the centre-tap simply 'because it is there', even when some other ground reference is already being used. A burnt-out resistor is a much cheaper way to discover a grounding conflict, rather than a burnt-out transformer winding.

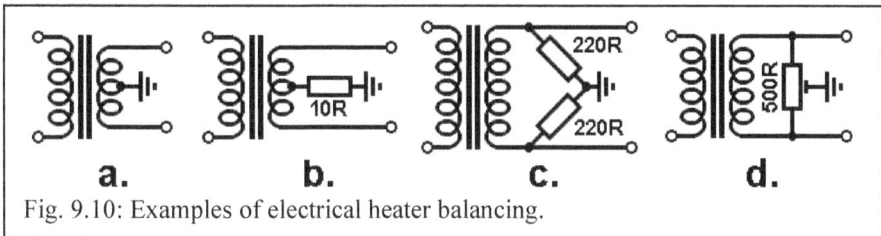

Fig. 9.10: Examples of electrical heater balancing.

If there is no centre tap then the heater supply can be balanced using a pair of resistors instead, as shown in fig. 9.10c. They should be small-valued to encourage the shunting of transformer leakage current to earth and to avoid adding unnecessarily to R_{hk}, so values of 100Ω $\frac{1}{2}$W to 220Ω $\frac{1}{4}$W are usual. There is no point in carefully matching these resistors since the coupling of hum into the audio circuit is itself not exactly balanced or predictable. This is particularly the case when rectifier-induced hum is the culprit, both because the rectifier clipping is never perfectly symmetrical, and the way in which it is coupled into the audio circuit is not symmetrical either, and will vary with every set of valves that is plugged in. In which case what we really need is an 'off-centre' tap! This is easily accomplished with a **humdinger**. This is simply a trimpot connected across the heater legs, with the wiper connected to ground, as shown in fig. 9.10d. Again, a small value is preferable; a 500Ω pot will dissipate less than 80mW at 6.3V. The pot can then be adjusted for minimum hum, which is not likely to occur at the exact centre setting.[*]

Electrical balancing can also be used on DC heater supplies, although the benefits may not be so obvious. A common beginner's error is to try to add a ground reference on both the AC and DC sides of the rectifier. This will short it out! Only one ground reference should be used (usually the AC side results in the least hum) as in fig. 9.11; the other side of the circuit will still receive this ground reference through the rectifier.

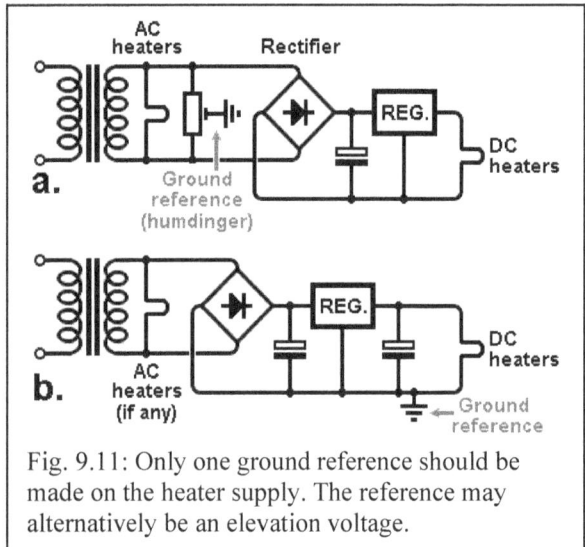

Fig. 9.11: Only one ground reference should be made on the heater supply. The reference may alternatively be an elevation voltage.

9.3.6: Heater Elevation

Heater Elevation means referencing the heater supply to a DC voltage other than ground or zero volts. The heaters still operate at 6.3V or whatever, but this floats on top of the elevation voltage. As mentioned earlier, some valve stages such as cathode followers require the heater supply to be elevated to avoid exceeding the heater-cathode voltage ($V_{hk(max)}$) rating, and this is the most common reason to use heater elevation. However, if hum persists in a circuit with good lead dress and a

---

[*]See also: Gilbert, G. R. (1963). Simple Hum-Bucking Circuit, *Electronics World*, August, p77.

humdinger, then the culprit may yet be heater-cathode leakage, in which case it is worth seeing if elevation will suppress it.

All that is needed for elevation is a potential divider connected to the HT (or other convenient DC supply). This has the natural advantage that the elevation voltage will track any changes in the HT. No current flows 'into' the heater supply from the divider; the heaters are simply jacked up by the elevation voltage. The DC voltage can be applied to a transformer centre tap, artificial centre tap, humdinger, or whatever reference

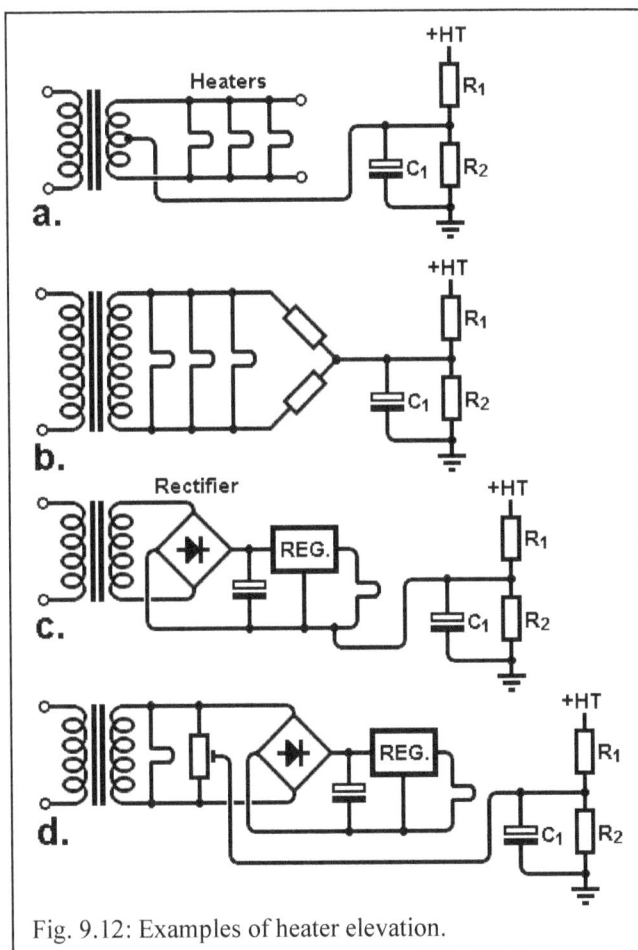

Fig. 9.12: Examples of heater elevation.

connection the heater supply would normally have. Fig. 9.12 shows some examples. As noted earlier, only one reference point should be used. The divider should have a fairly high resistance so as not to waste current, although R_2 should not be excessively large or $R_{hk(max)}$ may be grossly exceeded, so it is advisable not to make it greater than 100kΩ. C_1 provides decoupling/smoothing of the elevation voltage and can be arbitrarily large, say 10µF or more.

In a fixed-bias amp another possibility is to use the bias supply as the elevation (de-elevation?) voltage, as this is sometimes more effective at hum suppression than positive elevation.

9.4: DC Heater Supplies

In very sensitive circuits such as hi-fi preamps or high-gain guitar amps, AC heating may prove too noisy, in which case we resort to a DC supply instead. This should eliminate all heater noise since all EM coupling and leakage currents becomes

241

unvarying (0Hz). However, the DC *must* be well smoothed or even fully regulated; a dirty DC supply may prove noisier than an AC one, owing to the extra harmonics in the ripple voltage which are more easily coupled by stray capacitance and are more noticeable to the ear.

The audio input stage is invariably the most sensitive to picking up hum, while (indirectly heated) power valves are very insensitive to heater hum so almost never need a DC supply, which is a great relief considering how much more current they need. Therefore, we can often get away with supplying only the preamp with DC while the power valves receive ordinary AC. Even more frugal is to supply only the input valve heater with DC, which considerably eases the design requirements on the heater supply.

9.4.1: Simple DC Heater Supplies

The simplest DC heater supply is a rectifier and reservoir capacitor delivering unstabilised DC to the input valve heater,[7] leaving the rest of the valves operating on AC, as illustrated in fig. 9.13. The reservoir capacitance C_1 needs to be very large to bring the ripple voltage down to an acceptable level, typically 4700μF for every 300mA of heater current, and 10000μF would not be unwelcome. It is easy to see why this approach is normally used to supply only one heater. Furthermore, such a large reservoir results in a very poor power factor of around 0.5 on a 6.3V system. This means the DC heater looks twice as hungry from the point of view of the transformer, i.e. the transformer must be rated to handle 600mA for every 300mA heater being supplied with DC. Similarly, the capacitor needs a ripple current rating at least equal to this, too. Even with such a large reservoir the residual ripple voltage could still cause audible hum since we are dealing with the most sensitive valve in the amp, so a humdinger is included (this provides the ground reference for the entire heater system). Heater elevation could be used as well, of course, by returning the wiper to a voltage other than ground.

Also worth noting is that at very low voltages like 6.3V, the losses due to ordinary diode drop will generally result in the DC heater voltage being roughly equal to, or even slightly less than, the AC voltage. In other words, if the heater winding is loaded to its full capacity so the measured AC voltage before the rectifier really *is* 6.3V, the DC voltage can be expected to be close to 6.3V too. If it turns out too low then

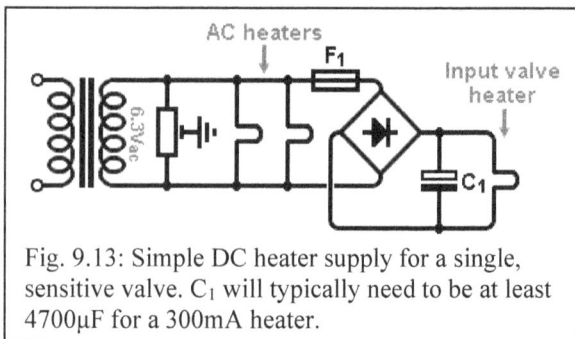

Fig. 9.13: Simple DC heater supply for a single, sensitive valve. C_1 will typically need to be at least 4700μF for a 300mA heater.

---

[7] French, G. A. (1967). Simple D.C. Heater Supply for A.F. Amplifiers, *The Radio Constructor*, August, pp17-18.

Schottky diodes can be used instead, as they have a smaller voltage drop than ordinary rectifier diodes. With higher-voltage systems the DC voltage will approach the peak AC voltage as usual, so any excess may need to be dropped with a resistor or additional series diode.

9.4.2: Common-Mode Filtering

An unexpected source of buzz in an otherwise clean heater supply is common-mode noise. This is caused by mains noise leaking across from other windings on the transformer, through stray capacitance, and appearing *equally* on both heater wires. This noise will be seeking a path to mains earth, e.g. through heater-cathode and heater-grid capacitances. But since the noise is common to both wires even a DC

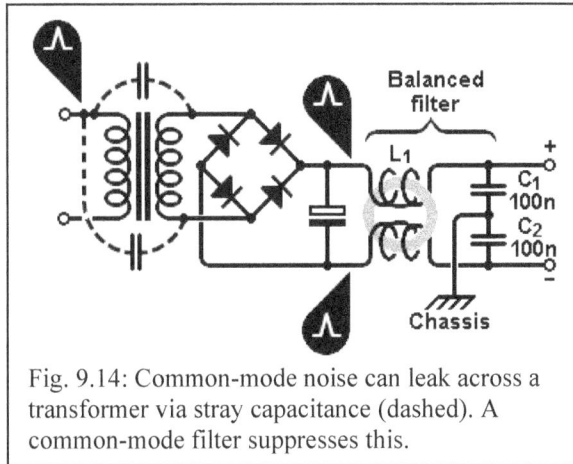

Fig. 9.14: Common-mode noise can leak across a transformer via stray capacitance (dashed). A common-mode filter suppresses this.

voltage regulator will be blind to it. This noise can be suppressed using a common-mode filter. Since heater current is usually quite large, a common-mode choke would normally be used, rather than series resistors. This is followed by a pair of capacitors connected to chassis (i.e. mains earth, not audio ground) as shown in fig. 9.14. The choke L_1 can be bought off the shelf or made from scratch by winding two wires a few times –and in the same direction– around a ferrite core. The values of inductance and shunt capacitors are not critical, though obviously the bigger the better. Such a filter could be used on an AC heater supply too, but any benefit would probably be negligible compared to the differential-mode hum and buzz that already come with an AC supply.

9.4.3: The Free Lunch

Another way to obtain a DC heater supply, at least in theory, is by borrowing the DC currents already used by other parts of the amplifier –most likely the power valves. For example, the heaters could form part of the cathode-bias resistance in a class-A output stage (class-A because the average current is constant, whereas in a class-AB amplifier the 'free' heater current would vary with signal level). Using existing DC currents in this way has come to be known as the 'free lunch' approach. The name is apt because, like the proverbial free lunch, there is really no such thing. Designing an amplifier that operates sensibly with enough current and voltage to supply some typical heaters usually leads to considerable compromises in every other area of the design. It's an interesting academic exercise, but results tend to look very contrived, like a solution that went in search of a problem.

The Bogen DB212 is a rare commercial example of the free lunch; the circuit is reproduced in fig. 9.15 (with measured voltages rather than the speculative ones indicated on the original Bogen schematic). The

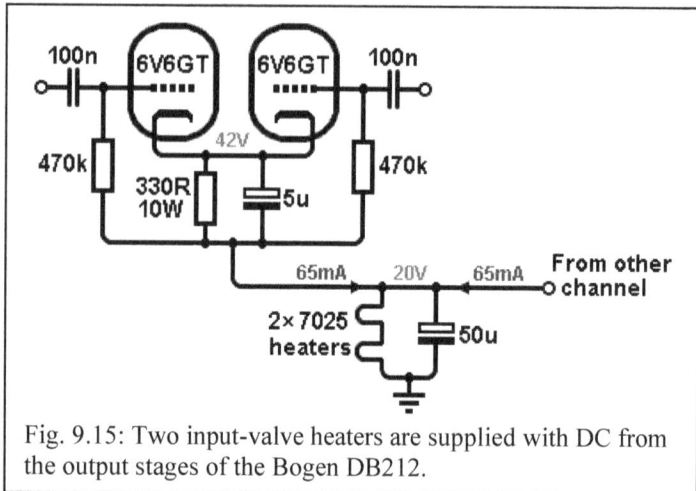

Fig. 9.15: Two input-valve heaters are supplied with DC from the output stages of the Bogen DB212.

cathode currents of its stereo output stages combine to supply the heaters of two 7025 phono input valves (other valves in the amplifier were supplied with AC). In this case the heaters are not even used for biasing the output valves –they are simply stacked on top of the heaters with a 50µF capacitor to decouple the channels from one another. There is a risk of overvoltage on this bypass capacitor if either of the 7025s is pulled from its socket, and the amplifier came with a warning label against this.

A more justifiable implementation of the free lunch would utilise higher-current output valves such as KT88s or 300Bs.[*] For example, a pair of KT88s could be operated at 75mA each and be cathode biased by one or two 150mA heaters, as illustrated in fig. 9.16. Any additional bias voltage could be furnished by individual bias

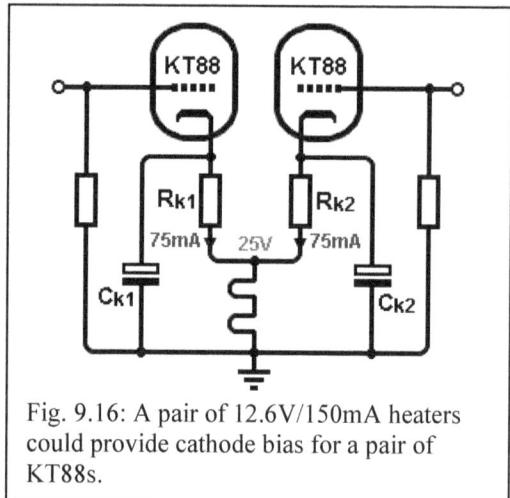

Fig. 9.16: A pair of 12.6V/150mA heaters could provide cathode bias for a pair of KT88s.

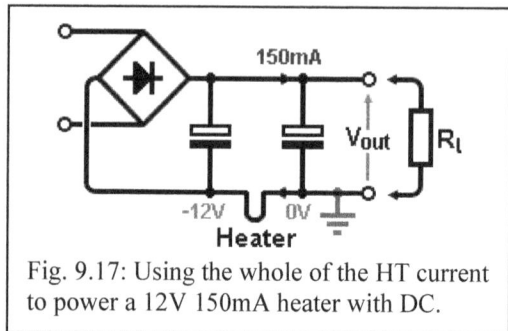

Fig. 9.17: Using the whole of the HT current to power a 12V 150mA heater with DC.

---

[*] See also the 1959 version of the Eico HF-87.

resistors R_{k1} and R_{k2} which would maintain some independence –and therefore better self-adjustment– between the output valves.

Still another free-lunch option is to use the HT current of the entire amplifier.[8] This is possible if the reservoir capacitor is not directly grounded, as illustrated in fig. 9.17. The voltage dropped across the heater is effectively subtracted from the HT available for the rest of the amplifier represented by R_l, so do we really get anything for 'free'? Maybe. With the right-hand end of the heater grounded, the left-hand end is negative and could therefore be used for biasing one or more valves in the set, i.e. back biasing (section 8.4.6). Only one heater is shown in the figure but more could be inserted. The Philips/Mullard 'U' series present some interesting possibilities since their heaters only require 100mA which is more in range of typical HT currents. Sadly, not many of the 'U' series are suited to hi-fi use. Also, be aware that there needs to be enough HT current flowing in the first place to get the free-lunch valve to heat up and add its own contribution to the final total. If you try to power *all* the heaters in the amplifier from their own HT current, it will not switch on in the first place!

9.4.4: Linear Regulators

If we need to supply more than one or two heaters with DC then a simple unstabilised supply may be impractical due to the extremely large reservoir capacitance that would be needed to keep ripple down to an acceptable level. We may therefore change tack and opt for a smaller reservoir (i.e. higher ripple voltage) and employ a voltage regulator to scrub away the hum. The 78xx series of three-pin regulator ICs can deliver up to 1A or 1.5A depending on the exact type, and a very low-cost circuit is shown in fig. 9.18. This uses a 5V fixed voltage regulator with a pair of diodes in series with the common pin to 'jack up' the voltage by a further 1.2V, to produce 6.2V at the output –ideal for 6.3V heaters.

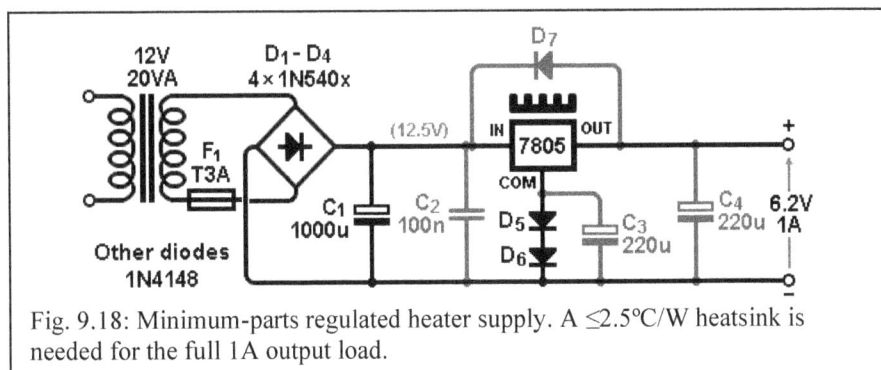

Fig. 9.18: Minimum-parts regulated heater supply. A $\leq2.5°C/W$ heatsink is needed for the full 1A output load.

---

[8] King, G. J. (1951). Hum Problems in Low-Level Amplifiers, *Practical Wireless*, July, pp317-18.

The IC requires at least 2V across itself to remain in regulation, so the *minimum* input voltage needed is 8.2V. To this we should add a further 10% to allow for mains brown-out, suggesting at least 9V input is needed. Immediately we know this cannot be supplied from a 6V transformer winding (except by voltage doubling), so the next most common standard of 12V is used here. Allowing 2V for two rectifier diode drops, at full load this should produce about $12V \times 1.3 - 2V = 13.6V_{pk}$. We cannot allow the valleys to fall below 9V, so the peak-to-peak ripple voltage must therefore not exceed $13.6 - 9 = 4.6V_{pp}$. This implies a reservoir capacitance of: $C = it/V = 1 \times 0.01/4.6 \approx 2200\mu F$. However, experiment proved that only $1000\mu F$ is necessary in practice.

From Ohm's law the total effective load resistance on the rectifier circuit is $11.3V/1A = 11.3\Omega$. The source impedance of a typical low-voltage transformer is likely to be about 1.2Ω. We can therefore estimate the ratio R_s/R_l to be about $1.2/11.3 = 0.11$, and from fig. 2.15 we can therefore predict the RMS ripple current to be about 1.6 times the DC load current. The transformer (and also the reservoir ripple current rating) must therefore be rated for at least 1.6A, or in other words a $1.6A \times 12V \approx 20VA$ transformer is needed.

For a TO-220 device, the junction-to-case thermal resistance is quoted as 5°C/W, to which we may add 0.5°C/W for thermal grease. Allowing a 60°C temperature rise above ambient, this requires a $60°C /7.5W - 5.5°C/W = 2.5°C/W$ heatsink, or better. Such a sink would typically be about the size of a deck of playing cards.

The audio-band output noise of this circuit was measured as 1mV (mostly 100Hz) at full load, which should be quiet enough for indirectly-heated valves. It can be reduced a little further to $640\mu V$ by adding C_{2-4}, shown faint, which in turn makes D_7 necessary. No ground reference is shown since this is at the discretion of the user.

Notice that the circuit in fig. 9.18 burns off quite a lot of heat and needs a 20VA transformer to supply a measly 6.2W of heater power. This is the price that must be paid for building low-voltage supplies using linear regulators with unimpressive

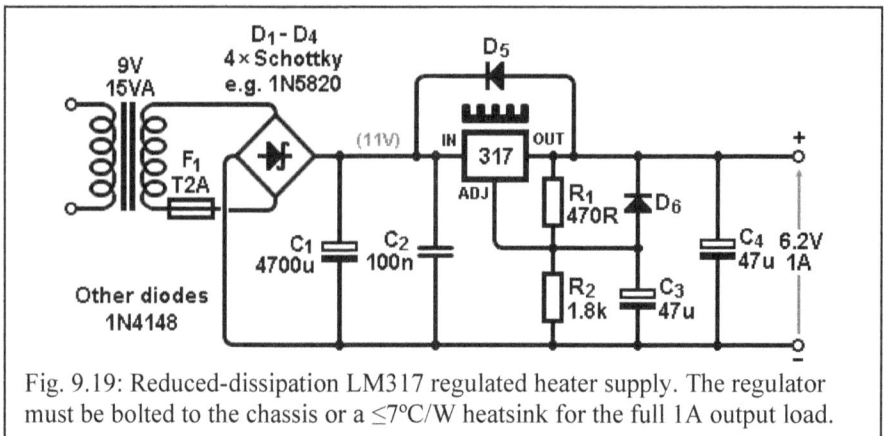

Fig. 9.19: Reduced-dissipation LM317 regulated heater supply. The regulator must be bolted to the chassis or a ≤7°C/W heatsink for the full 1A output load.

drop-out limitations. Slightly less power would be wasted by using a 9V, 15VA transformer, while also replacing D_{1-4} with Schottky diodes and increasing C_1 to 4700µF to keep the ripple valleys above the drop-out threshold. Such a circuit is shown in fig. 9.19; the regulator IC has also been replaced with an LM317 for interest. The dissipation is reduced by only 1.5W in the regulator IC, but this translates into a considerably cheaper 7°C/W heatsink –even an aluminium chassis with insulating hardware would suffice. The audio-band output noise was measured as 180µV –improved mainly because the ripple voltage is smaller.

What if we want to run a $6V_{dc}$ regulator from the $6.3V_{ac}$ winding that a conventional valve transformer provides? Sadly, after accounting for the various losses and ripple, an ordinary bridge rectifier will provide barely enough DC to run even a low drop-out regulator reliably (but see next section). Trying to use the 6.3V winding means we are forced to use a voltage doubler as in fig. 9.20, which means a big penalty in the available current. Notice that the transformer needs twice the current rating and the reservoir capacitors C_1 and C_2 are twice as large compared with fig. 9.18. The circuit is just as wasteful of power. When tested, the output noise was 570µV.

A possible problem with IC regulators operating close to their maximum current limit is that they may refuse to start if the inrush current to the cold heaters is too large. The author could not induce such a problem with the foregoing circuits, but nevertheless a textbook soft-start option is shown faint in fig. 9.20 as an example solution (a soft start may be desirable for other reasons too). When power is first applied, Q_1 will short out R_2, thereby pinning the regulator output to its default value of 1.25V. As C_6 then charges up through R_3 the transistor will gradually switch off and release R_2. With the values shown this creates a linear ramp-up over about ten seconds. Any general purpose PNP transistor will do.

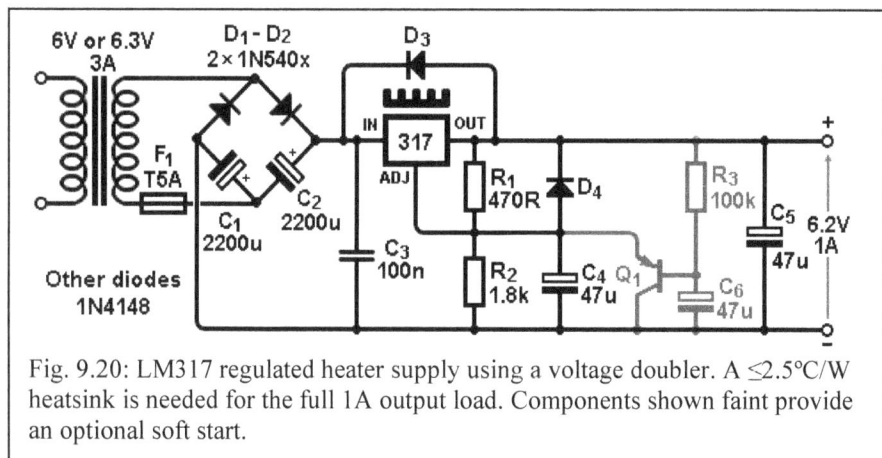

Fig. 9.20: LM317 regulated heater supply using a voltage doubler. A ≤2.5°C/W heatsink is needed for the full 1A output load. Components shown faint provide an optional soft start.

When a voltage-doubler is used as in the previous circuit, another possible way out of the cold-start-up problem is to add a preheating system, as illustrated in fig. 9.21. If the regulator refuses to start because the cold load is too heavy, the preheating diode D_1 will supply half the normal rectified voltage to the heaters to get them started. Once the regulator springs into life, D_2 becomes forward biased, leaving D_1 reverse biased and operation continues as normal.

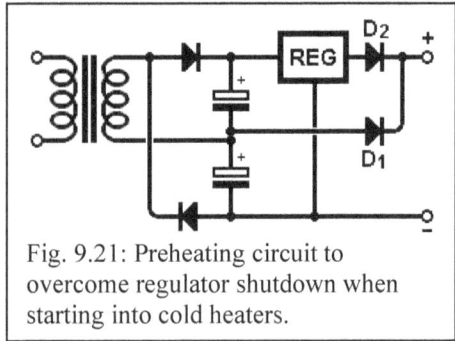

Fig. 9.21: Preheating circuit to overcome regulator shutdown when starting into cold heaters.

9.4.5: Switching Regulators

An alternative way to regulate a DC heater supply is to use a switching regulator module, also called a DC-DC converter. There are many kinds available, some of them extremely cheaply from China, though quality is variable. Such regulators can be obtained for boosting or bucking the voltage, and the latter typically have a drop-out voltage less than 0.25V, giving us great freedom over the raw input voltage. Note that such converters usually have positive and negative input and output terminals but are usually *non-isolating*, i.e. the negative terminals are actually connected together, so think of them like a three-terminal regulator.

Fig. 9.22 is a circuit tested by the author. It employed a buck converter module based on the XL4015; it had output voltage and current-limit trimmers and cost less than £2. The reservoir capacitor is relatively large and the output voltage was set to 6V rather than 6.3V to ensure comfortable headroom across the converter. The resulting audio-band output noise was 3mV at 1A load, mainly 100Hz. Although the converter

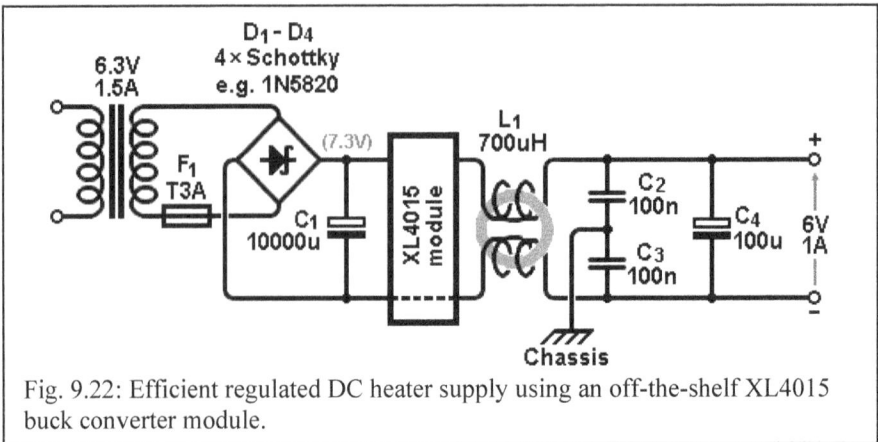

Fig. 9.22: Efficient regulated DC heater supply using an off-the-shelf XL4015 buck converter module.

could in theory handle more current,[*] transformer sag caused drop-out above 1.2A. Output switching noise was effectively eliminated by the output filter (the value of the common-mode choke is not critical), but this had negligible effect on the audio-band noise measurement since the switching frequency was well outside the audio range at 180kHz. The result is much noisier than a good linear regulator, but no doubt good enough for a guitar amp and many line-level hi-fi applications too.

The converter efficiency was found to be about 90%, meaning the input power was 6.7W for 6W of heaters. This means the average input current to the converter was actually less than the output current, and allowing for power factor the transformer need only be rated for 1.5A. Since no heatsink is required either, the cost saving compared with traditional linear regulator circuits is enormous! Indeed, the only reason a transformer –rather than an SMPS– is used here at all is for comparison with the previous circuits, and because a 6.3V winding is often available by default when using a traditional valve transformer.

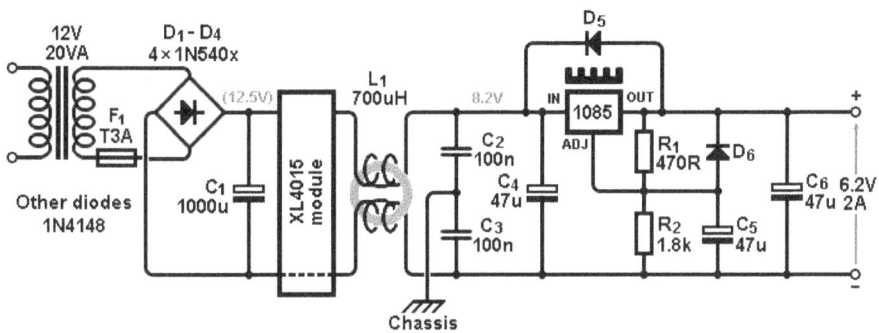

Fig. 9.23: Efficient regulated DC heater supply using an XL4015 buck converter module and low drop-out linear regulator. The LD1085 must be bolted to the chassis or a $\leq 10°C/W$ heat sink.

Given the previous circuit it would not be difficult to take another step and combine a switching regulator or SMPS with a linear regulator, to get the best of both worlds: efficiency and low noise. For example, the XL4015 module and an LM1085 low drop-out regulator have a combined drop-out voltage that is less than a single 7805, so it should be possible to use them to replace the regulator from fig. 9.18 without making any changes to the rectifier. Fig. 9.23 shows a circuit tested by the author. The switching module now serves as a pre-regulator, efficiently bucking the raw DC voltage down to a less-noisy 8.2V (the LD1085 delivers better line rejection with at least 2V across itself) which the linear regulator finally cleans up. The audio-band output noise was below 60μV at full load which, thanks to the superior efficiency, is now 2A. Again, no DC ground reference is shown since this is at the discretion of the user.

---

[*]It was advertised as a handling up to five amps, but experience suggests that such claims are inflated by at least a factor of two when it comes to anonymous, cheap sellers from China.

9.4.6: Constant-Current Regulators

A seen in previous sections, inefficiency in linear regulator circuits is caused by the rectifier and drop-out voltages being a significant fraction of the (usually low) output voltage. A different approach to improving efficiency is therefore to connect the heaters in series and thus use more voltage but less current. Sometimes this will alleviate problems with heater elevation too, since the valves with the highest cathode voltages can be supplied from the higher-voltage end of the heater chain.

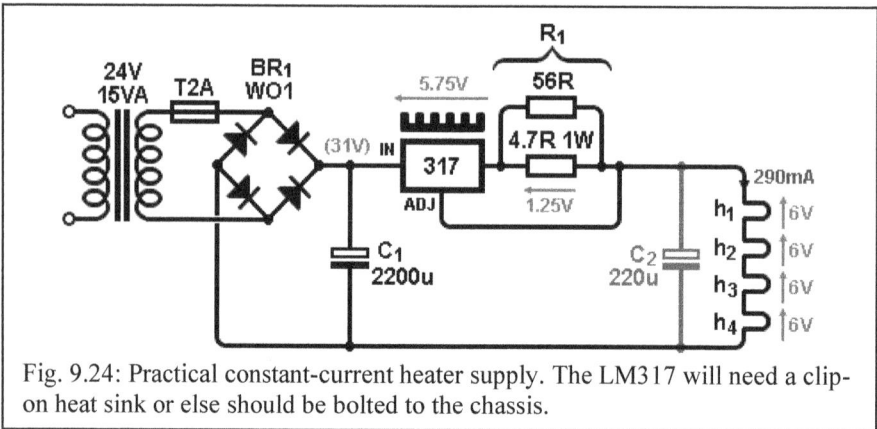

Fig. 9.24: Practical constant-current heater supply. The LM317 will need a clip-on heat sink or else should be bolted to the chassis.

Different heaters might not have matching warm-up times, so to avoid power-hogging during start-up, they should be supplied from a constant-current regulator or current limiter. Fig. 9.24 shows an example circuit supplying four 6.3V/300mA heaters, which are run deliberately low at 6V/290mA to allow a 24V/15VA transformer to be used. The current-programming resistance R_1 must be 1.25V/0.29A = 4.3Ω, which is created with 4.7Ω and 56Ω in parallel. As drawn, the regulator dissipates only 5.75V×0.29A = 1.7W so would be happy with a clip-on heatsink. A free bonus of this circuit is that the constant-current action eliminates heater inrush, providing an inherent soft start. The audio-band output noise was found to be 700μV without C_2, or 50μV with it included (remember, each of the four heaters will see only a quarter of this). Note that when power is first applied the heaters will be cold and the full supply voltage will be imposed across the regulator. The LM317 is rated for 40V, so a longer heater chain would need a different regulator such as the 125V-rated TL783. Unfortunately, this IC also has a much larger drop-out voltage –typically 10V when handling 300mA– which must be accommodated.

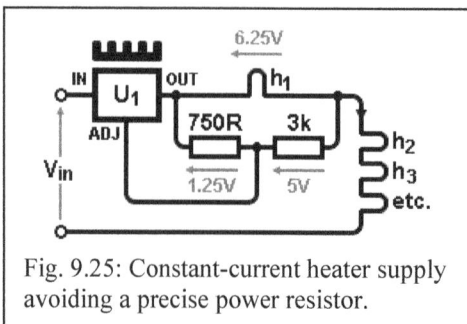

Fig. 9.25: Constant-current heater supply avoiding a precise power resistor.

In some applications the current-programming resistance in fig. 9.24 may be difficult to make using standard values, and may also waste excessive power. Why waste power when it can be used to feed a heater? Fig. 9.25 replaces this resistance with a heater itself. In this case a 6.3V heater (h_1) serves as 'master', and whatever current this requires will also feed the other heaters. However, this arrangement does not provide a soft start since the regulator will maintain the full voltage across h_1 from cold. Do not forget that all heater supplies must be referenced to ground somehow.

9.5: Substituting Different Valves

Valves with unpopular heater voltages are often much cheaper than their 6.3V equivalents, while being otherwise pin compatible. It would be quite attractive, therefore, if an amplifier could accept different versions of the same valve without having to turn on the soldering iron. Three methods for accomplishing this are described here, though all of them are somewhat wasteful of power. On the other hand, the freedom to use multiple valve types in the same circuit –so-called 'tube rolling'– might outweigh this disadvantage. The author has also used these tricks when building valve testers, so one socket can accept many valve types.

The first method works for AC and DC and can be used when two valves are completely pin compatible but have different heater voltages. It involves using a carefully chosen heater supply voltage, plus an equally carefully-chosen dropping resistor for each valve socket. With the right combination, two different heater types will receive the correct voltage (assuming one requires more voltage but less current than the other). To find the right combination, call the heater voltage of the first valve V_1, and its nominal heater resistance R_1. Likewise for the other valve type call its heater voltage V_2 and heater resistance R_2 (the nominal heater resistance is simply the rated heater voltage divided by the rated heater current, given on the datasheet). The unknown dropping resistor R_x must then be:

$$R_x = \frac{R_1 R_2 (V_1 - V_2)}{R_1 V_2 - R_2 V_1} \qquad (9.4)$$

And the heater supply voltage must be:

$$V_x = I_1 (R_1 + R_x) \qquad (9.5)$$

Where I_1 is the heater current of the first valve type.

For example, the ECC88/6DJ8 needs 6.3V/365mA so its heater resistance is $6.3/0.365 = 17.3\Omega$. The PCC88/7DJ8 needs 7V/300mA so its heater resistance is $7/0.3 = 23.3\Omega$. Applying these figures to the previous formulae:

$$R_x = \frac{17.3 \times 23.3 \times (6.3 - 7)}{17.3 \times 7 - 23.3 \times 6.3} = 11\Omega$$

And:

$$V_x = 0.365 \times (17.3 + 11) = 10.3V$$

But the close standards of 10V and 10Ω work well enough. The power dissipated by R_x can be found from I^2R, where I is the largest heater current. In this case it would amount to $0.365^2 \times 10 = 1.3\,W$. Since it will have to burn continuously, using a 2W resistor would be a little unkind, so we ought to use a 4W resistor or better, as shown in fig. 9.26. Sometimes more than two valves can be accommodated this way. For example, the 6SN7 (6.3V/600mA), 8SN7 (8.4V/450mA), and 12SN7/12SX7 (12.6V/300mA) would each receive the correct voltage within 4% using a 22Ω 15W dropping resistance and 19V supply voltage.

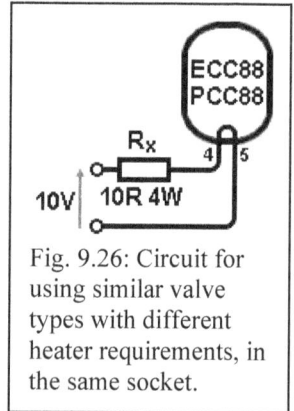

Fig. 9.26: Circuit for using similar valve types with different heater requirements, in the same socket.

The second method applies to 9-pin dual-triodes, most of which have one of the two common pin configurations shown in fig. 9.27. The difference is pin-9, which is either a tapping point between the two heaters to allow them to be operated on 12.6V or 6.3V, or else it is connected to an internal shield, in which case only 6.3V operation is possible, as shown in fig. 9.28. Normally these two families are incompatible in the same socket, because if a 9A-based valve is wired for 6.3V operation, trying to put a 9AJ-based valve in its place will result in no heater power at all. Conversely, if it is wired for 12.6V operation then the 9AJ-based valve in its place will suffer twice its

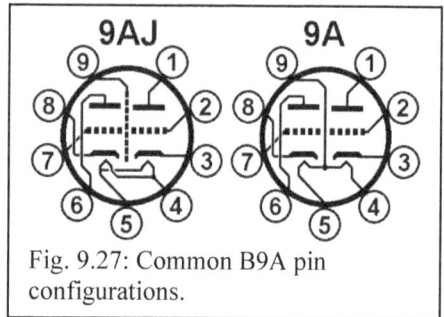

Fig. 9.27: Common B9A pin configurations.

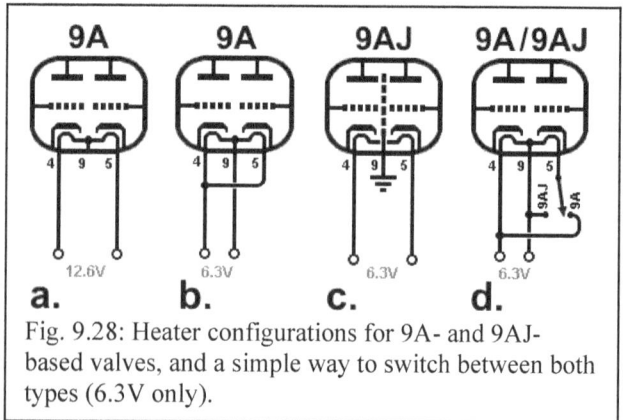

Fig. 9.28: Heater configurations for 9A- and 9AJ-based valves, and a simple way to switch between both types (6.3V only).

nominal heater voltage, which risks heater burn-out. On the other hand, if a 9AJ-based valve is plugged into a socket wired for a 9A-based valve, the heaters will run at half the normal voltage, which is equally unsatisfactory.

Table 9.1 gives a list of some of the valve types using these pin outs. Most of them are electrically unique, but guitarists may still want to make random substitutions to experiment with the different tonal possibilities (assuming other aspects of the audio circuit can tolerate this). A few of the bottles in the table contain *exactly the same*

9AJ base		9A base	
Type	**I$_h$ at 6.3V**	**Type**	**I$_h$ at 6.3V**
6AQ8 ECC85	435mA	12AD7	450mA
6BC8	400mA	12AT7 ECC81 ECC801 6060 CV4024 M8162	300mA
6BK7	450mA	12AU7 ECC82 ECC802 5814 CV4003 M8136	300mA
6BQ7	400mA	12AV7	450mA
6BZ7	400mA	12AX7 ECC83 ECC803 5751 7025 CV4004 M8137	300mA
6CG7	600mA	12AY7 6072	300mA
6DJ8 ECC88 E88CC 6922 7308 6N23P 6H23П	365mA	12AZ7	450mA
6DJ9 ECC189	365mA	12BH7	600mA
6FQ7	600mA	12BZ7	600mA
6GU7	600mA	12DF7	300mA
6N1P 6H1П	600mA	12DT7	300mA
6N2P 6H2П	340mA	12DW7	300mA
6N30 6H30	800mA	ECC99	800mA
6N6P 6H6П	800mA		

Table 9.1: Some pin-compatible dual triodes and their heater currents.

triodes, differing only in the base configuration and heater requirements, so these sorts of substitutions would be desirable for hi-fi purposes. Example pairs include the ECC81/12AT7 and ECC85/6AQ8; and the ECC83/12AX7 and 6N2P/6H2П. But the choice of plug-and-play substitutions is usually limited to *either* the 9AJ-based family (common in hi-fi), or the 9A-based family (common in guitar amps), but not both.

However, if you don't mind manual switching then this incompatibility can be overcome very simply with the arrangement in fig. 9.28c, which works for both AC and DC. Each valve socket will need its own switch. Note that the heater is never subjected to excess voltage; if the switch is set incorrectly then either one heater will light, or none, but in either case the mistake will soon be spotted with no harm done. Note that when a 9AJ-based valve is plugged in, the internal shield on pin-9 will be connected to the heater voltage, but since this is a low-impedance, low-voltage supply this does not significantly affect shielding or hum.

Fig. 9.29 shows a (DC only) circuit that accomplishes similar functionality to the above. It uses pin-9 as a sort of 'identity detection' pin; when a 9A-based valve is plugged in as shown, the base of Q_1 is connected to the heater common point and

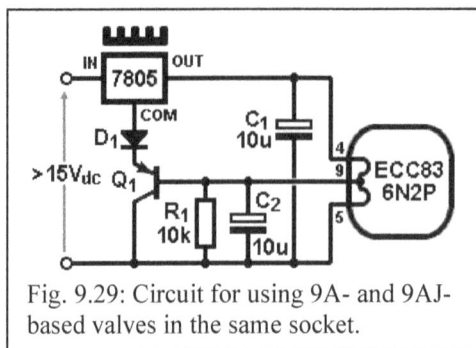

Fig. 9.29: Circuit for using 9A- and 9AJ-based valves in the same socket.

will act as an emitter follower. The regulator voltage will therefore be jacked up by the heater voltage of the 'lower' heater, plus V_{be}, plus the drop across D_1, which altogether work out close to 12.6V. Conversely, if a 9AJ-based valve is plugged in, pin-9 will be connected to nothing but the internal shield. The base of Q_1 will therefore be pulled down by R_1 so the regulator voltage will only be raised by V_{be} plus the diode drop, producing 6.3V. C_2 is necessary to stabilise the regulator for capacitive loads (e.g. C_1), and it decouples the internal shield too. D_1 and Q_1 can be any general purpose devices as they only pass the few milliamps flowing out of the adjust pin. The regulator will dissipate almost 4W depending on the valve type plugged in, so it will need a small heatsink or can be bolted to the chassis. The $\geq$15V supply voltage could be derived from a slightly under-loaded 12V_{ac} transformer.

9.6: Directly Heated Valves

Heating the filament of a directly-heated valve requires a little more thought than for an indirectly-heated valve, since the filament and cathode are now one and the same. The anode (audio) current therefore unavoidably shares some common wiring with the heating current. Nevertheless, provided we do not lose sight of the fact that they *are* two separate current paths, it is not difficult to design the heater circuit.

During operation, the voltage at one end of the filament wire is necessarily higher than at the other end, meaning the bias voltage between the grid and the filament is not the same along its length, but follows a gradient. Therefore, slightly more anode current will flow in the portion of the filament that sees the smaller bias voltage between grid and filament. This is illustrated in a simplified way in fig. 9.30 which shows a cathode-biased, directly-heated valve, with a battery supplying filament power. The grid is pulled to ground by R_g, making it zero volts, while R_k establishes a more positive bias voltage. But one end of the filament is even *more* positive because the filament battery voltage is added to it, meaning somewhat less

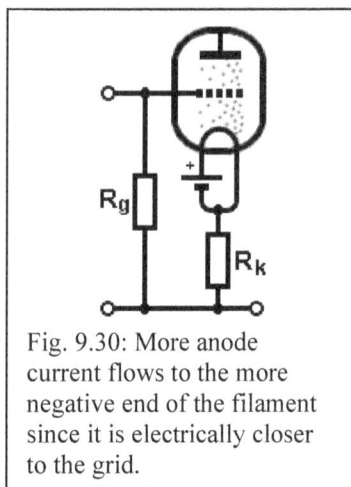

Fig. 9.30: More anode current flows to the more negative end of the filament since it is electrically closer to the grid.

anode current will flow to this end of the filament wire. In other words, the filament voltage gradient leads to a corresponding anode current density gradient along its length (possibly leading to premature ageing at the 'hot spot'). This phenomenon is sometimes demonstrated quite clearly in DC-heated vacuum fluorescent displays

254

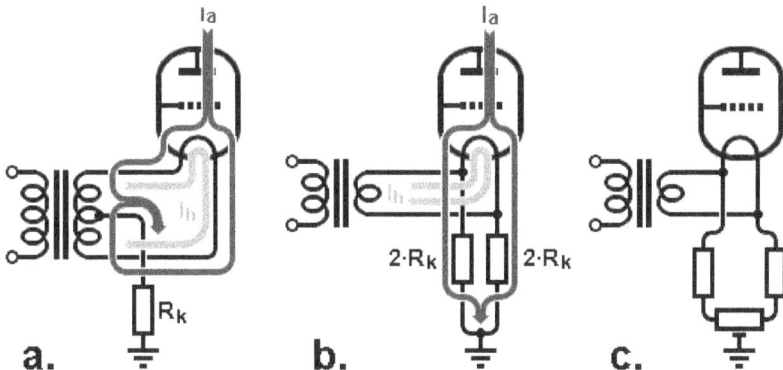

Fig. 9.31: Traditional AC filament supplies for directly-heated valves. The cathode resistors may be bypassed with capacitors in the usual way.

which may develop a brightness gradient from one end to the other along the line of the filament wire. The effect is not so great in audio valves since the grid and anode voltages are much larger, making a few volts of filament potential look small by comparison, but it is nevertheless something to ponder.

This effect disappears (or rather it averages out) if we use AC heating, since the potential gradient flips from one direction to the other with each cycle. An AC filament supply is also attractive for its simplicity, but instantly burdens us with mains-frequency hum. This must be tackled using the same principle of electrical balancing as explained in section 9.3.5 (plus neat, twisted lead dress, of course). If the filament transformer has a centre tap then a single bias resistor, R_k, may be used as in fig. 9.31a. However, this is deprecated because it inevitably means a long run of wiring to the transformer –carrying both audio and heater current– and because a centre tap does not necessarily result in the minimum hum anyway. Also, any imbalance in the split of the DC anode current will lead to DC flux in the transformer. A better method is to use two bias resistors of twice the required value, as in b. As indicated by the arrows, this separates the heating and audio current paths as far as possible. Better still is to incorporate a humdinger as in fig. 9.31c so the balancing can be trimmed a little either way to find the ultimate null. The two fixed resistors would be selected to provide most of the bias voltage, so an inexpensive 47Ω to 100Ω ½W trimpot may be usable, despite the full anode current flowing in it.

A secondary effect of AC heating is that the magnetic field around the filament collapses twice every mains cycle, leading to a corresponding disruption in the anode current electron stream at twice mains frequency (this effect is suppressed in indirectly heated valves by using a twisted or folded heater, which is not practical in filamentary valves). This source of residual hum is unaffected by the voltage balancing techniques above, but is usually small enough to pass unnoticed. Nevertheless, it can in theory be nulled out with feed-forward cancellation techniques.

A filament can alternatively be supplied with DC, which ought to eliminate any possible hum. Fig. 9.32a shows a first-principle approach using an ordinary regulator IC, probably a 1084, 1085, or 338 since filaments are often very hungry (the 338 boasts a time-lag current limiting feature that may be useful when coping with filament inrush). This is designed in the usual way, and common-mode filtering could also be added.

Alternatively, a constant-current supply might be used as in fig. 9.32b. This might be preferable for fragile thoriated-tungsten filaments, to eliminate the stress of inrush current. Notice that in either case the cathode resistor is

Fig. 9.32: First-principle DC filament supply for a directly-heated valve. **a:** Constant voltage; **b:** Constant current.

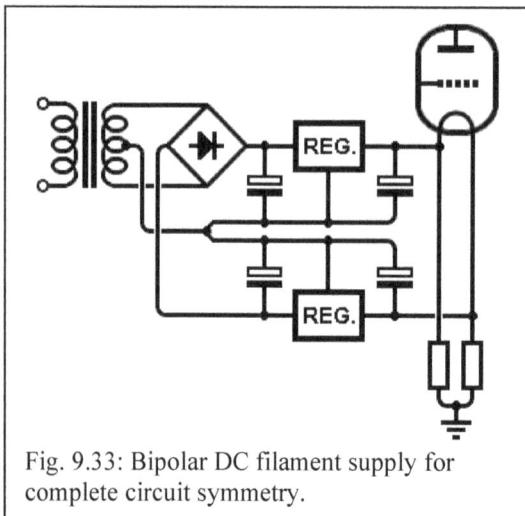

connected to the more positive side of the filament; this is preferred because but it means the extra anode current flowing in the negative end of the filament –as a result of the potential gradient described earlier– still has to travel along the full length of the filament, which helps to even out the current density along it.

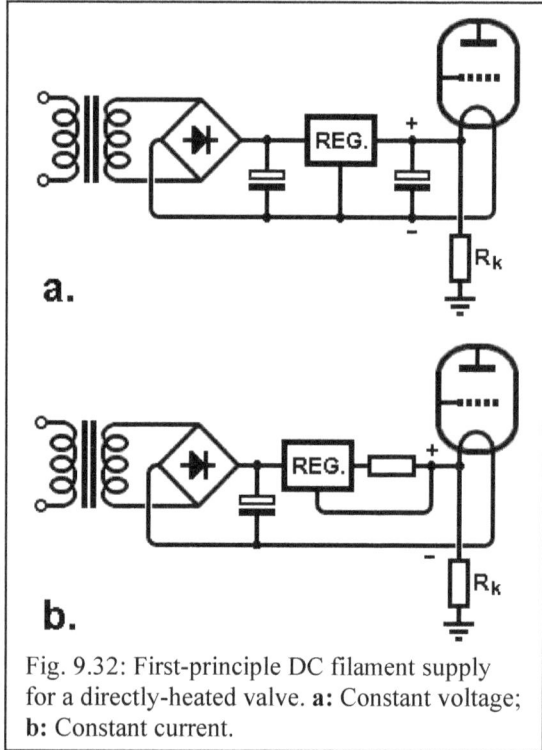

Fig. 9.33: Bipolar DC filament supply for complete circuit symmetry.

Other practitioners might prefer to mimic the appealing symmetry of traditional AC supplies by using a bipolar DC supply and a pair of cathode resistors, as in fig. 9.33. This also spreads the regulator dissipation, which may be advantageous. For example, an 805 filament needs 10V/3.25A which could be supplied by LM338 (positive) and LMS1585 (negative) regulators set to ±5V (not forgetting that the filament supply is actually floating on top of the cathode resistors). It might be tempting to connect a large bypass capacitor directly across the heater, i.e. from rail to rail, but the author has encountered regulator latch-up problems when trying this.

Chapter 10: Voltage Control

A common problem for designers is how to obtain a particular supply voltage from parts that are on hand or readily available, without necessarily using voltage regulators or resorting to custom-wound transformers. Typically, circumstances conspire to deliver higher voltages than we would like, rather than lower. This may be because a power transformer was designed for a lower historical wall voltage than exists today, or because it is more powerful than it really needs to be and the voltages are not sagging as much as expected, or because we are using a solid-state rectifier instead of valve.

Admittedly, some hobbyists obsess unnecessarily about the exact voltages measured in a circuit when compared with those quoted on a datasheet or schematic. Valve circuits are usually very forgiving; a 10% deviation from the design value is often of no consequence. Nevertheless, this chapter presents some practical methods for tweaking and controlling supply voltages, especially the HT voltage. This is of particular relevance to guitarists as it offers a way to control the output power and perceived loudness of an amplifier while maintaining its overdrive character, as discussed later.

10.1: Primary-Side Voltage Control

A classic way to control the mains wall voltage is with a variable autotransformer or variac.[*] An autotransformer has a single winding, and the desired output voltage may be tapped off any point along its length. This can be made continuously adjustable with a sliding contact as shown in fig. 10.1. It is similar to a big potentiometer, but being an inductor it dissipates negligible power, and it can step the voltage *up* as well as down.

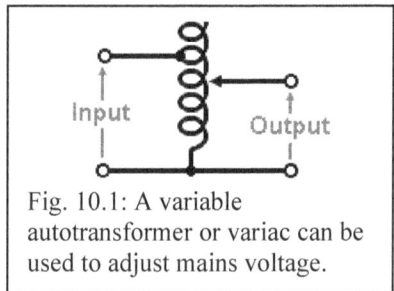

Fig. 10.1: A variable autotransformer or variac can be used to adjust mains voltage.

Most variacs allow the output voltage to be varied from zero up to about 10% higher than the input voltage, so appliances can be tested with the wall voltage both lower and higher than the nominal value.

It is important to remember that an autotransformer/variac does *not* provide galvanic isolation from the mains; it cannot safely be used alone as a make-shift power transformer. It should be regarded as an extension of the wall supply and just as dangerous.

---

[*]Variac is a shortening of *variable-AC* and was originally a trade-mark of the General Radio Corporation. It has since fallen into common use to describe all variable autotransformers used for adjusting the wall voltage.

Voltage Control

If we need to alter the mains voltage by a fixed amount then an ordinary power transformer can be used as an autotransformer, as shown in fig. 10.2. The secondary voltage either adds to (boosts), or subtracts from (bucks) the wall voltage, depending on the phase of the windings, as indicated by the dots. This is particularly useful for reducing the wall voltage to

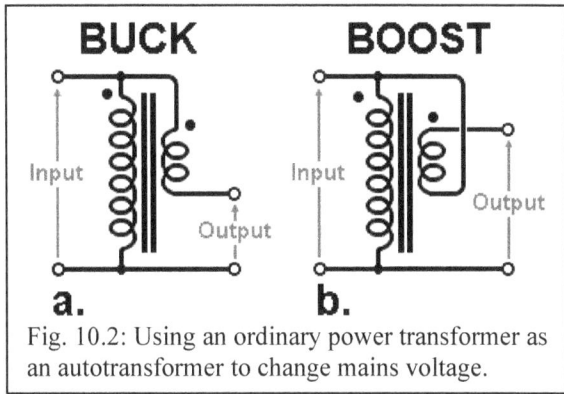

Fig. 10.2: Using an ordinary power transformer as an autotransformer to change mains voltage.

work with a vintage transformer that was wound for a lower standard. Similarly, cheap transformers from China are often wound for rather low wall voltages, and tend to run hot unless the voltage is bucked.

The secondary winding of the autotransformer carries the full load current of the following appliance, and must be rated accordingly. For example, if we wish to alter the wall voltage by 12V, and the maximum expected load current is 1A, then we need a 12V 1A (12VA) transformer, or better. This is the same whether bucking or boosting. This trick can also serve as a cheap substitute for a variac; for example, fig. 10.3 shows an

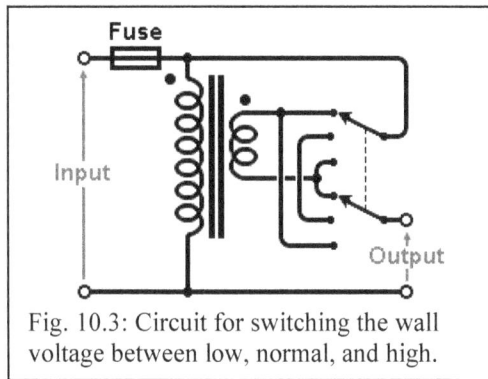

Fig. 10.3: Circuit for switching the wall voltage between low, normal, and high.

arrangement which could serve as test-bench equipment. By using a transformer with a secondary voltage which is about 10% of the normal wall voltage, the live output can be switched from 10% low, to normal, to 10% high, covering the full range of expected mains variation. A possible addition might be a filament lamp in series with the live input, to serve as a simple current limiter. Just to re-state the point, this does *not* provide galvanic isolation; it is a raw wall supply and just as dangerous.

10.2: Secondary-Side Voltage Control

Commonly we will be using a power transformer with several secondaries: HT, heaters, and so on. Changing the wall voltage with an autotransformer will therefore cause *all* these secondary voltages to change proportionately. What if we want to change just one of them? An auxiliary transformer can be used to buck or boost an individual secondary winding, while maintaining galvanic isolation from the mains. Fig. 10.4 shows the basic arrangement when a single secondary winding is used (a bridge-rectifier is shown simply as an example of a typical situation). The

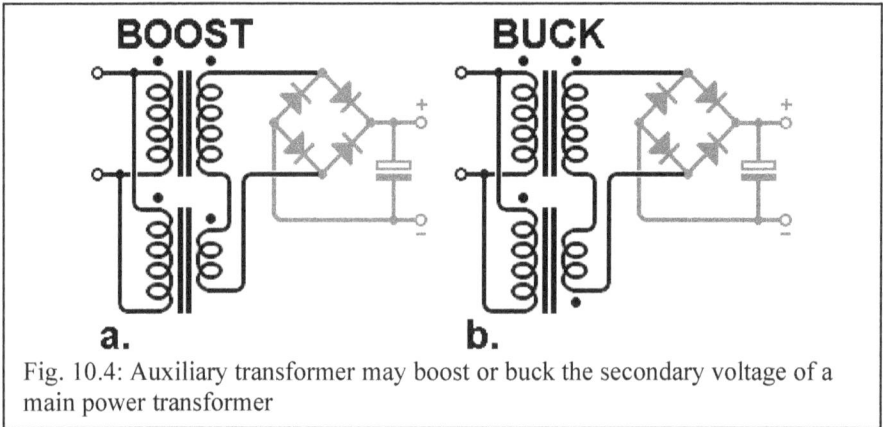

Fig. 10.4: Auxiliary transformer may boost or buck the secondary voltage of a main power transformer

boost/buck transformer must handle the same secondary current as the main power transformer of course, so its current rating should be equal to or greater than that of the main transformer.

To alter the voltage of a centre-tapped HT winding, two identical boost/buck windings would be needed. This is not necessarily a problem since many off-the-shelf transformers have identical dual secondaries, but it does place unusual voltage stress on the insulation of the auxiliary transformer. An alternative approach is to use stacked rectifiers as in fig. 10.5, although this can only provide boost, not buck. An optional 'low power' switch is also shown, allowing the main HT transformer to be turned off, reducing the HT to the lower value provided by the auxiliary transformer alone (note that current flows in two directions in the upper transformer, which defeats its inductance). For example, a 180-0-180V transformer combined with a 70V transformer would allow the HT to be switched between a conventional $350V_{dc}$ and an anaemic $100V_{dc}$ for bedroom practicing and recording.

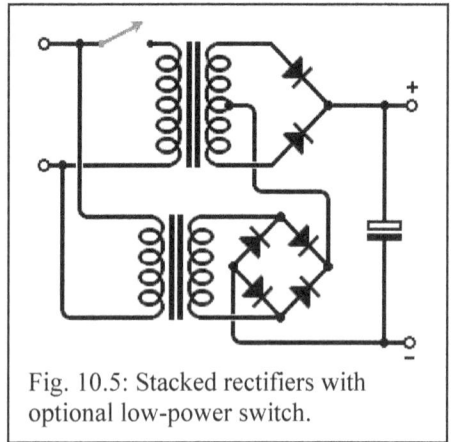

Fig. 10.5: Stacked rectifiers with optional low-power switch.

10.2.1: Zener Diodes

Another popular way to reduce HT voltage is with a Zener diode. This has the advantages of smaller size and lower cost than an auxiliary transformer, but the disadvantage that the Zener dissipates heat. Fig. 10.6 shows the arrangements frequently used by hobbyists. The HT voltage is reduced by the Zener voltage of course, and the dissipation will be equal to $V_z \times I_{dc}$. The device should be generously

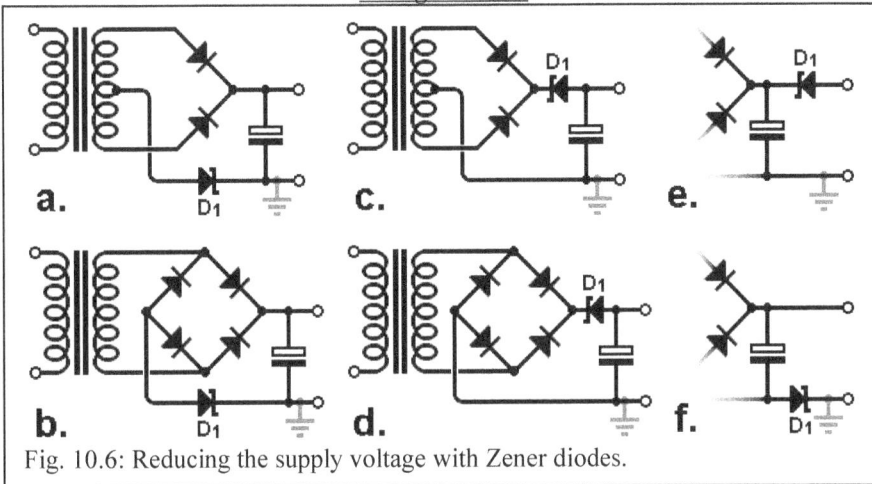

Fig. 10.6: Reducing the supply voltage with Zener diodes.

overrated to handle the power. A string of Zeners in series could alternatively be used to share the burden, of course.

Sometimes the Zener is placed in series with the negative side of the rectifier as in fig. 10.6a and b. If a power-Zener is used, and the casing is internally connected to the cathode, then it can in theory be bolted directly to the chassis for heatsinking. This of course will mean ripple current flows in the chassis, which may lead to hum being picked up elsewhere in the amplifier, which is not ideal. If a bias supply is obtained from the same HT winding, then placing the Zener in this position will also cause the bias voltage to increase (go more negative) by the same amount that the HT voltage falls. If we do not want to affect the bias voltage then the Zener could be placed immediately after the rectifier instead, as in fig. 10.6c or d.

However, owing to the shape of the Zener I/V curve, dissipation is minimised if it is placed *after* the reservoir so it does not handle ripple current, as in fig. 10.6e. If it is placed in the ground side as in e. then a power Zener can again be bolted to the chassis if it is the (more usual) casing-to-anode type, this time with less risk of hum. Placing the Zener after the reservoir also eliminates the stress of inrush current, so this option is preferable.

Power Zeners are expensive. Still lower dissipation can be achieved by dropping the voltage to the power valve/s alone, as in fig. 10.7, since this is usually the reason for wanting lower HT in the first place. Preamp voltage can always be dropped with series resistors in the usual way.

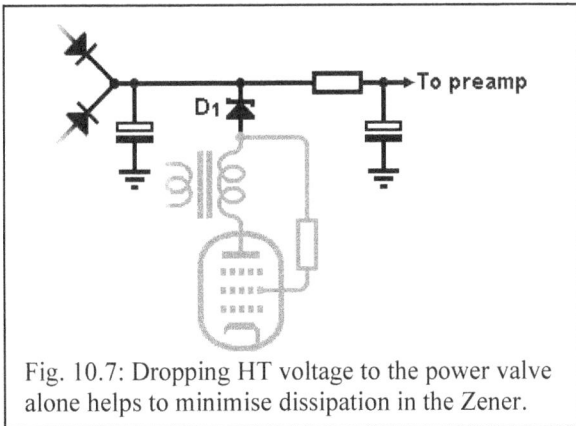

Fig. 10.7: Dropping HT voltage to the power valve alone helps to minimise dissipation in the Zener.

261

10.3: Scaling Output Power in Guitar Amps

The great appeal valve guitar amps is their ability to produce overdrive/distortion in a pleasing and musical way. However, a dilemma peculiar to guitarists is that the richest and most desirable distortion tones often come from overdriving the power valves, which necessarily means the amp is delivering maximum loudness. Guitar speakers usually have high sensitivity, so even a small amount of audio power can generate a surprisingly loud volume. Indeed, guitarists are notorious for subjecting themselves (and the audience) to high sound-pressure levels which can lead to permanent hearing damage, to say nothing of a poor live sound mix.

Some guitarists have grown up under the false impression that they *need* high power amps just to be heard above everyone else. This may have been true in the 1950s and 60s when bands had to play to dance halls with nothing but their own equipment, but that situation has changed. Now even very small venues usually have their own house PA system which can fill the whole place with whatever volume the sound-engineer deems necessary, since all the stage instruments will be mic'd or DI'd. The guitarist really only needs to be able to hear *himself* amongst the rest of the band, which does not usually demand a particularly loud amp (it may be taken care of by stage monitors anyway).

Sadly, some amateur guitarists still labour under the myth of needing high power, and a battle for control may ensue between the guitarist and the sound engineer. The result is usually a wall of noise through which almost no one can be heard clearly, and the audience retreats to the back of the room. The audience seems disinterested; maybe we're not loud enough? Aesop's fable of the competition between the north wind and the sun is apt.

Fortunately, this attitude does seem to be in decline. There is a growing demand for low-power tone amps; that is, amps which can reproduce the classic, fully-overdriven sound, but at much lower acoustic volumes. The most versatile of these allow the output power to be varied from full, down to almost a whisper, without seriously altering the tonal character of the amp. Since this is often achieved by controlling the power supply voltages, some discussion is deserved here.

10.3.1: Power and Perceived Loudness

A layman might expect a 100-watt amplifier to deliver twice the loudness as a 50-watt amplifier, and may be surprised to discover there is actually only a slight difference. This is because the human ear does not respond in a linear fashion to the actual **sound pressure level**, or **SPL**, acting on the ear drum. It actually takes quite large changes in SPL to cause significant changes in the perceived loudness of a sound.

The sound pressure level of an acoustic source is usually given in decibels,[*] relative to a standard of 20µPa (RMS) which is taken to be the threshold of human hearing. The relationship between SPL and perceived loudness is approximately given by:

$$x = 2^{\left(\frac{\Delta_{SPL}}{10}\right)} \qquad \text{or:} \qquad \Delta_{SPL} = 10 \times \log_2(x)$$

Where:
x = factor by which loudness is perceived to have changed;
Δ_{SPL} = change in sound pressure level, in decibels.

Notice the logarithm is to the base 2. For example, to double the perceived loudness of a sound requires an increase in SPL of about:

$$\Delta_{SPL} = 10 \times \log_2(2) = 10dB$$

The sound pressure level in dB produced by a loudspeaker is linearly related to the power pumped into it by the amplifier, also in dB. A change in output power is converted into decibels according to:

$$x = 10^{\left(\frac{\Delta_P}{10}\right)} \qquad \text{or:} \qquad \Delta_P = 10 \times \log_{10}(x)$$

Where:
x = factor by which power has changed;
Δ_P = change in power level, in decibels.

Notice the logarithm is base 10 this time. Therefore, to obtain a 10dB increase in SPL requires a 10dB increase in power, which is a factor of:

$$x = 10^{\left(\frac{10}{10}\right)} = 10 \text{ times}$$

This gives us the handy rule of thumb that doubling the loudness of a sound requires *ten times* more electrical power, not twice as much, as we might have supposed. Similarly, halving the loudness requires ten times less power.

There are further psychoacoustic complications which influence the perceived loudness, such as the frequency of the sound, its duration, and its harmonic content, but we will not dwell on them here. It is more important for the reader simply to realise that even quite small changes in perceived loudness require *much* larger changes in output power. To make this relationship a little more visual, if we were to take a typical volume knob on the front panel of an amp and label it with watts too, it would look something like fig. 10.8.

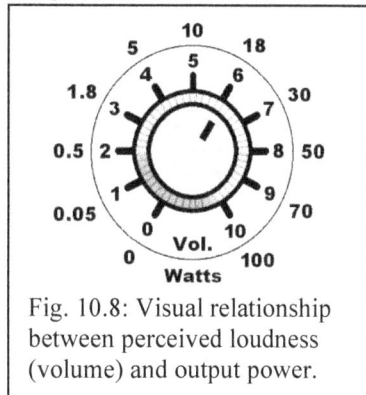

Fig. 10.8: Visual relationship between perceived loudness (volume) and output power.

---

[*] Specialised books may use **phons**, which are numerically equal to decibels SPL.

Most guitarists will already know that a 100-watt amplifier turned up half-way is already loud enough to drown out most drummers, but from fig. 10.8 we see it would be delivering an average power of only about 10W. When used for bedroom practicing, the same amp is unlikely to be turned above 1 on the volume scale, which is a mere 50mW of output power. It is also interesting to notice that everything from the popular Marshall *1974 18W*, to the Vox *AC30*, to any number of '50-watt' amplifiers, all fall with just two divisions of the volume knob of a 100W amp. Small differences indeed.*

SPL (dB)	Equivalent to (at 1m):	Consequence:
194	Loudest possible sound at atmospheric pressure.	Death.
140	Jet engine. Gunshot.	Permanent hearing damage within seconds.
120	Fast train. Loud rock concert.	Threshold of pain. Permanent hearing damage within 10 minutes.
100	Pneumatic road drill. Chain saw.	Permanent hearing damage within 10 hours.
80	Busy traffic. Vacuum cleaner.	
60	Conversation level. Washing machine.	
40	Moderate rainfall. Whisper.	
20	Soft breathing. Very quiet studio.	
0	Mosquito.	Threshold of hearing.

Table 10.1: Some noise levels and their consequences.

It is also useful to appreciate how a given SPL compares with everyday sounds, and table 10.1 gives some examples. A typical guitar loudspeaker will have a sensitivity of perhaps 100dB/1W/1m. This means that with 1W of input power it will produce an SPL of 100dB when measured at a distance of 1 metre away (hi-fi speakers are generally less sensitive, often around 86dB/1W/1m). Therefore, a guitarist standing close to the speaker and playing at a power level of just a few watts can suffer permanent hearing damage after a few hours of continuous exposure, even though the sound is not necessarily painful. It should be easy to understand why some guitarists develop unusually poor hearing over years of playing in live bands. The reader is duly warned.

---

* If Spinal Tap really wanted a 100-watt amp to go up to eleven it would need to be converted into a 137-watt amp.

10.2.2: Simple Power Reduction

One of the oldest power reduction 'tricks' is to switch the power valves from pentode/tetrode mode, to triode mode, as in fig. 10.9a. Unfortunately, this is probably the most pathetic means of power reduction since the available power drops by only about half, which in terms of loudness would be perceived as a 20% reduction. What's more, power triodes produce a very different tone from power pentodes, which some guitarists describe as dull, tending to harsh when overdriven, due to the higher levels of grid current. A triode/pentode switch is therefore effective as a tone modifier, but not very useful for loudness reduction.

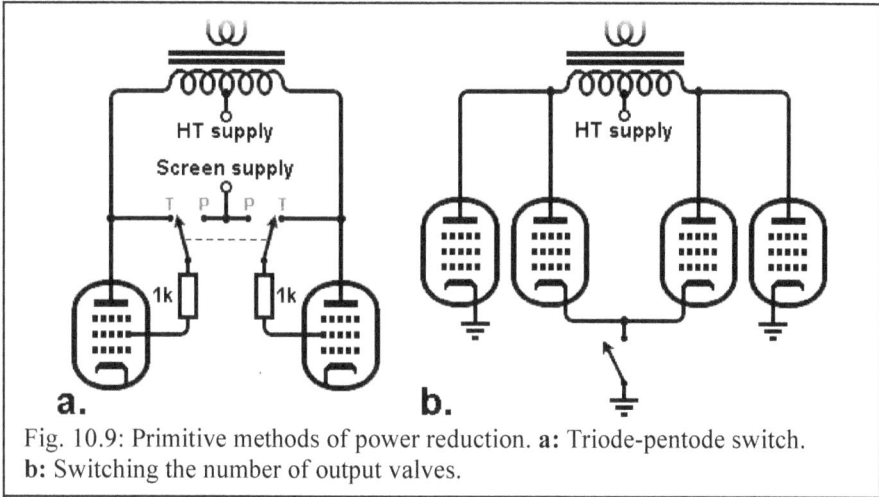

Fig. 10.9: Primitive methods of power reduction. **a:** Triode-pentode switch. **b:** Switching the number of output valves.

Another method, practiced by guitarists for decades, is to remove one or more pairs of power valves from the output stage (high-power amps are invariably push-pull), and some manufactures have made this switchable for convenience, as in fig. 10.9b. Although this does maintain good constancy of tone, it is not very effective for loudness reduction because the power is not reduced enough. Even starting with an amp with eight output valves, removing three pairs reduces the output power by three quarters, which is a loudness reduction of only about one third.

10.3.3: Proportional Voltage Scaling

Guitarists may be familiar with the apocryphal tale of Eddie van Halen using a variac to reduce the mains voltage to his Marshall amp, thereby reducing the power and creating the famous 'brown sound'. This it is not a practical solution for most players, not least because variacs are heavy and expensive. Moreover, *all* the voltages in the amp will be reduced; the preamp voltages, heater voltage and any other supply voltages for ancillary circuits, and at some moderately low setting the amp will cut out. Also, since the heater voltage is reduced, the characteristics of the valves will change at reduced settings, so the tone will also change –often for the worse– when compared with normal operation.

A more effective way to reduce power / loudness without significantly changing tone is to control proportionately the anode, screen, and bias voltages of the output valve/s. This technique has been pioneered and widely publicised by Kevin O'Connor of London Power, Ontario, since the 1990s. There it is marketed under the trademark *Power Scaling™*, though other manufactures now offer the same thing with their own imaginative names.* The beauty of this technique is that, by varying the three voltages: anode, screen, and grid (bias), in direct proportion, the overall transfer function of the amplifier remains substantially the same at any setting. It therefore retains the bulk of its non-linear characteristics, so the tone does not vary much even when the output power is reduced from a hundred watts to a hundred milliwatts, or even less.

At first, this direct proportionality might seem unlikely, given the rather non-linear nature of valves. Nevertheless, its utility can be illustrated with fig. 10.10. This shows the measured grid curves for a 6L6GC beam tetrode for three scaled settings, that is, HT, screen and bias voltages all controlled proportionately. Also shown is a simplified push-pull, class AB load line, corresponding to an anode-to-anode load of 5kΩ. The quiescent anode voltage is equal to the screen voltage, which is the case for most guitar amps, but the argument still holds for disparate anode and screen voltages too, and also for any operating class.

The upper image shows the usual case where all the voltages are at their maximum (400V HT) and the valve is biased at -35V resulting in 60mA idle anode current, indicated by the dot. The load line passes very slightly below the knee of the characteristics, and the peak clean output power is about $80W_{pk}$, or 40W average ('RMS').

Halving the anode, screen, and bias voltage produces the middle image. The load line now passes just above the knee, but on the whole the transfer characteristic has hardly changed. The average clean output power would now be around 7W.

Halving the voltages once more brings us to the bottom image. The load line still passes only slightly above the knee, and the transfer characteristic is nearly the same as before, yet the average output power would now be only 0.8W.

This demonstrates that to maintain substantially the same transfer function at reduced voltages (and therefore reduced output power), the bias voltage must track the screen voltage, and the screen voltage must track the anode voltage; all three varied in direct proportion, or close to it.

---

*In 2010 the author proposed the term *ayalodyne* as a name for the general principle, from the Greek *ayalogia* (proportion) and *dynamis* (power).

266

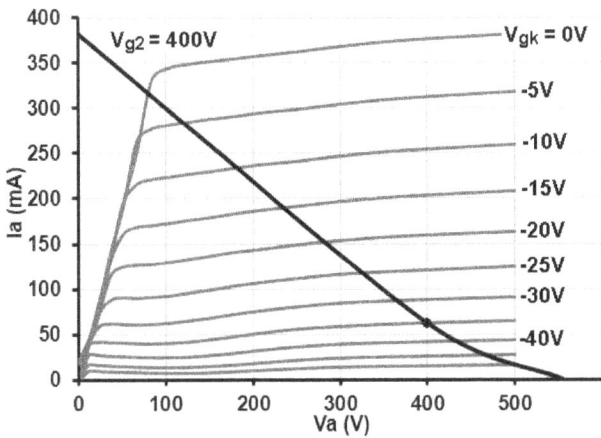

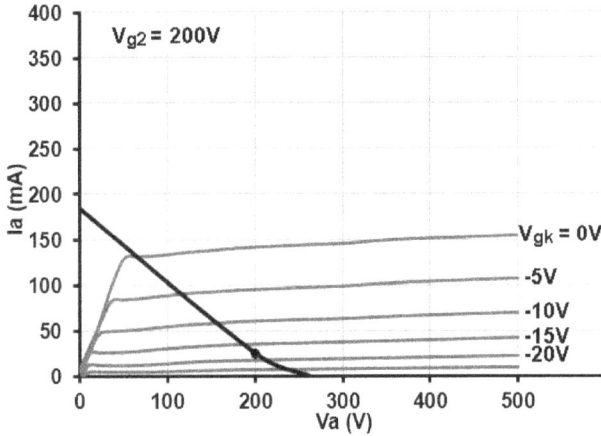

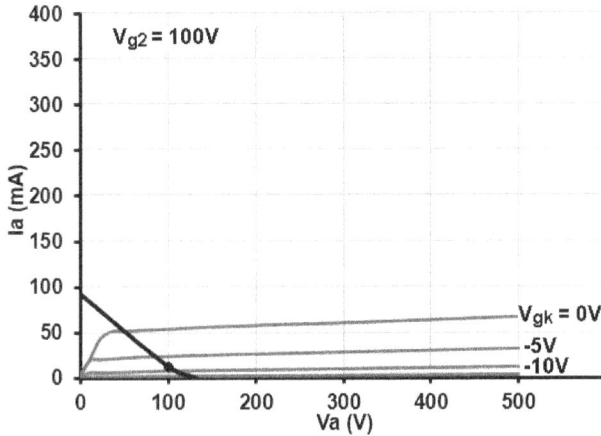

Fig. 10.10: Characteristics and typical push-pull loadline for a 6L6GT when all voltages are proportionally scaled.

There are some minor imperfections is the method, of course. Perhaps the most obvious is that if the amp has global negative feedback then the feedback factor will fall as power is reduced, which will alter the headroom of the driver stage and change its distortion character, which will also cause the usual presence control to become less effective (unless the feedback resistor is varied along with the power control). The knee of the anode characteristics begins to drop below the load line as voltages are reduced, so the harmonic distortion produced by the power valves, and the effects of sag and screen current compression, will not scale perfectly at all settings. In addition, the forward control-grid current will *increase* as the

screen voltage falls, so the audio signal will clip somewhat harder when the power valves are overdriven at low voltage settings. Finally, the bandwidth of the output

267

transformer changes slightly with flux density, and the distortion characteristics of the loudspeaker may also be different at low volumes. Nevertheless, these are fairly minor criticisms, and they do not cause such drastic changes in tone as one might expect. Overall it is a very effective method and has been met with almost universal satisfaction from guitarists, and is becoming ever more popular.

10.3.4: Power-Adjustment Topologies

So far it has been assumed that only the voltages around the power output stage will be adjustable, as in fig. 10.11. The preamp is supplied with the original, maximum voltage all the time by taking its HT directly from the reservoir capacitor. Consequently, the first dropping resistor may need to be be made larger than usual to bring the preamp voltages down to original design levels. An additional smoothing stage may also be desirable. Since the power valves become more easy to overdrive at

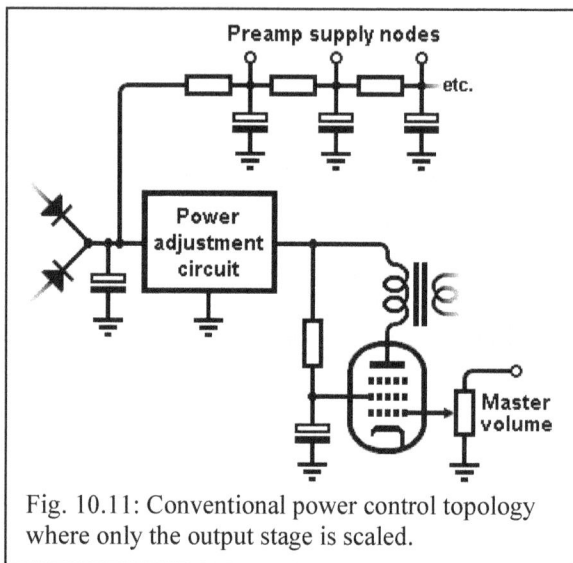

Fig. 10.11: Conventional power control topology where only the output stage is scaled.

reduced power settings, a master volume control immediately prior to the output stage is more-or-less obligatory when this type of power control is installed into an

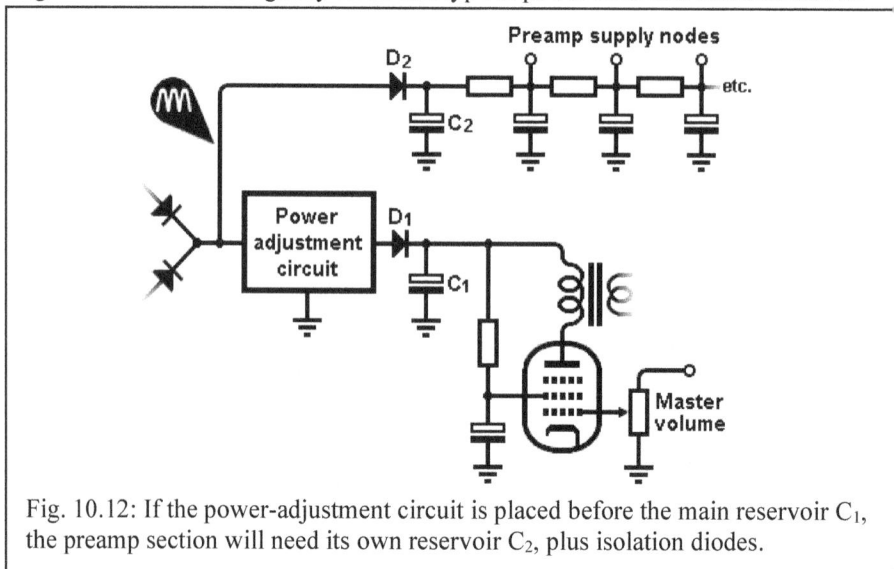

Fig. 10.12: If the power-adjustment circuit is placed before the main reservoir C_1, the preamp section will need its own reservoir C_2, plus isolation diodes.

existing amplifier. Note that if the power-adjustment circuit is placed *before* the main reservoir capacitor, then the preamp section will need its own reservoir, plus an isolation diode to prevent back-feeding, as illustrated in fig. 10.12.

In either case it is worth noting that as the power is dialled down, loading is reduced, so the raw HT voltage will un-sag and cause the preamp voltages to rise somewhat. Often this does not matter, but it can always be managed by regulating the preamp voltages with Zener diodes, amplified Zener, or a Zener follower to act as a voltage clamp.

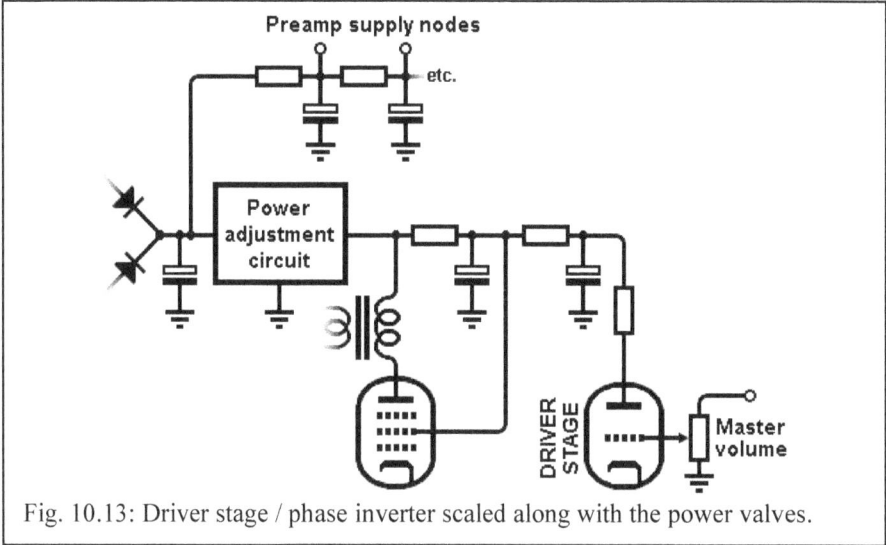

Fig. 10.13: Driver stage / phase inverter scaled along with the power valves.

The next-most popular power-adustment topology allows the supply voltage to the driver stage / phase inverter to scale along with the power valves, as shown in fig. 10.13. Since the driver stage will invariably be cathode biased it will reliably track the supply voltage. The phase inverter normally demands only a couple of milliamps of anode current, so additional dissipation in the pass device is negligible (next

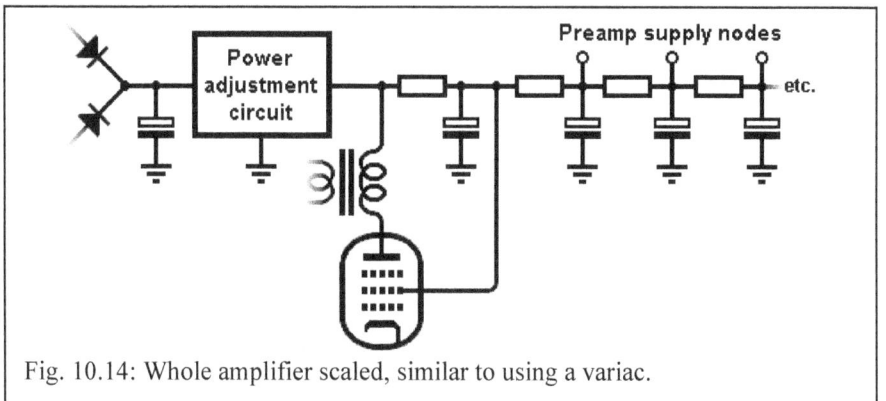

Fig. 10.14: Whole amplifier scaled, similar to using a variac.

section). With this approach the available drive voltage to the power valves reduces in proportion to everything else, so the master volume control can be moved to the input of the driver stage. Many conventional amplifiers already include a gain control in this position. On the other hand, as the power control is turned down the tone is likely to change more with this topology than with the previous one, but overall the results can still be satisfying. Again, the first dropping resistor to the preamp smoothing filters will need to be made larger than usual.

The most extreme option is to vary the supply voltage to the whole amplifier, as in fig. 10.14. This gives results closest to using a variac, except the heater voltage is unaffected. With this method any existing gain/volume controls should suffice. At very low-voltage settings the preamp distortion can suffer, approaching a dull fuzz, so this topology is probably the least appealing. On the other hand, it is the least invasive when retrofitted into an existing amp.

10.3.5: Design Considerations

The most obvious way to vary the HT voltage is by way of a voltage regulator or follower, such as those already discussed in chapter 7. Fig. 10.15 shows the simplest approach, reproduced from section 7.6.1. The voltage is varied by a pot which feeds a MOSFET source-follower that does the hard work of passing the load current and dissipating excess power.

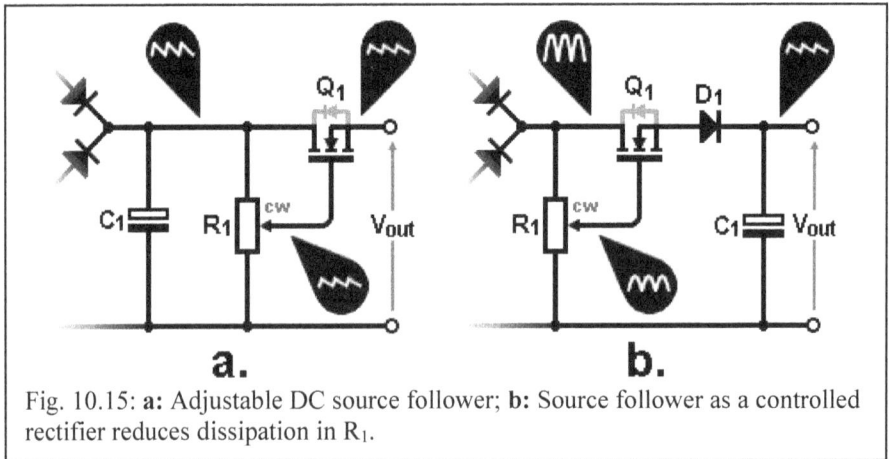

Fig. 10.15: **a:** Adjustable DC source follower; **b:** Source follower as a controlled rectifier reduces dissipation in R_1.

Assuming the load presented by the amplifier circuit is resistive –which is more-or-less true since we will be varying the screen and bias voltages proportionately– then since $P = V^2/R$ the load power will vary with the square of the HT voltage reduction, and the audio output power and SPL will follow suit. In other words, halving the voltage should reduce audio power to one quarter of its initial value, which is a loudness reduction of one third. In practice the power tends to drop at a slightly faster rate, as we saw from fig. 10.10, which works in our favour. Ideally though, for the smoothest 'feel' of loudness-versus-rotation we would like the power to drop by a factor of *ten* when the knob is turned down by half. This can be achieved with a pot whose taper is somewhere in between linear and logarithmic. A linear pot with a slugging resistance can achieve this quite well.

Power dissipation in the pass device varies as the voltage is reduced. The worst case occurs when the voltage is reduced by exactly half, at which point the dissipation in the pass device is equal to one-quarter of the maximum load power. This knowledge allows us to make some initial estimations about the choice of pass device. For example, suppose we have a typical push-pull guitar amplifier using two EL34s. The supply voltage is 400V and the output-transformer impedance is 3.5kΩ, so each valve sees one quarter of this or 875Ω. Estimating the anode voltage can swing as low as 50V, the maximum peak anode current per valve is $(400-50)/875 = 0.4A$.

Dividing the peak current by $\pi/2$ we get the average value of 250mA. Neglecting the screen-grid and preamp current which are negligible in comparison, the maximum average load power is found to be $400V \times 0.25A = 100W$. Therefore, we immediately know that our pass device will have to dissipate up to one quarter of this, or 25W, at the half-voltage setting when the amp is fully driven. At all other settings the dissipation will be less. Obviously a fairly large heat sink will be needed, and forced air cooling would also help. Incidentally, such an amplifier would probably be advertised as "50-watts". This leads to a rule of thumb: in a class-AB amplifier the worst-case dissipation in the pass device is usually about one half of the 'advertised' audio output power. In a class-A or single-ended amplifier the worst-case dissipation in the pass device is usually about equal to the 'advertised' audio output power.

In fig. 10.15 the pot is exposed to the full HT voltage. Ordinary 24mm pots are usually rated for 0.5W, which implies we could tolerate as much as 700V across a 1MΩ pot. However, this does not allow for insulation and arcing ratings; the actual voltage rating of the pot is likely to be less than 500V for linear types, while logarithmic pots usually have still lower ratings. An input voltage of 300V should be within the capability of most 24mm pots, but if a datasheet is available it should of course be consulted. Unsurprisingly, 16mm pots have lower voltage ratings, typically 200V for linear or 150V for logarithmic. Of course, there is nothing to stop us using a high-power or mil-spec pot with much higher ratings, other than the fact they cost at least ten times more than general-purpose pots. The alternative option is to use a circuit which does not place the full HT voltage across the pot, as shown later.

As explained in section 7.6, power dissipation in the pot can be minimised by placing it before the reservoir capacitor as in fig. 10.15b; an added advantage is that this also eliminates scratching sounds when the pot is rotated. A disadvantage of this arrangement is that the HT voltage will react immediately when we turn it up, but when we turn it down there will be some lag time as the reservoir discharges more slowly into the load. With a stiff pot that cannot be turned very quickly this is a minor inconvenience, and may even pass unnoticed.

271

10.3.6: Power-Adjustment for Low-Power Amps

For amplifiers operating at less than 350V, a direct implementation of fig. 10.15 can readily be used. Fig. 10.16 shows a practical example, which is very similar to the voltage reducer from section 7.6.1 and may be placed before or after the reservoir capacitor. As noted already, the main voltage limitation is imposed by the pot itself; an ordinary 24mm pot should be safe when the supply voltage is less than 350V, but a 16mm pot should not be used unless the datasheet allows. Much higher voltages could be accommodated using a special-quality pot and other suitably-rated parts, if cost is no object.

For low-power (i.e. less than 20W) amplifiers a fairly modest MOSFET like the IRF830 can be used for Q_1, though there is no reason why a more heavy-duty device cannot. Whatever the case, Q_1 will need generous heat sinking. For single-ended amplifiers up to 10W advertised

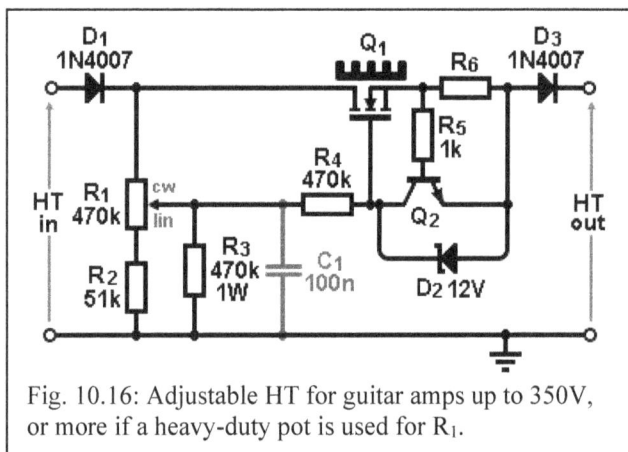

Fig. 10.16: Adjustable HT for guitar amps up to 350V, or more if a heavy-duty pot is used for R_1.

audio power, or 20W push-pull, an aluminium chassis should be sufficient for the heatsink. At higher power levels, or if the chassis is steel, a thicker slab of copper or aluminium, or a proper finned sink will be necessary, as discussed in section 7.3.

Q_2 provides a current limit equal to V_{be}/R_6 (as covered chapter 7) to protect the MOSFET from inrush or valve shorts. If the circuit is placed before the reservoir capacitor then a limit at least ten time higher than the maximum average operating current should be used; a value of 0.47Ω 2W would suit most circumstances. If the circuit is placed after the reservoir then a lower limit can be used; 1Ω 2W would be suitable for transparency. However, an even lower limit could be imposed to introduce a deliberate compressive sag effect. Indeed, this could be made switchable, but is left to the reader to experiment.

R_2 is an end-stop resistor that sets the minimum possible voltage to about one tenth of the maximum, which is about as low as is ever useful, though this could be adjusted of course. R_3 is a slugging resistor that bends the control taper, resulting in an extremely smooth loudness-versus-rotation action in most cases. It also provides a fail-safe if the wiper fails to make contact with the track. C_1 can be added to provide some smoothing, but some players insist that natural ripple on the power supply contributes to the amp's basic tone.

D_1 is not strictly necessary but is added for good measure to make the circuit highly 'self contained'. D_3 *is* necessary and is added to prevent the following smoothing capacitor from discharging through the body diode of the MOSFET, which might otherwise cause brief oscillations when turning P_1 from a high voltage to a low one, or if the circuit is placed before the reservoir capacitor.

10.3.7: Power-Adjustment for High-Power Amps

More powerful amplifiers generally use voltages too high to impose directly across an ordinary control pot. For these situations we can use a lower voltage reference and amplify it before sending to the gate of the pass device. Fig. 10.17 shows a circuit which can be used when the raw HT is up to $500V_{dc}$ (but only *after* the main reservoir capacitor).

The voltage across the pot is attenuated to less than 100V by R_1, and the end-stop resistor R_2 sets the minimum voltage. R_1 is comprised of two resistors because the total must be tweaked to suit the raw HT being supplied. Use the following as a guide:
HT = 500V, R_1 = 810kΩ
HT = 400V, R_1 = 700kΩ
HT = 300V, R_1 = 570kΩ
HT = 200V, R_1 = 400kΩ
The junction of the two R_1 resistors can be decoupled with a capacitor C_1 for superior smoothing, e.g. 10μF 450V, effectively making the circuit into a form of capacitor multiplier.

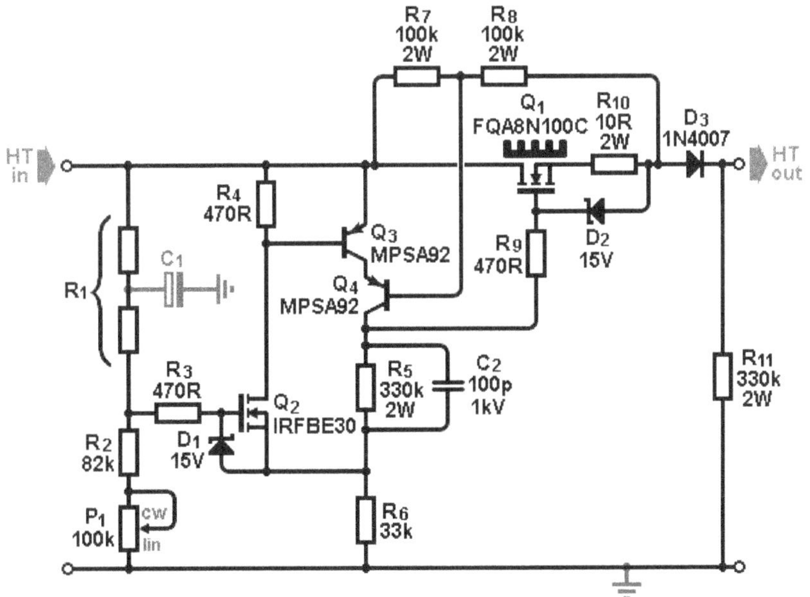

Fig. 10.17: Adjustable HT for guitar amps up to 500V, 100-watt. For amplifiers up to 50W, Q_1 should be bolted to a 3.3°C/W heat sink or better; for amplifiers up to 100W, Q_1 must be bolted to a 1.3°C/W heat sink or better.

Q_2 is a high-voltage MOSFET which forms a complimentary feedback pair (CFP, also sometimes called a Sziklai pair) with Q_3 and Q_4. A constant current of $V_{be}/R_4 = 1.3mA$ flows in Q_2, and Q_3–Q_4 are cascoded to increase voltage handling, the voltage across each always being half the voltage across the pass device, as determined by R_7 and R_8. The usual gate stoppers and protection Zeners are also included, and a 1A transient current limit is imposed by R_{10}.

At the maximum setting less than 10V is lost across the pass device, while the minimum output is close to one tenth of the raw HT. However, the circuit requires a minimum load of at least 3mA to achieve the minimum output voltage, so if testing without valves installed it may be necessary to use a dummy load.

Fig. 10.18 shows an alternative circuit that can again be used with HT voltages up to 500V. This is a simple feedback regulator using a vertical differential amplifier, so Q_2 and Q_3 share voltage burden. Notice that R_1 and R_2 are matched to P_1 and R_3, to form a $-6dB$ potential divider (R_2 might therefore need tweaking to match the actual pot resistance, given the poor tolerance of the pot which might be $250k\Omega$ in reality). This divider is mirrored by the feedback divider comprising R_8 and R_9. The circuit strives to make the voltage at the base of Q_2 almost equal to the voltage on the base of Q_3. In other words, at the maximum setting only half the total HT voltage ever appears across the pot or small transistors.

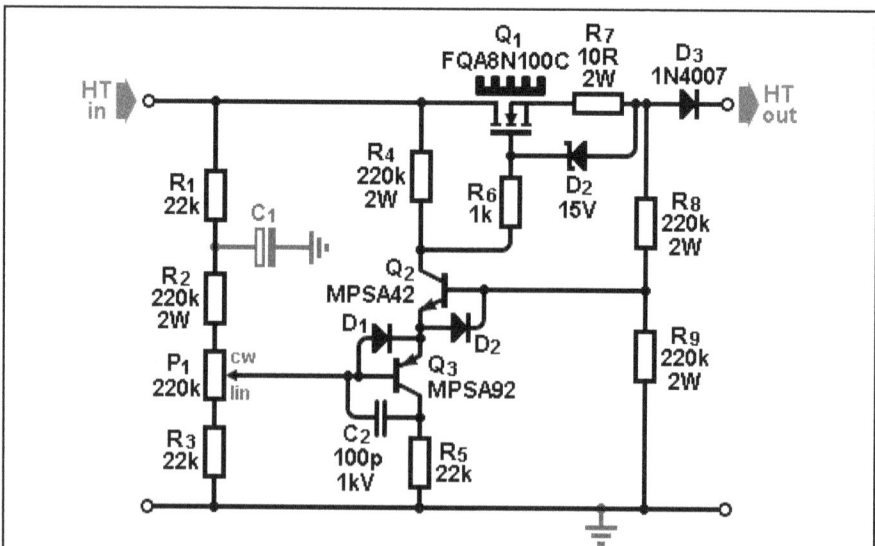

Fig. 10.18: Adjustable HT for guitar amps up to 500V, 100-watt. For amplifiers up to 50W, Q_1 should be bolted to a 3.3°C/W heat sink or better; for amplifiers up to 100W, Q_1 must be bolted to a 1.3°C/W heat sink or better.

D_1 and D_2 protect the base-emitter junctions and can be any diodes, e.g. 1N4148 or 1N400x. R_5 and C_2 limit the bandwidth of the amplifier in a way that is more consistent over the full range of operation than if a conventional feedback capacitor were used; remember, every capacitor in a high-voltage regulator hangs like a sword of Damocles. Similarly, C_1 can be included for superior smoothing, e.g. 10μF 450V, and putting it in this position –rather than hanging off the wiper of the pot– ensures that all charging or discharging transients that might otherwise damage the pot or the silicon are limited by R_2.

10.3.8: Screen Tracking

In most guitar amplifiers the screen grid is fed from an LC or RC filter which has fairly low DC resistance, so the screen voltage will quickly track a rising HT voltage. However, depending on the size of the screen decoupling capacitor, the screen current, and how fast we happen to turn down the HT voltage, it may take several seconds for the screen voltage to *fall* to a desired level. This will be more likely with beam tetrodes since they consume less screen current than pentodes. If there is any doubt about how fast the screen voltage can follow the HT, then it is simple enough to add a screen tracking circuit. A screen tracker is also preferred when an automatic bias tracker is being implemented (next section).

Fig. 10.19 shows what is required for screen tracking; a MOSFET follower replaces the traditional decoupling capacitor, eliminating any discharge delay, as well as providing a stable reference for any bias tracker that may be used. One power valve is shown but several could be supplied from a single tracker. As drawn, the idle screen voltage will be

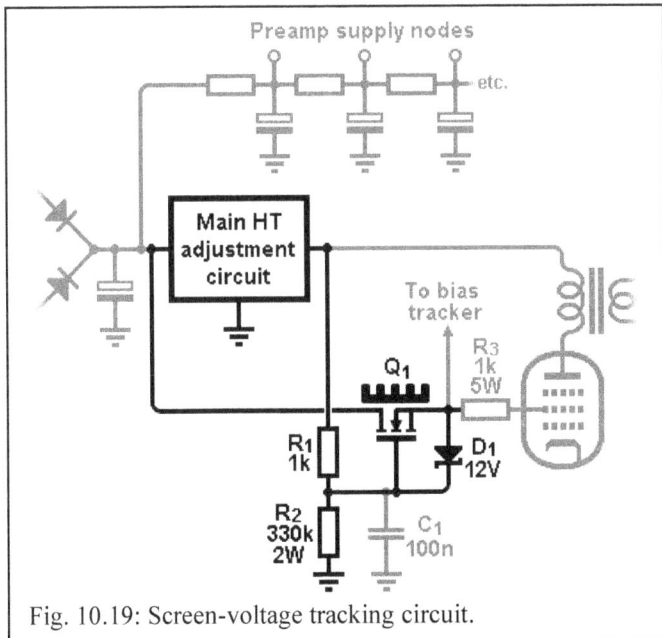

Fig. 10.19: Screen-voltage tracking circuit.

only a few volts (V_{GS}) lower than the anode, which is how most guitar amps are arranged. However, R_1 and R_2 form a potential divider that can be altered to obtain any lower screen voltage, if necessary. It is assumed that the output of the main adjustment circuit is already clean, but C_1 can be added to provide some additional smoothing for the screen, without seriously delaying the reaction time. Screen current is usually less than a fifth of anode current, so Q_1 should not need a great deal of heatsinking –the chassis should suffice– and the MOSFET could be a less heavy-duty device than used in the main adjustment circuit, such as a venerable IRF820. R_3 is the usual screen-stopper resistor, which is essential in any guitar amp both for protecting the screen grid against over dissipation, and for protecting the tracker against gross transients. If a bias tracker is also used then it should be fed from the point indicated.

10.3.9: Bias Tracking

To apply the ayalodyne technique to a fixed-biased amp we must force the bias voltage to track the screen voltage if constancy of tone is to be maintained. The general topology is illustrated in fig. 10.20. If the bias voltage was unable to track the screen, the valves would be biased colder and colder as the HT is reduced, quickly reaching cut-off, which is not what we want at all. Fortunately, the bias supply in a conventional guitar amp demands almost no current, so the bias tracking circuit can be quite simple and should not require any heat sink.

A first-principle approach is to use a dual-gang potentiometer for the power-adjust control, with the second gang being used to control the bias voltage, as illustrated in fig. 10.21 (it is assumed that the screen voltage tracks the HT by itself, or that a screen tracker will be implemented). In theory the bias could be taken directly from the wiper of R_{1B}, but since the pot is likely to be high resistance (necessary for the HT section) this might exceed $R_{gk(max)}$ of the power valves. A pot with dissimilar sections could be used, but they are rare. The problem is avoided by buffering the bias supply with Q_2. This can be any general-purpose PNP with sufficient voltage rating and $h_{FE} >$ 100. Note that the same end-stop and fail-safe resistors are used in the bias tracker to maintain the same control taper as the HT section (in fact, the input resistance of Q_2 bends the bias taper a little more, which works in our favour). A

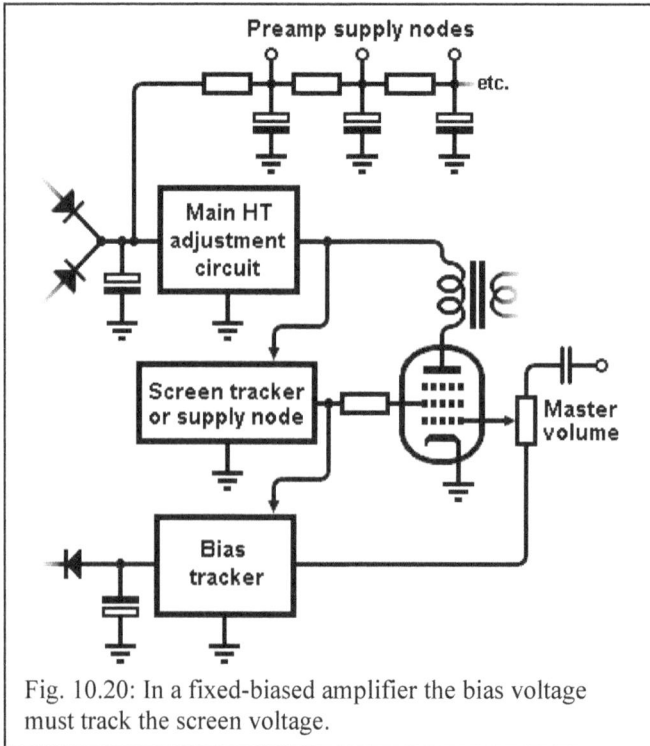

Fig. 10.20: In a fixed-biased amplifier the bias voltage must track the screen voltage.

typical bias-trimming circuit can be added either before or after the tracker, as shown faint. The bias should be trimmed with the power control at maximum, then left alone. No smoothing capacitors should be used to the right of the tracker, as it would then take too long for the bias voltage to respond to user changes (although a few nanofarads may be used in case of instability

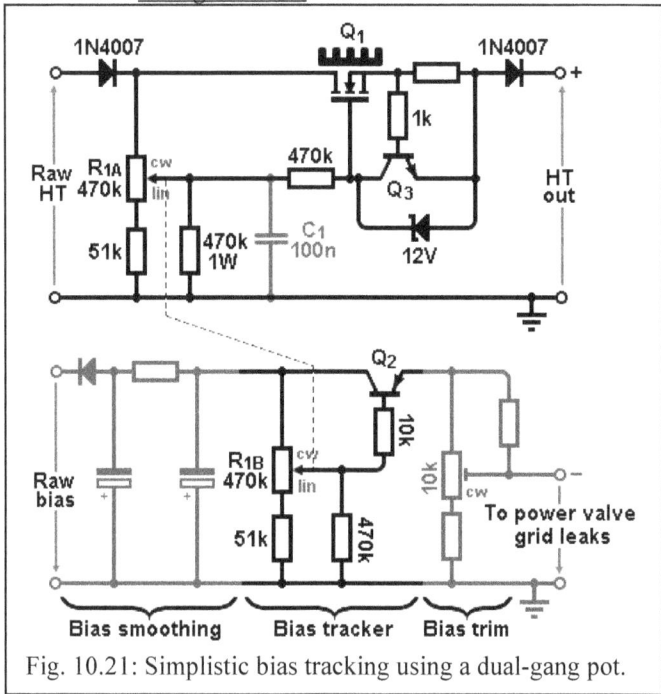

Fig. 10.21: Simplistic bias tracking using a dual-gang pot.

in Q_2). This approach to bias tracking is intuitive, but for amplifiers using voltages above about 300V where it becomes undesirable to place the full burden across the pot, an automatic bias tracking circuit is a better option.

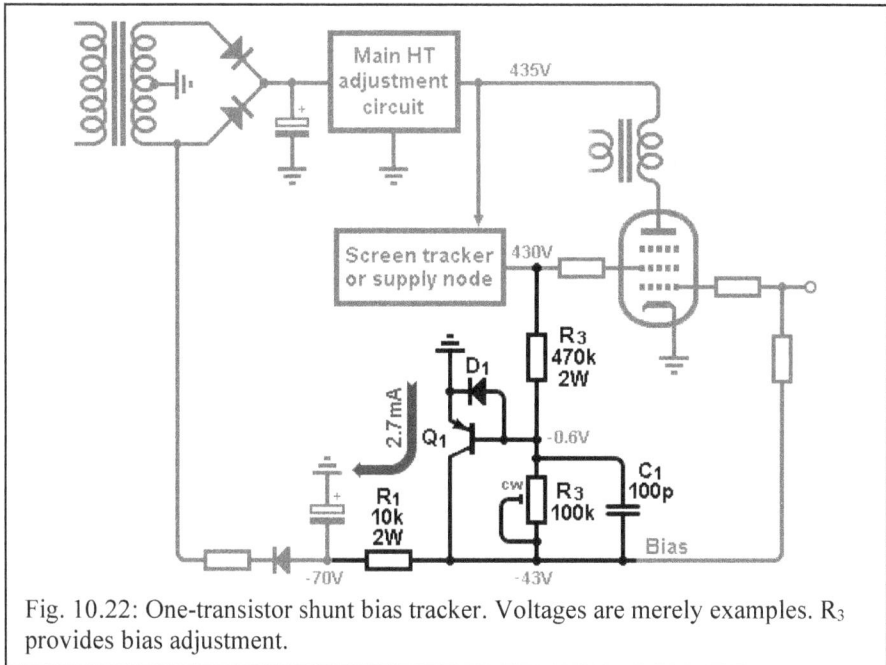

Fig. 10.22: One-transistor shunt bias tracker. Voltages are merely examples. R_3 provides bias adjustment.

277

Fig. 10.22 shows a simple one-transistor solution which automatically tracks the screen voltage. A typical amplifier architecture is shown faint, with example voltages such as might be found in a Bassman-derived design, but values can be adapted to suit other conditions of course. Here, Q_1 acts as a shunt regulator. The feedback divider R_2/R_3 sets the ratio of bias voltage to screen voltage; making R_2 a variable resistor allows the exact bias to be trimmed (with the power control at maximum). The junction of R_2/R_3 is always one V_{be} drop below ground (a virtual earth), and Q_1 will conduct whatever is necessary to make this so. For example, when R_2 is ten times smaller than R_3, the magnitude of the bias voltage will be one tenth of the screen voltage, as depicted in the figure. The value of R_1 should be selected to allow a couple of milliamps to flow in Q_1 under normal (maximum power) conditions. As the power control is turned down, Q_1 will conduct more current, pulling the bias voltage proportionally closer to ground. Q_1 can be any general purpose PNP with sufficient voltage rating to withstand the raw bias supply, e.g. 2N5401, MPSA92 etc. D_1 protects the base junction from reverse voltage and can be any diode, e.g. 1N4148. C_1 ensures stability and its value is not critical; it should also be rated to withstand the raw bias voltage.

The circuit in fig. 10.22 needs few parts but, being a shunt regulator, it is wasteful with current. Most guitar amp bias supplies are high impedance and would need the usual series dropping resistance to be reduced in value or even replaced with a pre-regulator to allow the shunt bias tracker to work. This can be avoided by using a two-transistor series tracker, as in fig. 10.23. This works essentially the same as before, except the large series resistor is replaced by Q_2, another general purpose PNP. There is now no need to specially select R_1 to suit different amplifier designs.

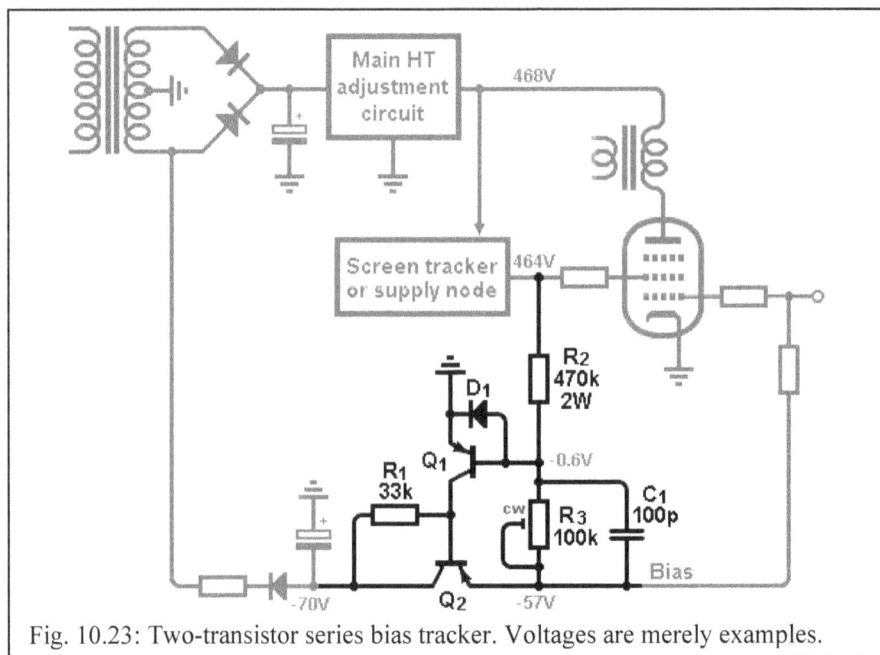

Fig. 10.23: Two-transistor series bias tracker. Voltages are merely examples.

Voltage Control

Since Q$_2$ is PNP, the bias cannot be pulled all the way to zero volts, but it is unlikely that we would ever need it to; the amplifier would become unusable before we reached that point.

10.4: Cheap and Dirty Power Supplies

For those who struggle to obtain a 'proper' valve transformer or who wish to experiment with small valve projects without getting too invested, it is possible to use low-voltage off-the-shelf transformers in a creative way to get both HT and heater power. Some options exploiting voltage multiplication were discussed in chapter 3, and the following can be added to that bag of tricks.

10.4.1: Back-to-Back Transformers

A popular way to make low-cost power supply for small valve projects is to use back-to-back transformers. In other words, use one transformer to step down to a low voltage ideal for heaters, then also use that low voltage to drive a second transformer 'backwards' to step back up again, producing a high voltage. This approach is quite lossy, however, so do not expect to be able to feed a big power amplifier this way. Back-to-back transformers tend to make the best use of parts when the first transformer is between 12VA and 50VA, and the voltage is stepped down as little as possible in between, e.g. avoid using 6V transformers if you can use 12V or more.

The first transformer must supply the heater power plus the power demanded by the second transformer, so the first one will often be a bigger device. On the other hand, sometimes it is easier to obtain two identical transformers than two different ones. Either way, the important thing is not to overload the first transformer. Small EI transformers (up to 50VA, say) will consume anywhere from 3 to 10VA of magnetising power. Small toroidal transformers demand less, up to perhaps 5VA. More powerful transformers are likely to require more magnetising power. This still applies when the transformer is being driven backwards, so the first transformer

must supply this magnetising power, thereby reducing its remaining 'useful' VA. As a rule of thumb, subtract 5VA from the first transformer's rating; whatever remains can be divided up between heater and HT power as required.

It is important not to be too optimistic about the high voltage that can be obtained with this approach, as

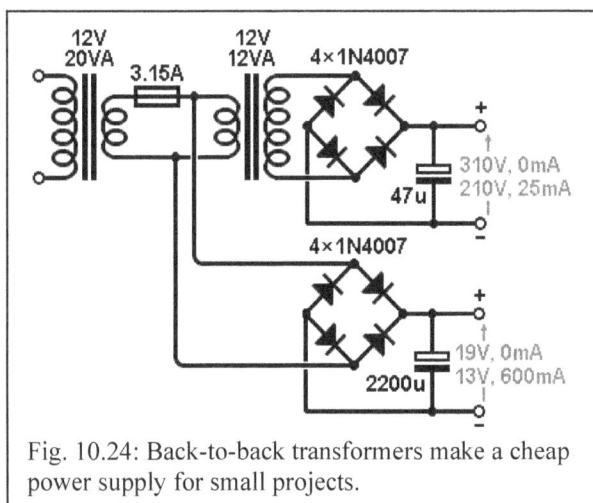

Fig. 10.24: Back-to-back transformers make a cheap power supply for small projects.

279

transformers are deliberately overwound to compensate for their regulation figure. In other words, a small 12V transformer is really a 14V transformer (or thereabouts) with unavoidable resistance. Assuming it is optimised for 230V mains, driving it backwards with $12V_{ac}$ will therefore produce closer to $12/14 \times 230 = 197V_{ac}$ rather than the full $230V_{ac}$. We also have to put up with twice the normal source impedance, so the voltage will sag more under load. The author has used back-to-back EI transformers many times and it tends to yield about $200V_{dc}$ to $230V_{dc}$ under load (this could be voltage doubled of course). Toroidal transformers may yield a little more, but nowhere near the $300V_{dc}$ figure that beginners might anticipate. Incidentally, do not be tempted to cheat nature by driving the second transformer with a higher-than-rated voltage, as this will saturate it.

Fig. 10.24 shows a practical supply using a couple of 12V transformers. Measured output voltages are shown, and the loading has been arranged to result in full use of the first (20VA) transformer. Note the monumentally poor regulation of the HT! Of course, this is of little consequence for a preamp or flea-power 'saggy' guitar practice amp. The low-voltage supply might serve a 6V regulator, or a 12V low-dropout regulator if slightly under loaded. Alternatively, up to 900mA of heaters can be supplied directly from the 12V AC if the rectifier is omitted.

10.4.2: Dual-Primary Misuse

A different way to make a cheap and dirty power supply is possible where the wall voltage is 120V and you can obtain a universal dual-primary transformer. Normally the primaries would be connected in parallel for 120V operation, but by driving only one of them, the other, being identical, will *produce* $120V_{ac}$. After rectification this will yield close to $160V_{dc}$. The 'proper' secondary winding can be used for heaters or another purpose. An ordinary 6V transformer could therefore provide everything we need for a small project.

Once again this is a wasteful trick because only one half of the intended primary is used, meaning the available VA of the transformer is *also halved*; demanding more may overload the lone primary winding. A further concern is that the insulation between the two 120V windings might not be as robust as between a 'proper' primary and secondary, so use at your own risk. Voltage doubling the HT is not recommended since it would further stress this insulation.

Fig. 10.25 shows a practical example using a 6V 20VA transformer. Since only half this figure is safely available,

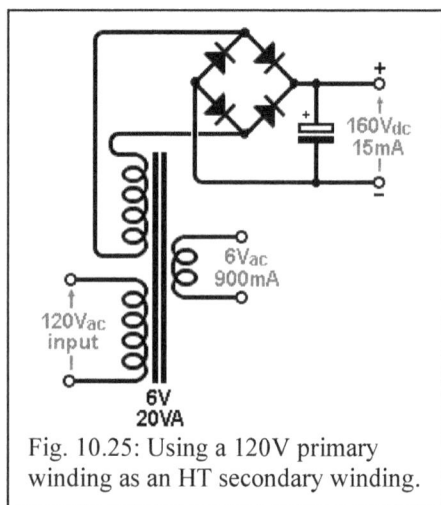

Fig. 10.25: Using a 120V primary winding as an HT secondary winding.

5.4VA has been apportioned to the heaters leaving 3.6VA for the HT winding. Allowing for a typical power factor this amounts to about 2.4W of actual HT power, enough for a preamp project.

10.5: External Power Supplies

Very sensitive equipment may require a separate power supply unit to achieve maximum separation from the (noisy) power transformer and rectifier. If this is the case then it is important to ensure that all the reservoir capacitance is contained within the power supply box. Fig. 10.26a shows a violation of this rule where a smoothing capacitor has been added to the amplifier, with nothing separating it from the reservoir capacitor. The second capacitor is therefore directly in parallel with – and hence part of– the reservoir capacitance. A portion of the ripple current will flow down the cable and into the amplifier box, which can pollute the quiet amplifier ground, leading to hum and buzz. To avoid this a dropping resistor, smoothing filter or voltage regulator should be added to the output of the power supply unit to keep the ripple current contained inside it. An example is shown in fig. 10.26b where a common-mode smoothing filter has been added to serve the dual purpose of containing the ripple current and shunting common-mode hash to earth before it reaches the amplifier box. A common-mode choke where power enters the amplifier box is also favourable addition, especially when using an SMPS. The power supply chassis (if it is metal) must be connected to mains earth for safety, but in theory the amplifier chassis does not since mains does not enter it. Nevertheless, the amplifier chassis should in practice be earthed too for shielding reasons (the extra safety is a bonus).

To link the two boxes we will presumably need some connectors, but finding ones that are officially rated for high-voltage DC is very difficult, even though the currents involved in valve projects are usually small. However, since the voltages

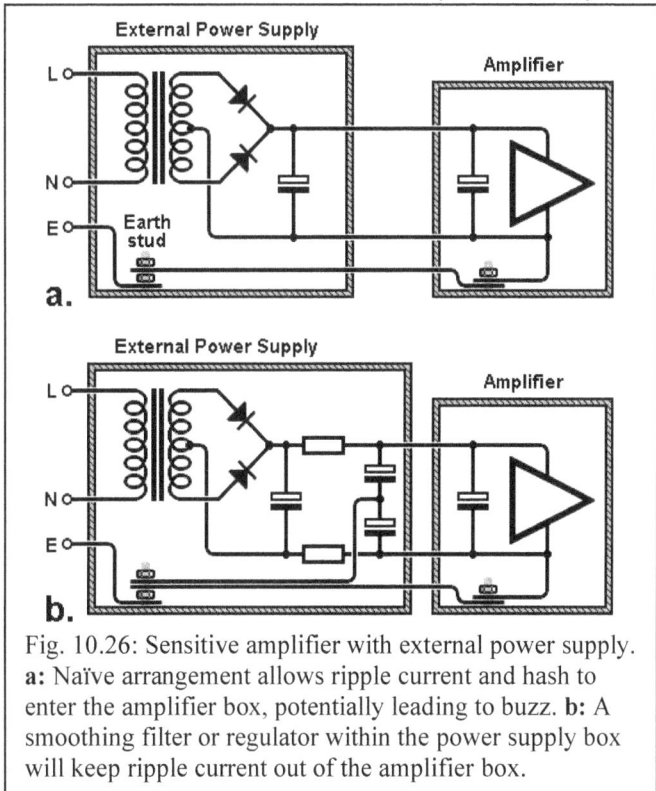

Fig. 10.26: Sensitive amplifier with external power supply. **a:** Naïve arrangement allows ripple current and hash to enter the amplifier box, potentially leading to buzz. **b:** A smoothing filter or regulator within the power supply box will keep ripple current out of the amplifier box.

281

being handled here are isolated from the mains, the immediate danger to the user is much less than with mains voltage itself. Most builders will therefore resort to sensible-looking, heavy-duty connectors, even though they may not have 'official' high-voltage ratings (but to avoid dangerous confusion never use a standard mains connector –such as an IEC socket– for anything other than actual mains).

The author has used *unusual* miniature mains connectors such as those made by Bulgin. The 'Cliffcon' ZC (miniature multi-pole), and the S (loudspeaker) series, made by Cliff, are also likely candidates.

Index

www.ingramcontent.com/pod-product-compliance
Lightning Source LLC
Chambersburg PA
CBHW031404180326

41458CB00043B/6617/J

* 9 7 8 0 9 5 6 1 5 4 5 4 5 *